Artificial Intelligence: Foundations, Theory, and Algorithms

Series Editors

Barry O'Sullivan, Department of Computer Science, University College Cork, Cork, Ireland

Michael Wooldridge, Department of Computer Science, University of Oxford, Oxford, UK

Artificial Intelligence: Foundations, Theory and Algorithms fosters the dissemination of knowledge, technologies and methodologies that advance developments in artificial intelligence (AI) and its broad applications. It brings together the latest developments in all areas of this multidisciplinary topic, ranging from theories and algorithms to various important applications. The intended readership includes research students and researchers in computer science, computer engineering, electrical engineering, data science, and related areas seeking a convenient way to track the latest findings on the foundations, methodologies, and key applications of artificial intelligence.

This series provides a publication and communication platform for all AI topics, including but not limited to:

- Knowledge representation
- Automated reasoning and inference
- Reasoning under uncertainty
- Planning, scheduling, and problem solving
- Cognition and AI
- Search
- Diagnosis
- Constraint processing
- Multi-agent systems
- Game theory in AI
- Machine learning
- Deep learning
- Reinforcement learning
- Data mining
- Natural language processing
- Computer vision
- Human interfaces
- Intelligent robotics
- Explanation generation
- Ethics in AI
- Fairness, accountability, and transparency in AI

This series includes monographs, introductory and advanced textbooks, state-of-the-art collections, and handbooks. Furthermore, it supports Open Access publication mode.

More information about this series at https://link.springer.com/bookseries/13900

Chuan Shi • Xiao Wang • Philip S. Yu

Heterogeneous Graph Representation Learning and Applications

 Springer

Chuan Shi
School of Computer Science
Beijing University of Posts
and Telecommunications
Beijing, China

Xiao Wang
School of Computer Science
Beijing University of Posts
and Telecommunications
Beijing, China

Philip S. Yu
Department of Computer Science
University of Illinois at Chicago
Chicago, IL, USA

ISSN 2365-3051   ISSN 2365-306X   (electronic)
Artificial Intelligence: Foundations, Theory, and Algorithms
ISBN 978-981-16-6165-5   ISBN 978-981-16-6166-2   (eBook)
https://doi.org/10.1007/978-981-16-6166-2

This Springer imprint is published by the registered company Springer Nature Singapore Pte Ltd.
The registered company address is: 152 Beach Road, #21-01/04 Gateway East, Singapore 189721,
Singapore

Foreword

Graphs and networks are ubiquitous in today's interconnected world. Among complex network models, a specific one, called *heterogeneous network* (or *heterogeneous graph*), which models real-world systems as interactions among a massive set of multi-modal and multi-typed objects, is of particular importance because the explicit modeling of the inherent structure of complex networks facilitates powerful, in-depth network analysis. In recent years, *representation learning* (which is also known as *embedding learning*), which represents high-dimensional data with lower dimensional distributions by various deep learning or embedding methods, has been rapidly developed as a powerful tool for high-dimensional data analysis. Similarly, *graph representation learning* (also called *network embedding*), which learns representations of nodes/edges in a lower-dimensional space, has demonstrated its effectiveness for various graph mining and graph analysis tasks.

This book is the first book dedicated to *heterogeneous graph representation learning*, which learns node/edge representations in a lower dimensional space while preserving the heterogeneous structures and semantics for downstream tasks (e.g., node/graph classification and link prediction). Heterogeneous graph representation learning has become a powerful, realistic, and general network modeling tool in recent years and has attracted increasing attention in both academia and industry.

This book serves as a comprehensive and extensive introduction to heterogeneous graph representation learning and its applications, including a survey of current developments and the state of the art in this booming field. It not only extensively introduces the mainstream techniques and models, including structure-preserved, attribute-assisted, and dynamic graph, but also presents wide applications in recommendation, text mining, and industry. In addition, the book provides a platform and practice of heterogeneous graph representation learning. As the first book on the theme, it summarizes the latest developments and presents cutting-edge research on heterogeneous graph representation learning. It may have double benefits: (1) providing researchers with an understanding of the fundamental issues and a good entry point for working in this rapidly expanding field, and (2) presenting the latest research on applying heterogeneous graphs to model real systems and learning structural features of interaction systems.

The authors of this book have done substantial research on heterogeneous graph representation learning and the related themes. Philip S. Yu is one of the leading experts on data mining and heterogeneous information networks. Chuan Shi is a long-term collator with Philip on research into heterogeneous information networks. Chuan has systematically studied the recommendation and representation learning based on heterogeneous graphs, applied heterogeneous information network modeling to e-commerce and text mining, and dived recently into heterogeneous graph representation learning. Wang Xiao is a rising-star scholar in network embedding community. The book systematically summarizes their contributions in the direction of heterogeneous graph representation learning. This book can be used not only as a guidebook for academia and industry but also as a textbook for undergraduate and graduate students. I hope you enjoy reading it.

Michael Aiken Chair Professor   Jiawei Han
University of Illinois at Urbana-Champaign

Preface

Heterogeneous graph, containing different types of nodes and links, is ubiquitous in the real world, ranging from bibliographic networks and social networks to recommendation systems. Currently, heterogeneous graph representation learning, which learns node/edge representations in a lower dimensional space while preserving the heterogeneous structures and semantics for downstream tasks (e.g., node/graph classification and link prediction), has attracted considerable attentions, and we have witnessed the impressive performance of heterogeneous graph representation learning methods on various real-world applications (e.g., recommender systems). The increasing number of works on heterogeneous graph representation learning indicates a global trend in both academic and industrial communities. Thus, there is a pressing demand for comprehensively summarizing and discussing heterogeneous graph representation learning methods.

Compared with homogeneous graph representation learning, heterogeneous graph representation learning shows different challenges because of the heterogeneity. For example, heterogeneous graph has more complex structures caused by multiple relations, where the node attributes are also heterogeneous. The heterogeneous graph representation learning is highly related with real-world applications from the heterogeneous graph construction to learning, which may need more advanced domain knowledge. All these factors heavily affect the performance of heterogeneous graph representation learning, which should be carefully considered. Therefore, researches on heterogeneous graph representation learning are of great scientific and application value.

This book serves the interests of specific reader groups. Generally, the book is intended for anyone who wishes to understand the fundamental problems, techniques, and applications of heterogeneous graph representation learning. In particular, we hope that students, researchers, and engineers will find this book inspiring.

This book is divided into four parts, and the readers are able to quickly understand this field through the first part, deeply study the techniques and applications with the second and third parts, and perform typical algorithms with a platform in the fourth part.

- In the first part, we first present an overview of recent heterogeneous graph representation learning methods from different aspects, including both of the methodology and technique levels. Open sources are also summarized so as to facilitate future research and applications in this area. This part will help readers rapidly understand the overall development of this field. In particular, in Chap. 1, the basic concepts and definitions, as well as the background of homogeneous and heterogeneous graph representation learning, will be introduced. The method taxonomy and open sources will be summarized in Chap. 2.
- In the second part, we then provide an in-depth and detailed introduction of representative heterogeneous graph representation learning techniques. This part will help readers understand the fundamental problems in this field and illustrate how to design the state-of-the-art heterogeneous graph representation learning methods for these problems. In particular, the structure-preserved heterogeneous graph representation learning methods are discussed in Chap. 3, including the meta-path structure and network schema structure. In Chap. 4, the heterogeneous graph representation learning methods with attributes are presented, mainly focusing on the heterogeneous graph neural networks. After that, we introduce the dynamic heterogeneous representation learning methods in Chap. 5, which consider the incremental learning, sequence information, and temporal interaction. Then, in Chap. 6, we discuss some emerging topics of heterogeneous graph representation learning, covering the adversarial learning, sampling, and hyperbolic representation learning.
- In the third part, we summarize the real-world applications based on heterogeneous graph representation learning. This part enables readers to know the successful applications of heterogenous graph representation learning and the way of applying the advanced techniques to the real-world scenarios. Specifically, in Chap. 7, we show how the heterogeneous graph representation learning improves different recommender systems, e.g., the top-N recommendation, cold-start recommendation, and author-set recommendation. The application on text mining is introduced in Chap. 8, focusing on the short text classification and the news recommendation scenarios. In Chap. 9, we present the heterogeneous graph representation learning in industry applications, e.g., the cash-out user detection, intent recommendation, share recommendation, and friend-enhanced recommendation.
- In the fourth part, we introduce a platform of heterogeneous graph representation learning and conclude this book. Considering the importance of deep learning platforms, we introduce the foundation platforms on graph machine learning, especially the platform of heterogeneous graph representation learning in Chap. 10. Also, we take three representative heterogeneous graph neural networks as examples, showing how we can perform them using the platform. Finally, the future research directions and open problems are discussed in Chap. 11.

Writing a book always involves more people than just the authors. We would like to express our sincere thanks to all those who worked with us on this book. They are Deyu Bo, Jiawei Liu, Ruijia Wang, Yugang Ji, Houye Ji, Yiding Zhang, Mengmei

Zhang, Tianchi Yang, Shaohua Fan, Chunchen Wang, Hui Han, Qi Cui, Qi Zhang, Nian Liu, Yuanxin Zhuang, Zhenyi Wang, Guanyi Chu, Hongrui Liu, Chen Li, Tianyu Zhao, Xinlong Zhai, Donglin Xia, and Fengqi Liang. We also give our thanks to many students of Prof. Philip S. Yu for their careful proofreading. They are Yuwei Cao, Yingtong Dou, Ziwei Fan, He Huang, Xiaohan Li, Zhiwei Liu, and Congying Xia. In addition, the work is supported by the National Natural Science Foundation of China (No. U20B2045, U1936220, 61772082, 61702296, 62002029, 62172052). It is also supported in part by NSF under grants III-1763325, III-1909323, III-2106758, and SaTC-1930941. We also thank the supports of these grants. Finally, we thank our families for their wholehearted support throughout this book.

About the Book

Representation learning in heterogeneous graphs (HGs) is intended to provide a meaningful vector representation for each node so as to facilitate downstream applications, such as link prediction, personalized recommendation, node classification, etc. This task, however, is challenging not only because of the need to incorporate heterogeneous structural (graph) information consisting of multiple types of node and edge but also because of the need to consider heterogeneous attributes or types of content (e.g., text or image) associated with each node. Although considerable advances have been made in homogeneous (and heterogeneous) graph embedding, attributed graph embedding, and graph neural networks, few are capable of simultaneously and effectively taking into account the heterogeneous structural (graph) information as well as the heterogeneous content information of each node.

In this book, we provide a comprehensive survey of the current developments in HG representation learning. More importantly, we present the state-of-the-art in this field, including theoretical models and real applications that have been showcased at the top conferences and journals, such as TKDE, KDD, WWW, IJCAI, and AAAI. This book has two major objectives: (1) to provide researchers with an understanding of the fundamental issues and a good point of departure for working in this rapidly expanding field and (2) to present the latest research on applying heterogeneous graphs to model real systems and learning structural features of interaction systems. To the best of our knowledge, it is the first book to summarize the latest developments and present cutting-edge research on heterogeneous graph representation learning. To gain the most from it, readers should have a basic grasp of computer science, data mining, and machine learning.

Contents

About the Authors

Chuan Shi is a professor in the School of Computer Sciences of Beijing University of Posts and Telecommunications and the deputy director of the Beijing Key Lab of Intelligent Telecommunication Software and Multimedia. His main research interests include data mining, machine learning, artificial intelligence, and big data analysis. He has published more than 100 refereed papers, including top journals and conferences in data mining, such as IEEE TKDE, ACM TKDD, KDD, WWW, NeurIPS, AAAI, and IJCAI. In the meanwhile, his first monograph about heterogeneous information networks has been published by Springer. He has been honored with the best paper award in ADMA 2011 and ADMA 2018 and has guided students to the world championship in the IJCAI Contest 2015, the premier international data mining competition. He is also the recipient of "the Youth Talent Plan" and "the Pioneer of Teacher's Ethics" in Beijing.

Xiao Wang is the associate professor in the School of Computer Sciences of Beijing University of Posts and Telecommunications. He was a postdoc in the Department of Computer Science and Technology at Tsinghua University. He got his Ph.D. in the School of Computer Science and Technology at Tianjin University and a joint-training Ph.D. at the Washington University in St. Louis. His main research interests include data mining, machine learning, artificial intelligence, and big data analysis. He has published more than 70 refereed papers, including top journals and conferences in data mining, such as IEEE TKDE, KDD, NeurIPS, AAAI, IJCAI, and WWW. He also serves as SPC/PC member and Reviewer of several high-level international conferences, e.g., KDD, AAAI, IJCAI, and journals, e.g., IEEE TKDE.

Philip S. Yu's main research interests include big data, data mining (especially on graph or network mining), social network, privacy-preserving data publishing, data stream, database systems, and Internet applications and technologies. Dr. Yu is a distinguished professor in the Department of Computer Science at UIC and also holds the Wexler Chair in Information and Technology. Before joining UIC, he was with the IBM Thomas J. Watson Research Center, where he was the manager of the Software Tools and Techniques department. He has published more than 1,300 papers in refereed journals and conferences with more than 149,000 citations and an H-index of 176. He holds or has applied for more than 300 U.S. patents. He is a fellow of the ACM and the IEEE. He is a recipient of the ACM SIGKDD 2016 Innovation Award and the IEEE Computer Society's 2013 Technical Achievement Award.

About the Authors

Chuan Shi is a professor in the School of Computer Sciences of Beijing University of Posts and Telecommunications and the deputy director of the Beijing Key Lab of Intelligent Telecommunication Software and Multimedia. His main research interests include data mining, machine learning, artificial intelligence, and big data analysis. He has published more than 100 refereed papers, including top journals and conferences in data mining, such as IEEE TKDE, ACM TKDD, KDD, WWW, NeurIPS, AAAI, and IJCAI. In the meanwhile, his first monograph about heterogeneous information networks has been published by Springer. He has been honored with the best paper award in ADMA 2011 and ADMA 2018 and has guided students to the world championship in the IJCAI Contest 2015, the premier international data mining competition. He is also the recipient of "the Youth Talent Plan" and "the Pioneer of Teacher's Ethics" in Beijing.

Xiao Wang is the associate professor in the School of Computer Sciences of Beijing University of Posts and Telecommunications. He was a postdoc in the Department of Computer Science and Technology at Tsinghua University. He got his Ph.D. in the School of Computer Science and Technology at Tianjin University and a joint-training Ph.D. at the Washington University in St. Louis. His main research interests include data mining, machine learning, artificial intelligence, and big data analysis. He has published more than 70 refereed papers, including top journals and conferences in data mining, such as IEEE TKDE, KDD, NeurIPS, AAAI, IJCAI, and WWW. He also serves as SPC/PC member and Reviewer of several high-level international conferences, e.g., KDD, AAAI, IJCAI, and journals, e.g., IEEE TKDE.

Philip S. Yu's main research interests include big data, data mining (especially on graph or network mining), social network, privacy-preserving data publishing, data stream, database systems, and Internet applications and technologies. Dr. Yu is a distinguished professor in the Department of Computer Science at UIC and also holds the Wexler Chair in Information and Technology. Before joining UIC, he was with the IBM Thomas J. Watson Research Center, where he was the manager of the Software Tools and Techniques department. He has published more than 1,300 papers in refereed journals and conferences with more than 149,000 citations and an H-index of 176. He holds or has applied for more than 300 U.S. patents. He is a fellow of the ACM and the IEEE. He is a recipient of the ACM SIGKDD 2016 Innovation Award and the IEEE Computer Society's 2013 Technical Achievement Award.

Chapter 1
Introduction

Abstract Networks (or graphs) are ubiquitous in the real world, such as social networks, academic networks, biological networks, and so on. Heterogeneous information network (HIN), a.k.a., heterogeneous graph (HG), is an important type of network, which contains multiple types of nodes and edges. To date, the research of HG has attracted extensive attention, the most important of which is the heterogeneous graph representation (HGR), a.k.a., heterogeneous network embedding. In this chapter, we first introduce some basic concepts and definitions in HG and emphasize the importance of graph representation learning in the field of data mining. Then, we analyze the unique challenges of HGR compared with homogeneous network. In the end, we briefly introduce the organization of this book.

1.1 Basic Concepts and Definitions

Before introducing HGR, we first give some basic definitions in HG. The first is information network, which is a template of the real-world networks. Specially, both homogeneous network and heterogeneous network can be seen as special cases of information network. We formally define them as follows:

Definition 1 Information Network [15]. An information network is defined as a graph $\mathcal{G} = \{\mathcal{V}, \mathcal{E}\}$, in which $\mathcal{V}$ and $\mathcal{E}$ represent the node set and the link set, respectively. Each node $v \in \mathcal{V}$ and link $e \in \mathcal{E}$ is associated with their mapping functions $\phi(v) : \mathcal{V} \to \mathcal{A}$ and $\varphi(e) : \mathcal{E} \to \mathcal{R}$, where $\mathcal{A}$ and $\mathcal{R}$ denote the node types and link types, respectively. **Homogeneous Network (or Homogeneous Graph)** is an instance of the information network, with $|\mathcal{A}| = |\mathcal{R}| = 1$. **Heterogeneous Network (or Heterogeneous Graph)** requires $|\mathcal{A}| + |\mathcal{R}| > 2$, i.e., it contains different types of nodes and links.

Compared with homogeneous graph, heterogeneous graph not only has stronger expressive power but also is more complex. An example of heterogeneous academic graph is illustrated in Fig. 1.1a, which consists of four node types (Author, Paper,

© The Author(s), under exclusive license to Springer Nature Singapore Pte Ltd. 2022
C. Shi et al., *Heterogeneous Graph Representation Learning and Applications*,
Artificial Intelligence: Foundations, Theory, and Algorithms,
https://doi.org/10.1007/978-981-16-6166-2_1

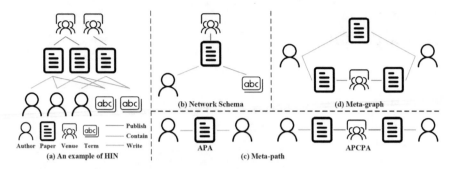

Fig. 1.1 A heterogeneous academic graph, including (**a**) four types of nodes (i.e., Author, Paper, Venue, and Term) and three types of link (i.e., Publish, Contain, and Write), (**b**) network schema, (**c**) meta-paths (i.e., Author–Paper–Author and Paper–Term–Paper), and (**d**) meta-graph

Venue, and Term) and three link types (Author–Write–Paper, Paper–Contain–Term, and Conference–Publish–Paper). In the next, we will introduce some unique definitions in HG, including network schema (Fig. 1.1b), meta-paths (Fig. 1.1c), and meta-graph (Fig. 1.1d). Finally, we will give the definition of graph representation learning.

Since an HG contains multiple node types and link types, to understand the whole structure of it, it is necessary to provide a meta-level (or schema-level) description of the graph. Therefore, as the blueprint of HG, network schema is proposed to give an abstraction of the graph:

Definition 2 Network Schema of $\mathcal{G}$ is a directed graph $\mathcal{S} = (\mathcal{A}, \mathcal{R})$, which can be seen as a meta template of an HG with the node type mapping function $\phi(v) : \mathcal{V} \rightarrow \mathcal{A}$ and the link type mapping function $\varphi(e) : \mathcal{E} \rightarrow \mathcal{R}$. Figure 1.1b illustrates the network schema of the academic graph.

Network schema describes the associations between different types of nodes. Based on it, we can further mine the higher-level semantics of the data. Therefore, meta-path [16] is further proposed to capture the higher-order relationships, i.e., semantics, between nodes. The definition of meta-path is given below:

Definition 3 Meta-Path [16]. A meta-path m is based on a network schema $\mathcal{S}$, which is denoted as $m = A_1 \xrightarrow{R_1} A_2 \xrightarrow{R_2} \cdots \xrightarrow{R_l} A_{l+1}$ (simplified to $A_1 A_2 \cdots A_{l+1}$) with node types $A_1, A_2, \cdots, A_{l+1} \in \mathcal{A}$ and link types $R_1, R_2, \cdots R_l \in \mathcal{R}$.

Different meta-paths capture the semantic relationships from different views. For example, the meta-path of "APA" indicates the co-author relationship and "APCPA" represents the co-conference relation. Both of them can be used to formulate the proximity over authors. Although meta-path can be used to depict the connections over nodes, it fails to capture a more complex relationship, such as motifs [10]. To address this challenge, meta-graph [6] is proposed to use a directed acyclic graph of node and link types to capture more complex relationship between two HG nodes.

Definition 4   Meta-Graph [6]. A meta-graph $\mathcal{T}$ can be seen as a directed acyclic graph (DAG) composed of multiple meta-paths with common nodes. Formally, meta-graph is defined as $\mathcal{T} = (V_\mathcal{T}, E_\mathcal{T})$, where $V_\mathcal{T}$ is a set of nodes and $E_\mathcal{T}$ is a set of links. For any node $v \in V_\mathcal{T}, \phi(v) \in \mathcal{A}$; for any link $e \in E_\mathcal{T}, \varphi(e) \in \mathcal{R}$.

An example of meta-graph is shown in Fig. 1.1d, which can be regarded as the combination of meta-path "APA" and "APCPA," reflecting a high-order similarity of two nodes. Note that a meta-graph can be symmetric or asymmetric [30].

The research of graph is always an important topic in machine learning. However, due to the non-Euclidean property, traditional heuristic methods suffer from high computational cost and low parallelizability [2], which cannot be used for real applications. Therefore, a critical challenge for this field is to find effective data representation. Through graph representation learning, the nodes are projected into vectors and can be incorporated with the advanced machine learning technologies and tasks. We formalize the problem of graph representation learning as follows.

Definition 5   Graph Representation Learning [2], also known as network embedding, aims to learn a function $\Phi : \mathcal{V} \to \mathbb{R}^d$ that embeds the nodes $v \in \mathcal{V}$ in a graph into a low-dimensional Euclidean space where $d \ll |\mathcal{V}|$.

A toy example of graph representation learning is shown in Fig. 1.2. Through graph representation learning, the complex network in the non-Euclidean space is projected into a low-dimensional Euclidean space. Therefore, the high computational cost and low parallelizability issues are well-solved. In the next, we will give a brief review of the recent development of graph representation learning.

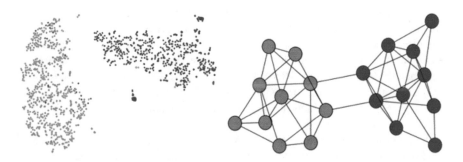

Fig. 1.2 A toy example of graph representation learning. (**a**) Input graph. (**b**) Node representations

1.2   Graph Representation Learning

In the aforementioned chapter, we refer that the analysis of graphs suffers from the high computational complexity and low parallelizability issues. To deal with these problems, graph representation learning is proposed and rapidly becomes the major tool in network analysis [2, 23].

Previous graph representation learning methods focus on preserving the structural information of the graph. For example, DeepWalk [11] uses random walk to generate node sequences and then employs the skip-gram model to learn the co-occurrence of nodes within a window, thus capturing the local structures. LINE [17] preserves both the first- and second-order structure similarities, node2vec [5] extends DeepWalk to global structures with a Depth-First Sampling (DFS) and Breadth-First Sampling (BFS), M-NMF [19] learns the community structures, and AROPE [27] preserves arbitrary order proximity through Singular Value Decomposition (SVD). In particular, Qiu et al. [12] proved that most existing graph representation learning methods can be unified into a matrix factorization framework.

Further, some methods begin to incorporate the rich node or edge attributes in node representation. TADW [24] jointly factorizes the adjacency matrix and text matrix to fuse the structural and attribute information. DANE [4] enforces the structural representations and attribute representations to be consistent, so that the learned representations can capture these two kinds of information at the same time. ANRL [28] designs a neighbor enhancement autoencoder. It aims to reconstruct the target neighbors and attributes to model the structural and attribute information.

With the development of deep learning, the emerging graph neural networks (GNNs) show powerful capability in combining the network structures and node attributes. Graph convolutional network (GCN) [7] is one of the most representative works, which designs a convolutional operator in spatial domain to filter the node attributes by network structures. Graph Attention Network (GAT) [18] uses self-attention to learn the importance of nodes in fusing the attributes of neighbors. Klicpera et al. proposed Predict then Propagate (PPNP) [8], which incorporates personalized PageRank into GNNs and alleviates the over-smoothing problem. SGC [22] simplifies the design of GCN through decoupling the transformation step and aggregation step, which not only reduces the parameters of GNNs but also accelerates the training process.

In addition to the structural and attribute information, recently, some researchers tend to explore the semantic information from the multiple node or edge types in graphs [13], leading to the research of HGR.

1.3 Heterogeneous Graph Representation Learning and Challenges

Different from homogeneous graph representation learning that mainly needs to preserve the structural information, heterogeneous graph representation learning aims to preserve the structural and semantic information simultaneously. However, due to the heterogeneity of HG, HGR imposes more challenges to this problem, which are illustrated below.

- **Complex structure** (the complex HG structure caused by multiple types of nodes and edges). In a homogeneous graph, the fundamental structure can be considered as the so-called first-order, second-order, and even higher-order structure [11, 17, 19]. All these structures are well-defined and have good intuition. However, the structure in HG will dramatically change depending on the selected relations. Let us still take the bibliographic network in Fig. 1.1a as an example: the neighbors of one paper will be authors with the "write" relation, while with "contain" relation, the neighbors become terms. Complicating things further, the combination of these relations, which can be considered as a higher-order structure in HG, will result in different and more complicated structures. How to efficiently and effectively preserve these complex structures is thereby an urgent need but it is still a significant challenge in HGR, and current efforts have been made towards the meta-path structure [3] and meta-graph structure [26], etc.
- **Heterogeneous attributes** (the fusion problem caused by the heterogeneity of attributes). Since the nodes and edges in a homogeneous graph have the same type, each dimension of the node or edge attributes has the same meaning. In this situation, node can directly fuse the attributes of its neighbors. However, in heterogeneous graph, the attributes of different types of nodes and edges may have different meanings [20, 29]. For example, the attributes of author can be the research fields, while paper may use keywords as attributes. Therefore, how to overcome the heterogeneity of attributes and effectively fuse the attributes of neighbors is an important challenge in HGR.
- **Application dependent** (the domain knowledge hidden in HG structures and attributes). HG is closely related to the real-world applications, while many practical problems remain unsolved. For example, constructing an appropriate HG may require sufficient domain knowledge in a real-world application. Also, meta-path and/or meta-graph are widely used to capture the structure of HG. However, unlike homogeneous graph, where the structure, e.g., the first-order and second-order structure, is well-defined, meta-path selection may also need prior knowledge. Furthermore, to better facilitate the real-world applications, we usually need to elaborately encode the side information, e.g., node attributes [20, 21, 25, 29] or more advanced domain knowledge [1, 9, 14] to the HGR process.

1.4   Organization of the Book

This book is written to comprehensively review the development of HGR and introduce the state-of-the-art methods. We first summarize existing works from two perspectives: method and technique, and then we introduce some open sources of this field. Then we introduce the state-of-the-art models of each category in detail. Part one focuses on the four main HGR models. Part two introduces the development of HGR on real-world industrial scene. After that, we introduce the platform and practice of heterogeneous graph representation. Finally, we discuss the future research direction of HGR and summarize the content of this book.

The remainder of this book is organized as follows: In Chap. 2, we summarize the developments of HGR, including taxonomy, technique, and open sources. In Chaps. 3–6, we categorize existing HGR methods into four categories, including structure-preserved HGR, attribute-assisted HGR, dynamic HGR, and some other emerging topics. In each chapter, we will introduce their unique challenges and designs in detail. In Chaps. 7–9, we further explore the transformativeness of existing HGR methods that have been successfully deployed in real-world applications, e.g., recommendation, text mining, cash-out user detection, etc. In Chap. 11, we forecast the future research directions in this field.

References

1. Chen, T., Sun, Y.: Task-guided and path-augmented heterogeneous network embedding for author identification. In: Proceedings of the Tenth ACM International Conference on Web Search and Data Mining, pp. 295–304. ACM, New York (2017)
2. Cui, P., Wang, X., Pei, J., Zhu, W.: A survey on network embedding. IEEE Trans. Knowl. Data Eng. 31(5), 833–852 (2018)
3. Dong, Y., Chawla, N.V., Swami, A.: metapath2vec: scalable representation learning for heterogeneous networks. In: Proceedings of the 23rd ACM SIGKDD International Conference on Knowledge Discovery and Data Mining, pp. 135–144. ACM, New York (2017)
4. Gao, H., Huang, H.: Deep attributed network embedding. In: Proceedings of the Twenty-Seventh International Joint Conference on Artificial Intelligence, pp. 3364–3370. ijcai.org (2018)
5. Grover, A., Leskovec, J.: node2vec: scalable feature learning for networks. In: Proceedings of the 22nd ACM SIGKDD International Conference on Knowledge Discovery and Data Mining, pp. 855–864. ACM, New York (2016)
6. Huang, Z., Zheng, Y., Cheng, R., Sun, Y., Mamoulis, N., Li, X.: Meta structure: computing relevance in large heterogeneous information networks. In: Proceedings of the 22nd ACM SIGKDD International Conference on Knowledge Discovery and Data Mining, pp. 1595–1604. ACM, New York (2016)
7. Kipf, T.N., Welling, M.: Semi-supervised classification with graph convolutional networks. In: Published as a Conference Paper at ICLR (2017)
8. Klicpera, J., Bojchevski, A., Günnemann, S.: Predict then propagate: graph neural networks meet personalized pagerank. In: Published as a Conference Paper at ICLR 2019 (Poster). OpenReview.net (2019)

9. Liu, Z., Zheng, V.W., Zhao, Z., Li, Z., Yang, H., Wu, M., Ying, J.: Interactive paths embedding for semantic proximity search on heterogeneous graphs. In: Proceedings of the 24th ACM SIGKDD International Conference on Knowledge Discovery & Data Mining, pp. 1860–1869. ACM, New York (2018)
10. Milo, R., Shen-Orr, S., Itzkovitz, S., Kashtan, N., Chklovskii, D., Alon, U.: Network motifs: simple building blocks of complex networks. Science **298**(5594), 824–827 (2002)
11. Perozzi, B., Al-Rfou, R., Skiena, S.: Deepwalk: Online learning of social representations. In: Proceedings of the 20th ACM SIGKDD International Conference on Knowledge Discovery and Data Mining, pp. 701–710 (2014)
12. Qiu, J., Dong, Y., Ma, H., Li, J., Wang, K., Tang, J.: Network embedding as matrix factorization: unifying deepwalk, LINE, PTE, and node2vec. In: WSDM '18: Proceedings of the Eleventh ACM International Conference on Web Search and Data Mining, pp. 459–467. ACM, New York (2018)
13. Shi, C., Li, Y., Zhang, J., Sun, Y., Yu, P.S.: A survey of heterogeneous information network analysis. IEEE Trans. Knowl. Data Eng. **29**(1), 17–37 (2017)
14. Shi, C., Hu, B., Zhao, W.X., Yu, P.S.: Heterogeneous information network embedding for recommendation. IEEE Trans. Knowl. Data Eng. **31**(2), 357–370 (2018)
15. Sun, Y., Han, J.: Mining heterogeneous information networks: a structural analysis approach. SIGKDD Explorat. **14**(2), 20–28 (2012)
16. Sun, Y., Han, J., Yan, X., Yu, P.S., Wu, T.: Pathsim: meta path-based top-k similarity search in heterogeneous information networks. Proc. VLDB Endowment **4**(11), 992–1003 (2011)
17. Tang, J., Qu, M., Wang, M., Zhang, M., Yan, J., Mei, Q.: Line: Large-scale information network embedding. In: WWW '15: Proceedings of the 24th International Conference on World Wide Web, pp. 1067–1077 (2015)
18. Veličković, P., Cucurull, G., Casanova, A., Romero, A., Lio, P., Bengio, Y.: Graph attention networks. ICLR 2018 Conference (2018)
19. Wang, X., Cui, P., Wang, J., Pei, J., Zhu, W., Yang, S.: Community preserving network embedding. In: AAAI'17: Proceedings of the Thirty-First AAAI Conference on Artificial Intelligence (2017)
20. Wang, X., Ji, H., Shi, C., Wang, B., Ye, Y., Cui, P., Yu, P.S.: Heterogeneous graph attention network. In: The World Wide Web Conference, pp. 2022–2032. ACM, New York (2019)
21. Wang, X., Lu, Y., Shi, C., Wang, R., Cui, P., Mou, S.: Dynamic heterogeneous information network embedding with meta-path based proximity. IEEE Trans. Knowl. Data Eng. (2020)
22. Wu, F., Jr., A.H.S., Zhang, T., Fifty, C., Yu, T., Weinberger, K.Q.: Simplifying graph convolutional networks. In: ICML, Proceedings of Machine Learning Research, vol. 97, pp. 6861–6871. PMLR (2019)
23. Wu, Z., Pan, S., Chen, F., Long, G., Zhang, C., Yu, P.S.: A comprehensive survey on graph neural networks. IEEE Trans. Neural Netw. Learn. Syst. **32**(1), 4–24 (2021)
24. Yang, C., Liu, Z., Zhao, D., Sun, M., Chang, E.Y.: Network representation learning with rich text information. In: IJCAI'15: Proceedings of the 24th International Conference on Artificial Intelligence, pp. 2111–2117. AAAI Press, Palo Alto (2015)
25. Yang, L., Xiao, Z., Jiang, W., Wei, Y., Hu, Y., Wang, H.: Dynamic heterogeneous graph embedding using hierarchical attentions. In: ECIR, Lecture Notes in Computer Science, vol. 12036, pp. 425–432. Springer, Berlin (2020)
26. Zhang, D., Yin, J., Zhu, X., Zhang, C.: Metagraph2vec: complex semantic path augmented heterogeneous network embedding. In: Pacific-Asia Conference on Knowledge Discovery and Data Mining 2018, pp. 196–208. Springer, Berlin (2018)
27. Zhang, Z., Cui, P., Wang, X., Pei, J., Yao, X., Zhu, W.: Arbitrary-order proximity preserved network embedding. In: KDD '18: Proceedings of the 24th ACM SIGKDD International Conference on Knowledge Discovery & Data Mining, pp. 2778–2786. ACM, New York (2018)
28. Zhang, Z., Yang, H., Bu, J., Zhou, S., Yu, P., Zhang, J., Ester, M., Wang, C.: ANRL: attributed network representation learning via deep neural networks. In: International Joint Conferences on Artificial Intelligence Organization, pp. 3155–3161. ijcai.org (2018)

29. Zhang, C., Song, D., Huang, C., Swami, A., Chawla, N.V.: Heterogeneous graph neural network. In: KDD '19: Proceedings of the 25th ACM SIGKDD International Conference on Knowledge Discovery & Data Mining, pp. 793–803. ACM, New York (2019)
30. Zhang, W., Fang, Y., Liu, Z., Wu, M., Zhang, X.: mg2vec: learning relationship-preserving heterogeneous graph representations via metagraph embedding. IEEE Trans. Knowl. Data Eng. (2020)

Chapter 2
The State-of-the-Art of Heterogeneous Graph Representation

Abstract In this chapter, we give a comprehensive review of the recent development on heterogeneous graph representation (HGR) methods and techniques. In the method aspect, according to the information used in HGR, existing works are divided into four categories, i.e., structure-preserved HGR, attribute-assisted HGR, dynamic HGR, and application-oriented HGR. In the technique aspect, we summarize five commonly used techniques in HGR and categorize them into shallow model and deep model. In addition, we also provide some public sources, e.g., benchmark datasets, source code, and available tools.

2.1 Method Taxonomy

Various types of nodes and links in HG bring complex graph structures and rich attributes, i.e., the heterogeneity of HG. To make the node representation capture the heterogeneity, we need to consider the information of different aspects, including graph structures, attributes, specific domain knowledge, and so on. In this chapter, we categorize the existing methods into four categories based on the information they used in HGR. An overview of existing HGR methods explored in this book is shown in Fig. 2.1.

2.1.1 Structure-Preserved Representation

One basic requirement of graph representation is to preserve the graph structures properly [7]. For example, in the homogeneous graph representation, existing works consider a lot of graph structures, e.g., first-order structure [37], second-order structure [48, 50], high-order structure [1, 67], and community structure [51]. Due to the heterogeneity of HG, the graph structures become more complex and even have semantic information, e.g., the co-author relationship. Therefore, an important direction of HGR is to learn both the structural and semantic information

© The Author(s), under exclusive license to Springer Nature Singapore Pte Ltd. 2022   9
C. Shi et al., *Heterogeneous Graph Representation Learning and Applications*,
Artificial Intelligence: Foundations, Theory, and Algorithms,
https://doi.org/10.1007/978-981-16-6166-2_2

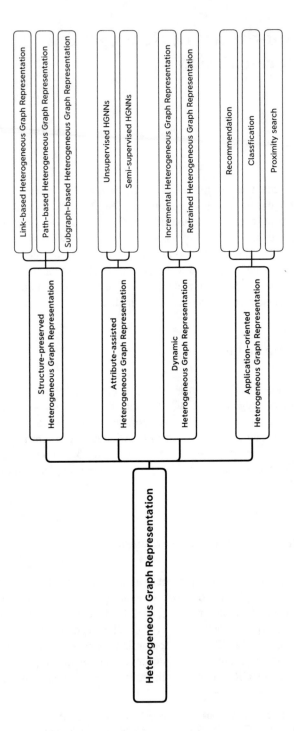

Fig. 2.1 An overview of heterogeneous graph representation methods

simultaneously. In this chapter, we review the typical structure-preserved HGR methods. Each of them considers different structures in HG, including link/edge, meta-path, and subgraph.

A basic requirement of HGR is to preserve the multiple relationships, i.e., links, in node representations. Different from homogeneous graph, links in HG have different types and semantics. To distinguish various types of links, one classical idea is to project them into different metric spaces, rather than a unified metric space. A representative work of this idea is PME [5], which treats each link type as a relation and uses a relation-specific matrix to transform the nodes into different metric spaces. In this way, nodes connected by different types of links can be close to each other in different metric spaces, thus capturing the heterogeneity of the graph. Different from PME, EOE [59] and HeGAN [22] use the relation-specific matrix to calculate the similarity between two nodes. AspEM [43] and HEER [44] aim to maximize the probability of existing links. Generally, the key point of designing link-based methods is to find a proper heterogeneous similarity function to preserve the proximity between nodes.

Link-based methods can only capture the local structures of HG, i.e., the first-order information. In fact, the higher-order relation, describing more complex semantic information (e.g., the co-author relationship), is also critical for HGR. Meta-path is a commonly used tool in modeling the high-order relationship of HG. A representative work is metapath2vec [8], which uses meta-path-guided random walk to generate heterogeneous node sequences with rich semantics; and then it designs a heterogeneous skip-gram technique to preserve the proximity between nodes and their context nodes. Based on metapath2vec, a series of variants have been proposed. For example, Spacey [17] designs a heterogeneous spacey random walk to unify different meta-paths into a second-order hyper-matrix. JUST [26] proposes a random walk method with Jump and Stay strategies, which can flexibly choose to change or maintain the type of the next node in the random walk without meta-path. BHIN2vec [29] designs an extended skip-gram technique to balance the various types of relations. HHNE [54] conducts the meta-path-guided random walk in hyperbolic spaces [18]. Besides, HEAD [57] separates the node representations into intrinsic representations and meta-path specific representations, so that the highly coupled representations can be well-disentangled and become more robust.

Subgraph represents a more complex structure of HG. Incorporating subgraphs into graph representation can significantly improve the ability of capturing complex structural relationship. Zhang et al. proposed metagraph2vec [65], which uses a meta-graph-guided random walk to generate heterogeneous node sequence. Then a heterogeneous skip-gram technique [8] is employed to learn the node representations. Based on this strategy, metagraph2vec can capture the high-order similarity and rich semantic information between nodes. DHNE [49] is a hyperedge-based HGR method. Specifically, it designs a novel deep model to produce a non-linear tuple-wise similarity function while capturing the local and global structures of a given HG.

Compared with link and meta-path, subgraph usually contains more higher-order structural and semantic information. However, one obstacle of subgraph-based

HGR methods is the high complexity of subgraph. Therefore, how to balance the effectiveness and efficiency is required for a practical subgraph-based HGR method, which is worthy of further exploration.

2.1.2   Attribute-Assisted Representation

In addition to the graph structures, another important component of HGR is the rich attributes. Attribute-assisted HGR methods, i.e., heterogeneous graph neural networks (HGNNs), aim to encode the complex structures and rich attributes together to learn node representations. Different from graph neural networks (GNNs) that can directly fuse the attributes of neighbors to update node representations, HGNNs need to overcome the heterogeneity of node or edge attributes and design effective fusion mechanisms to utilize the neighborhood information, which is more challenging. In this chapter, we divide HGNNs into unsupervised and semi-supervised settings and discuss their own pros and cons.

The goal of unsupervised HGNNs is to learn node representations that benefit downstream tasks in an unsupervised manner. To this end, they usually utilize the interactions between different types of nodes or edges to capture the potential proximity, so that the learned representation can have good generalization.

HetGNN [68] is the representative work of unsupervised HGNNs. It consists of three parts: content aggregation, neighbor aggregation, and type aggregation. Content aggregation is designed to fuse the multiple attributes in a node, e.g., a film can have image and text attributes simultaneously. Neighbor aggregation aims to aggregate the nodes with the same type. And type aggregation uses an attention mechanism to mix the representations of different types and produces the final node representations. Through these three components, HetGNN can preserve the heterogeneity of both graph structures and node attributes. Some other unsupervised methods can be regarded as special cases of HetGNN because they capture either the heterogeneity of node attributes or the heterogeneity of graph structures. HNE [3] is proposed to learn representations for the cross-model data in HG, but it ignores the various types of links. SHNE [69] focuses on capturing the semantic information of nodes by designing a deep semantic encoder with gated recurrent units (GRUs) [6]. Although it uses heterogeneous skip-gram to preserve the heterogeneity of graph, SHNE is designed only for text data.

Besides, GATNE [2] aims to learn node representations in multiplex graph, i.e., a heterogeneous graph with different types of edges. Therefore, it pays more attention to distinguish different link relationships. HeCo [58] uses self-supervised learning, i.e., contrastive learning, to generate supervised signals. It designs a novel co-contrastive mechanism to capture the meta-path information and network schema information simultaneously.

It can be seen that the purpose of unsupervised HGNNs is to save as much information as possible. For example, HetGNN uses three types of aggregation functions to learn the information of content, neighbor, and node type separately.

HeCo captures the information of meta-path and network schema. The reason is that the representations learned by unsupervised HGNNs need to be used for downstream tasks, so that it should cover the information of different aspects.

Different from unsupervised HGNNs, semi-supervised HGNNs aim to learn task-specific node representations. Therefore, they prefer to use attention mechanism to capture the most relevant structural and attribute information. Wang et al. [53] proposed heterogeneous graph attention network (HAN), which uses a hierarchical attention mechanism to capture both node and semantic importance. Then a series of attention-based HGNNs were proposed [12, 13, 19, 25]. MAGNN [13] designs intra-meta-path aggregation and inter-meta-path aggregation. HetSANN [19] and HGT [25] use self-attention mechanism, which treats one type of node as query to calculate the importance of other types of nodes around it. MEGAE [12] uses meta-paths are used as virtual edges to enhance the performance of graph attention operator.

Compared with structure-preserved HGR methods, HGNNs have an obvious advantage that they have the ability of inductive learning, i.e., learning representations for the out-of-sample nodes [20]. Besides, HGNNs need less memory space because they only need to store model parameters. These two reasons are important for the real-world applications. However, they still suffer from the huge time costing in inference and retraining.

2.1.3   Dynamic Representation

The real-world graphs are constantly changing over time. For example, in the social platform, people follow and unfollow others daily. Therefore, capturing the temporal information of HG is an important research direction. In this chapter, we introduce the typical dynamic HGR methods, which can be divided into two categories: incremental update and retrained update methods. The former learns the representation of new node in the next timestamp by utilize existing node representations, while the latter retrains the models in each timestamp.

DyHNE [56] is an incremental update method based on the theory of matrix perturbation, which aims to learn node representations and considers both the heterogeneity and evolution of HG at the same time. DyHNE first preserves the meta-path-based first- and second-order proximity. Then it uses the perturbation of meta-path augmented adjacency matrices to capture the changes of graph. Besides, some methods try to use GNNs to learn node or edge representations in each timestamp and then design some advanced neural network, e.g., RNN or attention mechanism, to capture the temporal information of HG. DyHATR [61] aims to capture the temporal information through the changes of node's representations in different timestamps. To this end, it first designs a hierarchical attention mechanism (HAT), which contains a node- and edge-level attention, to learn node representations by fusing the attributes of neighbors.

It can be seen that the incremental update methods are efficient, but they can only capture the short-term temporal information, i.e., the last timestamp [61]. Besides, incremental update methods focus on utilizing non-linear model and lack expressive power. On the contrary, the retrained update methods employ neural networks to capture the long-term temporal information. However, they suffer from the high computational cost. Therefore, how to combine the advantages of these two kinds of models is an important problem.

2.1.4   Application-Oriented Representation

HGR can be integrated with some specific applications. In this situation, one usually needs to consider two factors: one is how to construct an HG for a specific application and the other is what information, i.e., domain knowledge, should be incorporated into HGR. Here, we discuss three common types of applications: recommendation, classification, and proximity search.

Recommendation can be naturally modeled as a link prediction task on HG, where there are at least two types of nodes to represent users and items separately, and links represent the interaction between them. Therefore, HGR is widely used in the recommendation scenario [41]. Besides, other types of information, such as the social relationships, can be easily applied to HG [40], so applying HGR to recommendation application is an important research field.

HERec [42] aims to learn the representations of users and items under different meta-paths and fuse them for recommendation. It first finds the co-occurrence of users and items based on the meta-path-guided random walks on user–item HG. Then it uses node2vec [14] to learn preliminary representation from node sequences of users and items. Because the representations under different meta-paths contain different semantic information, for better recommendation performance, HERec designs a fusion function to unify the multiple representations. Apart from random walk, some methods try to use matrix factorization to learn user and item representations. HeteRec [62] considers the implicit user feedback in HG. HeteroMF [27] designs a heterogeneous matrix factorization technique to consider the context dependence of different types of nodes. FMG [72] incorporates meta-graph into HGR, which can capture some special patterns between users and items.

Previous methods mainly use non-linear model to learn the representations of users and items, which cannot fully capture users' preferences. Therefore, some neural network-based methods are proposed. One of the most important techniques is attention mechanism. MCRec [21] designs a neural co-attention mechanism to capture the relationship between user, item, and meta-path. NeuACF [16] and HueRec [55] first calculate multiple meta-path-based user–item proximity matrices. Then an attention mechanism is designed to learn the importance of different proximity matrices, which learns the importance of users' preferences.

Another type of methods is to apply HGNNs to recommendation. PGCN [60] converts the user–item interaction sequences into item–item graph, user–item graph, and user–sequence graph. Then it designs an HGNN to aggregate the information of user and item in the three graphs, thus capturing the collaborative filtering signals. SHCF [30] uses HGNNs to capture both the high-order heterogeneous collaborative signals and sequential information simultaneously. GNewsRec [23] and GNUD [24] are designed for news recommendation. They consider both the content information of news and the collaborative information between users and news.

Classification is a fundamental task in machine learning. Here we mainly introduce two types of classification tasks that require models to capture the heterogeneity of HG: author identification [4, 36, 64] and user identification [10, 70, 71].

Author identification aims to find the potential authors for an anonymous paper in the academic network. Camel [64] is designed to consider both the content information, e.g., the text of papers, and context information, e.g., the co-occurrence of paper, author, and conference. PAHNE [4] uses meta-paths to augment the pair-wise relations between paper and author. TaPEm [36] further maximizes the proximity between the paper–author pair and the context path around them.

User identification requires the model to make use of the heterogeneity of HG to learn discriminating user representations with weak supervision information. Player2vec [71], AHIN2vec [10], and Vendor2vec [70] are the principal methods. They can be summarized as a general framework: first, some advanced neural networks are used to learn preliminary node representations from the input features. Then the representations will be propagated on the constructed HG to capture the heterogeneity of HG. Finally, a semi-supervised loss function is used to make the node representations contain application-specific information. Under the guidance of partially labeled nodes, the node representations can distinguish special users from the ordinary users in the graph, which can be used for user identification.

Proximity search aims to find the nodes that are closest to the target node by using structural and semantic information of HG. Some earlier studies have dealt with this problem in homogeneous graphs, for example, web search [28]. Recently, some methods try to utilize HG in proximity search [39, 45]. However, these methods only use some statistical information, e.g., the number of connected meta-paths, to measure the similarity of two nodes in HG, which lack flexibility. With the development of deep learning, some graph representation methods are proposed. IPE [31] considers the interactions among different meta-path instances and proposes an interactive-path structure to improve the performance of HGR. SPE [32] proposes a subgraph-augmented HGR method, which uses a stacked autoencoder to learn the subgraph representation so as to enhance the effect of semantic proximity search. D2AGE [33] explores the directed acyclic graph (DAG) structure for better measuring the similarity between two nodes and designs a DAG-LSTM to learn node representations.

In summary, incorporating HGR into specific applications usually needs to consider the domain knowledge. For example, in recommendation, meta-path "user–item–user" can be used to capture the user-based collaborative filtering, while "item–user–item" represents the item-based collaborative filtering; in proximity search, the methods use meta-paths to capture the semantic relationships between nodes, thus enhancing the performance. Therefore, utilizing HG to capture the application-specific domain knowledge is essential for application-oriented HGR.

2.2 Technique Summary

In this chapter, from the technical perspective, we summarize the widely used techniques (or models) in HGR, which can be generally divided into two categories: shallow model and deep model.

2.2.1 Shallow Model

Early HGR methods focus on employing shallow model. They first initialize the node representations randomly and then learn the node representations through optimizing some well-designed objective functions. We divide the shallow model into two categories: random walk based and decomposition based.

Random Walk Based In homogeneous graph, random walk, which generates some node sequences in a graph, is usually used to capture the local structure of a graph [14]. While in heterogeneous graph, the node sequence should contain not only the structural information but also the semantic information. Therefore, a series of semantic-aware random walk techniques are proposed [8, 17, 26, 29, 42, 54, 66]. For example, metapath2vec [8] uses meta-path-guided random walk to capture the semantic information of two nodes, e.g., the co-author relationship in academic graph. Spacey [17] and metagraph2vec [65] design meta-graph-guided random walks, which preserve a more complex similarity between two nodes.

Decomposition Based Decomposition-based techniques aim to decompose HG into several subgraphs and preserve the proximity of nodes in each subgraph [5, 15, 35, 43, 44, 47, 59]. PME [5] decomposes the heterogeneous graph into some bipartite graphs according to the types of links and projects each bipartite graph into a relation-specific semantic space. PTE [47] divides the documents into word–word graph, word–document graph, and word–label graph. Then it uses LINE [48] to learn the shared node representations for each subgraph. HEBE [15] samples a series of subgraphs from an HG and preserves the proximity between the center node and its subgraph.

2.2.2 Deep Model

Deep model aims to use advanced neural networks to learn representation from the node attributes or the interactions among nodes, which can be roughly divided into three categories: message passing based, encoder–decoder based, and adversarial based.

Message Passing Based   The idea of message passing is to send the node representation to its neighbors, which is always used in GNNs. The key component of message passing-based techniques is to design a suitable aggregation function, which can capture the semantic information of HG [2, 13, 19, 38, 53, 63, 68, 75, 77]. For example, HAN [53] designs a hierarchical attention mechanism to learn the importance of different nodes and meta-paths, which captures both structural information and semantic information of HG. HetGNN [68] uses bi-LSTM to aggregate the representation of neighbors so as to learn the deep interactions among heterogeneous nodes. GTN [63] designs an aggregation function, which can find the suitable meta-paths automatically during the process of message passing.

Encoder–Decoder Based   Encoder–decoder-based techniques aim to employ some neural networks as encoder to learn representation from node attributes and design a decoder to preserve some properties of the graphs [3, 4, 36, 49, 64, 69]. For example, HNE [3] focuses on multi-modal heterogeneous graph. It uses CNN and autoencoder to learn representation from images and texts, respectively. Then it uses the representation to predict whether there is a link between the images and texts. Camel [64] uses GRU as encoder to learn paper representation from the abstracts. A skip-gram objective function is used to preserve the local structures of the graphs. DHNE [49] uses autoencoder to learn representation for the nodes in a hyperedge. Then it designs a binary classification loss to preserve the indecomposability of the hyper-graph.

Adversarial Based   Adversarial-based techniques utilize the game between generator and discriminator to learn robust node representation. In homogeneous graph, the adversarial-based techniques only consider the structural information; for example, GraphGAN [52] uses Breadth First Search when generating virtual nodes. In a heterogeneous graph, the discriminator and generator are designed to be relation aware, which captures the rich semantics on HGs. HeGAN [21] is the first to use GAN in HGR. It incorporates the multiple relations into the generator and discriminator, so that the heterogeneity of a given graph can be considered. MV-ACM [76] uses GAN to generate the complementary views by computing the similarity of nodes in different views.

2.3   Open Sources

In this chapter, we summarize the commonly used datasets of HGR. In addition, we will introduce some useful resources and open-source tools about HGR.

2.3.1 Benchmark Datasets

High-quality datasets are essential for academic research. Here, we introduce some popular real-world HG datasets, which can be divided into three categories: academic networks, business networks, and film networks.

- **DBLP**[1] This is a network that reflects the relationship between authors and papers. There are four types of nodes: author, paper, term, and venue.
- **Aminer**[2] This academic network is similar to DBLP, but with two additional node types: keyword and conference.
- **Yelp**[3] This is a social media network, including five types of nodes: user, business, compliment, city, and category.
- **Amazon**[4] This is an E-commercial network, which records the interactive information between users and products, including co-viewing, co-purchasing, etc.
- **IMDB**[5] This is a film rating network, recording the preferences of users on different films. Each film contains its directors, actors, and genre.
- **Douban**[6] This network is similar to IMDB, but it contains more user information, such as the group and location of the users.

2.3.2 Open-Source Code

Source code is important for researchers to reproduce the corresponding method. In Table 2.1, we refer to the related papers of the datasets. Furthermore, we collect the source code of the related papers and list them in Table 2.1. Besides, we provide some commonly used website about graph representation.

- Stanford Network Analysis Project (SNAP): It is a network analysis and graph mining library, which contains different types of networks and multiple network analysis tools. The address is http://snap.stanford.edu/.
- ArnetMiner (AMiner) [46]: In the early days, it was an academic network used for data mining. Now it has become a comprehensive academic system that provides a variety of academic resources. The address is https://www.aminer.cn/.

---

[1] http://dblp.uni-trier.de.

[2] https://www.aminer.cn.

[3] http://www.yelp.com/datasetchallenge/.

[4] http://jmcauley.ucsd.edu/data/amazon.

[5] https://grouplens.org/datasets/movielens/100k/.

[6] http://movie.douban.com/.

Table 2.1 Source code of related papers

Method	Source code	Programing platform
metapath2vec [8]	https://github.com/apple2373/metapath2vec	Tensorflow
metagraph2vec [65]	https://github.com/daokunzhang/MetaGraph2Vec	C++
AspEM [43]	https://github.com/ysyushi/aspem	Python
HEER [44]	https://github.com/GentleZhu/HEER	Python
HEBE [15]	https://github.com/olittle/Hebe	C++
JUST [26]	https://github.com/eXascaleInfolab/JUST	Python
HIN2vec [11]	https://github.com/csiesheep/hin2vec	Python & C++
BHIN2vec [29]	https://github.com/sh0416/BHIN2VEC	Pytorch
HHNE [54]	https://github.com/ydzhang-stormstout/HHNE	C++
HeRec [42]	https://github.com/librahu/HERec	Python
MNE [66]	https://github.com/HKUST-KnowComp/MNE	Python
PTE [47]	https://github.com/mnqu/PTE	C++
RHINE [34]	https://github.com/rootlu/RHINE	Pytorch
HAN [53]	https://github.com/Jhy1993/HAN	Tensorflow
MAGNN [13]	https://github.com/cynricfu/MAGNN	Pytorch
HetSANN [19]	https://github.com/didi/hetsann	Tensorflow
HGT [25]	https://github.com/acbull/pyHGT	Pytorch
HetGNN [68]	https://github.com/chuxuzhang/KDD2019_HetGNN	Pytorch
GATNE [2]	https://github.com/THUDM/GATNE	Pytorch
RSHN [77]	https://github.com/CheriseZhu/RSHN	Pytorch
RGCN [38]	https://github.com/tkipf/relational-gcn	Tensorflow
IntentGC [74]	https://github.com/peter14121/intentgc-models	Python
MEIRec [9]	https://github.com/googlebaba/KDD2019-MEIRec	Tensorflow
GNUD [24]	https://github.com/siyongxu/GNUD	Tensorflow
FMG [73]	https://github.com/HKUST-KnowComp/FMG	Python & C++
HeteRec [62]	https://github.com/mukulg17/HeteRec	R
DHNE [49]	https://github.com/tadpole/DHNE	Tensorflow
SHNE [69]	https://github.com/chuxuzhang/WSDM2019_SHNE	Pytorch
NSHE [75]	https://github.com/Andy-Border/NSHE	Pytorch
PAHNE [4]	https://github.com/chentingpc/GuidedHeteEmbedding	C++
Camel [64]	https://github.com/chuxuzhang/WWW2018_Camel	Tensorflow
TaPEm [36]	https://github.com/pcy1302/TapEM	Python
HeGAN [22]	https://github.com/librahu/HeGAN	Tensorflow
DyHNE [56]	https://github.com/rootlu/DyHNE	Python & Matlab

- Open Academic Society (OAS): It is an open and expanding knowledge graph for research and education, contributed by Microsoft Research and AMiner. It publishes Open Academic Graph (OAG), which unifies two billion-scale academic graphs. The address is https://www.openacademic.ai/.
- HG Resources: It is a website focusing on heterogeneous graphs, which collects a series of papers on HG and divides them into different categories, including

classification, clustering, and embedding. Code and datasets of the popular methods are also provided. The address is http://shichuan.org/.

2.3.3 Available Tools

Open-source platforms and toolkits can help researchers build the workflow of graph representation quickly and easily. Generally, there are many toolkits designed for homogeneous graph. For example, OpenNE[7] and CogDL.[8] However, the toolkits and platforms for heterogeneous graph are rarely mentioned. To bring this gap, we summarize the popular toolkits and platforms that are suitable for heterogeneous graph.

- AliGraph: It is an industrial-grade machine learning platform for graph data, supporting the calculation of hundreds of millions of nodes and edges. Besides, it considers the characteristics of real-world industrial graph data, i.e., large-scale, heterogeneous, attributed, and dynamic, and makes special optimizations. One instance can be found in https://www.aliyun.com/product/bigdata/product.
- Deep Graph Library (DGL): It is an open-source deep learning platform for graph data, which designs its own data structures and implements many popular methods. Specifically, it provides independent Application Programming Interfaces (APIs) for homogeneous graph, heterogeneous graph, and knowledge graph. One instance can be found in https://www.dgl.ai/.
- Pytorch Geometric: It is a geometric deep learning extension library for Pytorch. Specifically, it focuses on the methods for deep learning on graphs and other irregular structures. Same as DGL, it also has its own data structures and operators. One instance can be found in https://pytorch-geometric.readthedocs. io/en/latest/.
- OpenHINE: It is an open-source toolkit for HGR, which implements many popular HGR methods with a unified data interface. One instance can be found in https://github.com/BUPT-GAMMA/OpenHINE.
- OpenHGNN: It is an open-source toolkit for Heterogeneous Graph Neural Network (HGNN) based on Deep Graph Library (DGL) and PyTorch. It integrates many SOTA models of HGNN and offers easy-to-use APIs, which can be found in https://github.com/BUPT-GAMMA/OpenHGNN.

---

[7] https://github.com/thunlp/OpenNE.
[8] https://github.com/THUDM/cogdl.

References

1. Cao, S., Lu, W., Xu, Q.: Grarep: Learning graph representations with global structural information. In: CIKM '15: Proceedings of the 24th ACM International on Conference on Information and Knowledge Management, pp. 891–900. ACM, New York (2015)
2. Cen, Y., Zou, X., Zhang, J., Yang, H., Zhou, J., Tang, J.: Representation learning for attributed multiplex heterogeneous network. In: The 25th ACM SIGKDD Conference on Knowledge Discovery and Data Mining (KDD '19). ACM, New York (2019)
3. Chang, S., Han, W., Tang, J., Qi, G.J., Aggarwal, C.C., Huang, T.S.: Heterogeneous network embedding via deep architectures. In: KDD '15: Proceedings of the 21th ACM SIGKDD International Conference on Knowledge Discovery and Data Mining, pp. 119–128. ACM, New York (2015)
4. Chen, T., Sun, Y.: Task-guided and path-augmented heterogeneous network embedding for author identification. In: WSDM '17: Proceedings of the Tenth ACM International Conference on Web Search and Data Mining, pp. 295–304. ACM, New York (2017)
5. Chen, H., Yin, H., Wang, W., Wang, H., Nguyen, Q.V.H., Li, X.: Pme: projected metric embedding on heterogeneous networks for link prediction. In: ACM International Conference on Knowledge Discovery and Data Mining 2018, pp. 1177–1186. ACM, New York (2018)
6. Chung, J., Gulcehre, C., Cho, K., Bengio, Y.: Empirical evaluation of gated recurrent neural networks on sequence modeling. arXiv preprint arXiv:1412.3555 (2014)
7. Cui, P., Wang, X., Pei, J., Zhu, W.: A survey on network embedding. IEEE Trans. Knowl. Data Eng. 31(5), 833–852 (2018)
8. Dong, Y., Chawla, N.V., Swami, A.: metapath2vec: scalable representation learning for heterogeneous networks. In: Proceedings of the 23rd ACM SIGKDD International Conference on Knowledge Discovery and Data Mining, pp. 135–144. ACM, New York (2017)
9. Fan, S., Zhu, J., Han, X., Shi, C., Hu, L., Ma, B., Li, Y.: Metapath-guided heterogeneous graph neural network for intent recommendation. In: KDD '19: Proceedings of the 25th ACM SIGKDD International Conference on Knowledge Discovery & Data Mining, pp. 2478–2486 (2019)
10. Fan, Y., Zhang, Y., Hou, S., Chen, L., Ye, Y., Shi, C., Zhao, L., Xu, S.: iDev: enhancing social coding security by cross-platform user identification between github and stack overflow. In: Proceedings of the Twenty-Eighth International Joint Conference on Artificial Intelligence, pp. 2272–2278 (2019)
11. Fu, T.Y., Lee, W.C., Lei, Z.: Hin2vec: explore meta-paths in heterogeneous information networks for representation learning. In: CIKM '17: Proceedings of the 2017 ACM on Conference on Information and Knowledge Management, pp. 1797–1806. ACM, New York (2017)
12. Fu, Y., Xiong, Y., Yu, P.S., Tao, T., Zhu, Y.: Metapath enhanced graph attention encoder for hins representation learning. In: BigData, pp. 1103–1110. IEEE, Piscataway (2019)
13. Fu, X., Zhang, J., Meng, Z., King, I.: MAGNN: Metapath aggregated graph neural network forheterogeneous graph embedding. In: WWW '20: Proceedings of The Web Conference 2020 (2020)
14. Grover, A., Leskovec, J.: node2vec: Scalable feature learning for networks. In: KDD, pp. 855–864. ACM, New York (2016)
15. Gui, H., Liu, J., Tao, F., Jiang, M., Norick, B., Han, J.: Large-scale embedding learning in heterogeneous event data. In: ICDM, pp. 907–912. IEEE, Piscataway (2016)
16. Han, X., Shi, C., Wang, S., Yu, P.S., Song, L.: Aspect-level deep collaborative filtering via heterogeneous information networks. In: Proceedings of the Twenty-Seventh International Joint Conference on Artificial Intelligence, pp. 3393–3399 (2018)
17. He, Y., Song, Y., Li, J., Ji, C., Peng, J., Peng, H.: Hetespaceywalk: a heterogeneous spacey random walk for heterogeneous information network embedding. In: 28th ACM International Conference on Information and Knowledge Management, CIKM, pp. 639–648. ACM, New York (2019)

18. Helgason, S.: Differential Geometry, Lie Groups, and Symmetric Spaces. Academic Press, Cambridge (1979)
19. Hong, H., Guo, H., Lin, Y., Yang, X., Li, Z., Ye, J.: An attention-based graph neural network for heterogeneous structural learning. In: Proceedings of AAAI Conference (AAAI'20) (2020)
20. Hou, S., Fan, Y., Zhang, Y., Ye, Y., Lei, J., Wan, W., Wang, J., Xiong, Q., Shao, F.: αcyber: enhancing robustness of android malware detection system against adversarial attacks on heterogeneous graph based model. In: CIKM '19: Proceedings of the 28th ACM International Conference on Information and Knowledge Management, pp. 609–618 (2019)
21. Hu, B., Shi, C., Zhao, W.X., Yu, P.S.: Leveraging meta-path based context for top-n recommendation with a neural co-attention model. In: KDD '18: Proceedings of the 24th ACM SIGKDD International Conference on Knowledge Discovery & Data Mining, pp. 1531–1540. ACM, New York (2018)
22. Hu, B., Fang, Y., Shi, C.: Adversarial learning on heterogeneous information networks. In: KDD '19: Proceedings of the 25th ACM SIGKDD Conference On Knowledge Discovery and Data Mining, pp. 120–129. ACM, New York (2019)
23. Hu, L., Li, C., Shi, C., Yang, C., Shao, C.: Graph neural news recommendation with long-term and short-term interest modeling. Inf. Process. Manag. **57**(2), 102,142 (2020)
24. Hu, L., Xu, S., Li, C., Yang, C., Shi, C., Duan, N., Xie, X., Zhou, M.: Graph neural news recommendation with unsupervised preference disentanglement. In: Proceedings of the 58th Annual Meeting of the Association for Computational Linguistics (2020)
25. Hu, Z., Dong, Y., Wang, K., Sun, Y.: Heterogeneous graph transformer. In: Proceedings of The Web Conference 2020 (2020)
26. Hussein, R., Yang, D., Cudré-Mauroux, P.: Are meta-paths necessary?: revisiting heterogeneous graph embeddings. In: Proceedings of the 27th ACM International Conference on Information and Knowledge Management, pp. 437–446 (2018)
27. Jamali, M., Lakshmanan, L.V.S.: Heteromf: recommendation in heterogeneous information networks using context dependent factor models. In: International World Wide Web Conferences Steering Committee, pp. 643–654. ACM, New York (2013)
28. Jeh, G., Widom, J.: Scaling personalized web search. In: WWW '03: Proceedings of the 12th International Conference on World Wide Web, pp. 271–279 (2003)
29. Lee, S., Park, C., Yu, H.: Bhin2vec: balancing the type of relation in heterogeneous information network. In: CIKM '19: Proceedings of the 28th ACM International Conference on Information and Knowledge Management, pp. 619–628 (2019)
30. Li, C., Hu, L., Shi, C., Song, G., Lu, Y.: Sequence-aware heterogeneous graph neural collaborative filtering. In: Proceedings of the 2021 SIAM International Conference on Data Mining (SDM) (2021)
31. Liu, Z., Zheng, V.W., Zhao, Z., Li, Z., Yang, H., Wu, M., Ying, J.: Interactive paths embedding for semantic proximity search on heterogeneous graphs. In: Proceedings of the 24th ACM SIGKDD International Conference on Knowledge Discovery & Data Mining, pp. 1860–1869. ACM, New York (2018)
32. Liu, Z., Zheng, V.W., Zhao, Z., Yang, H., Chang, K.C., Wu, M., Ying, J.: Subgraph-augmented path embedding for semantic user search on heterogeneous social network. In: 27th International World Wide Web, WWW 2018, pp. 1613–1622. ACM, New York (2018)
33. Liu, Z., Zheng, V.W., Zhao, Z., Zhu, F., Chang, K.C.C., Wu, M., Ying, J.: Distance-aware dag embedding for proximity search on heterogeneous graphs. In: 32nd AAAI Conference on Artificial Intelligence, AAAI 2018 (2018)
34. Lu, Y., Shi, C., Hu, L., Liu, Z.: Relation structure-aware heterogeneous information network embedding. In: Proceedings of the AAAI Conference on Artificial Intelligence (2019)
35. Matsuno, R., Murata, T.: MELL: effective embedding method for multiplex networks. In: WWW '18: Proceedings of The Web Conference 2018, pp. 1261–1268. ACM, New York (2018)
36. Park, C., Kim, D., Zhu, Q., Han, J., Yu, H.: Task-guided pair embedding in heterogeneous network. In: Proceedings of the 28th ACM International Conference on Information and Knowledge Management, pp. 489–498 (2019)

37. Perozzi, B., Al-Rfou, R., Skiena, S.: Deepwalk: online learning of social representations. In: Proceedings of the 20th ACM SIGKDD International Conference on Knowledge Discovery and Data Mining, pp. 701–710 (2014)
38. Schlichtkrull, M., Kipf, T.N., Bloem, P., Van Den Berg, R., Titov, I., Welling, M.: Modeling relational data with graph convolutional networks. In: European Semantic Web Conference, pp. 593–607. Springer, Berlin (2018)
39. Shi, C., Kong, X., Huang, Y., Yu, P.S., Wu, B.: Hetesim: a general framework for relevance measure in heterogeneous networks. IEEE Trans. Knowl. Data Eng. **26**(10), 2479–2492 (2014)
40. Shi, C., Zhang, Z., Luo, P., Yu, P.S., Yue, Y., Wu, B.: Semantic path based personalized recommendation on weighted heterogeneous information networks. In: CIKM '15: Proceedings of the 24th ACM International on Conference on Information and Knowledge Management, pp. 453–462. ACM, New York (2015)
41. Shi, C., Li, Y., Zhang, J., Sun, Y., Yu, P.S.: A survey of heterogeneous information network analysis. IEEE Trans. Knowl. Data Eng. **29**(1), 17–37 (2017)
42. Shi, C., Hu, B., Zhao, W.X., Yu, P.S.: Heterogeneous information network embedding for recommendation. IEEE Trans. Knowl. Data Eng. **31**(2), 357–370 (2018)
43. Shi, Y., Gui, H., Zhu, Q., Kaplan, L., Han, J.: Aspem: Embedding learning by aspects in heterogeneous information networks. In: 2018 SIAM International Conference on Data Mining, SDM 2018, pp. 144–152. SIAM, Philadelphia (2018)
44. Shi, Y., Zhu, Q., Guo, F., Zhang, C., Han, J.: Easing embedding learning by comprehensive transcription of heterogeneous information networks. In: Proceedings of the 24th ACM SIGKDD International Conference on Knowledge Discovery and Data Mining, pp. 2190–2199. ACM, New York (2018)
45. Sun, Y., Han, J., Yan, X., Yu, P.S., Wu, T.: Pathsim: Meta path-based top-k similarity search in heterogeneous information networks. Proc. VLDB Endow. **4**(11), 992–1003 (2011)
46. Tang, J., Zhang, J., Yao, L., Li, J., Zhang, L., Su, Z.: Arnetminer: extraction and mining of academic social networks. In: Proceedings of the 14th ACM SIGKDD International Conference on Knowledge Discovery and Data Mining, pp. 990–998. ACM, New York (2008)
47. Tang, J., Qu, M., Mei, Q.: PTE: predictive text embedding through large-scale heterogeneous text networks. In: Proceedings of the 21th ACM SIGKDD International Conference on Knowledge Discovery and Data Mining, pp. 1165–1174. ACM, New York (2015)
48. Tang, J., Qu, M., Wang, M., Zhang, M., Yan, J., Mei, Q.: Line: large-scale information network embedding. In: Proceedings of the 24th International Conference on World Wide Web, pp. 1067–1077 (2015)
49. Tu, K., Cui, P., Wang, X., Wang, F., Zhu, W.: Structural deep embedding for hyper-networks. In: Thirty-Second AAAI Conference on Artificial Intelligence (2018)
50. Wang, D., Cui, P., Zhu, W.: Structural deep network embedding. In: Proceedings of the 22nd ACM SIGKDD International Conference on Knowledge Discovery and Data Mining, pp. 1225–1234 (2016)
51. Wang, X., Cui, P., Wang, J., Pei, J., Zhu, W., Yang, S.: Community preserving network embedding. In: Thirty-First AAAI Conference on Artificial Intelligence (2017)
52. Wang, H., Wang, J., Wang, J., Zhao, M., Zhang, W., Zhang, F., Xie, X., Guo, M.: Graphgan: graph representation learning with generative adversarial nets. In: Proceedings of the AAAI Conference on Artificial Intelligence (2018)
53. Wang, X., Ji, H., Shi, C., Wang, B., Ye, Y., Cui, P., Yu, P.S.: Heterogeneous graph attention network. In: The World Wide Web Conference, pp. 2022–2032. ACM, New York (2019)
54. Wang, X., Zhang, Y., Shi, C.: Hyperbolic heterogeneous information network embedding. In: Proceedings of the AAAI Conference on Artificial Intelligence (2019)
55. Wang, Z., Liu, H., Du, Y., Wu, Z., Zhang, X.: Unified embedding model over heterogeneous information network for personalized recommendation. In: Proceedings of the 28th International Joint Conference on Artificial Intelligence, pp. 3813–3819. AAAI Press, Palo Alto (2019)
56. Wang, X., Lu, Y., Shi, C., Wang, R., Cui, P., Mou, S.: Dynamic heterogeneous information network embedding with meta-path based proximity. IEEE Trans. Knowl. Data Eng. (2020)

57. Wang, R., Shi, C., Zhao, T., Wang, X., Ye, F.Y.: Heterogeneous information network embedding with adversarial disentangler. IEEE Trans. Knowl. Data Eng. (2021)
58. Wang, X., Liu, N., Han, H., Shi, C.: Self-supervised heterogeneous graph neural network with co-contrastive learning. In: Proceedings of the 27th ACM SIGKDD Conference on Knowledge Discovery & Data Mining (2021)
59. Xu, L., Wei, X., Cao, J., Yu, P.S.: Embedding of embedding (EOE): joint embedding for coupled heterogeneous networks. In: WSDM '17: Proceedings of the Tenth ACM International Conference on Web Search and Data Mining, pp. 741–749. ACM, New York (2017)
60. Xu, Y., Zhu, Y., Shen, Y., Yu, J.: Learning shared vertex representation in heterogeneous graphs with convolutional networks for recommendation. In: Proceedings of the Twenty-Eighth International Joint Conference on Artificial Intelligence, pp. 4620–4626 (2019)
61. Xue, H., Yang, L., Jiang, W., Wei, Y., Hu, Y., Lin, Y.: Modeling dynamic heterogeneous network for link prediction using hierarchical attention with temporal RNN. Preprint. arXiv:2004.01024 (2020)
62. Yu, X., Ren, X., Sun, Y., Sturt, B., Khandelwal, U., Gu, Q., Norick, B., Han, J.: Recommendation in heterogeneous information networks with implicit user feedback. In: RecSys, pp. 347–350. ACM, New York (2013)
63. Yun, S., Jeong, M., Kim, R., Kang, J., Kim, H.J.: Graph transformer networks. In: Advances in Neural Information Processing Systems, pp. 11960–11970 (2019)
64. Zhang, C., Huang, C., Yu, L., Zhang, X., Chawla, N.V.: Camel: Content-aware and meta-path augmented metric learning for author identification. In: Proceedings of the 2018 World Wide Web Conference, pp. 709–718 (2018)
65. Zhang, D., Yin, J., Zhu, X., Zhang, C.: MetaGraph2Vec: complex semantic path augmented heterogeneous network embedding. In: Pacific-Asia Conference on Knowledge Discovery and Data Mining 2018, pp. 196–208. Springer, Berlin (2018)
66. Zhang, H., Qiu, L., Yi, L., Song, Y.: Scalable multiplex network embedding. In: Proceedings of the Twenty-Seventh International Joint Conference on Artificial Intelligence, vol. 18, pp. 3082–3088 (2018)
67. Zhang, Z., Cui, P., Wang, X., Pei, J., Yao, X., Zhu, W.: Arbitrary-order proximity preserved network embedding. In: Proceedings of the 24th ACM SIGKDD International Conference on Knowledge Discovery & Data Mining, pp. 2778–2786. ACM, New York (2018)
68. Zhang, C., Song, D., Huang, C., Swami, A., Chawla, N.V.: Heterogeneous graph neural network. In: Proceedings of the 25th ACM SIGKDD International Conference on Knowledge Discovery & Data Mining, pp. 793–803. ACM, New York (2019)
69. Zhang, C., Swami, A., Chawla, N.V.: Shne: Representation learning for semantic-associated heterogeneous networks. In: Proceedings of the Twelfth ACM International Conference on Web Search and Data Mining, pp. 690–698 (2019)
70. Zhang, Y., Fan, Y., Song, W., Hou, S., Ye, Y., Li, X., Zhao, L., Shi, C., Wang, J., Xiong, Q.: Your style your identity: Leveraging writing and photography styles for drug trafficker identification in darknet markets over attributed heterogeneous information network. In: WWW '19: The World Wide Web Conference, pp. 3448–3454 (2019)
71. Zhang, Y., Fan, Y., Ye, Y., Zhao, L., Shi, C.: Key player identification in underground forums over attributed heterogeneous information network embedding framework. In: CIKM '19: Proceedings of the 28th ACM International Conference on Information and Knowledge Management, pp. 549–558 (2019)
72. Zhao, H., Yao, Q., Li, J., Song, Y., Lee, D.L.: Meta-graph based recommendation fusion over heterogeneous information networks. In: Proceedings of the 23rd ACM SIGKDD International Conference on Knowledge Discovery and Data Mining, pp. 635–644. ACM, New York (2017)
73. Zhao, H., Zhou, Y., Song, Y., Lee, D.L.: Motif enhanced recommendation over heterogeneous information network. In: Proceedings of the 28th ACM International Conference on Information and Knowledge Management, pp. 2189–2192 (2019)

74. Zhao, J., Zhou, Z., Guan, Z., Zhao, W., Ning, W., Qiu, G., He, X.: IntentGC: a scalable graph convolution framework fusing heterogeneous information for recommendation. In: Proceedings of the 25th ACM SIGKDD International Conference on Knowledge Discovery & Data Mining, pp. 2347–2357 (2019)
75. Zhao, J., Wang, X., Shi, C., Liu, Z., Ye, Y.: Network schema preserving heterogeneous information network embedding. In: 29th International Joint Conference on Artificial Intelligence (2020)
76. Zhao, K., Bai, T., Wu, B., Wang, B., Zhang, Y., Yang, Y., Nie, J.: Deep adversarial completion for sparse heterogeneous information network embedding. In: Proceedings of the Web Conference 2020, pp. 508–518. ACM/IW3C2, New York (2020)
77. Zhu, S., Zhou, C., Pan, S., Zhu, X., Wang, B.: Relation structure-aware heterogeneous graph neural network. In: IEEE International Conference On Data Mining (2019)

Part I
Techniques

Chapter 3
Structure-Preserved Heterogeneous Graph Representation

Abstract Heterogeneous graph (HG) contains various types of nodes or links, which are highly correlated and present intricate structures due to different links. These structures reflect the crucial factors of topology. Therefore encoding meaningful structures is a basic requirement to obtain node representations with high quality. So far, some representative structures have been studied in an HG, from one-hop edges to high-order local structures, such as meta-paths and network schema. In this chapter, we will introduce several works focusing on structure preservation. By capturing respective structures, they successfully depict the rich semantics and complex heterogeneity, and effectively support downstream tasks.

3.1 Introduction

One basic requirement of graph representation learning is to preserve the graph structure properly. A lot of early works focus on the structural preservation of homogeneous graph representation learning, which focus on how to maintain the second-order [44], higher-order [3], and community structure [47] in homogeneous graphs, and therefore have achieved good results in downstream tasks such as node classification and link prediction. Compared with homogeneous graphs, there are more challenges in maintaining structural information for heterogeneous graphs (HGs), because the latter contain multiple types of nodes and multiple relations among nodes.

Traditional heterogeneous graph modeling often uses meta-paths [39] to measure the structural similarity of nodes. However, these meta-path based methods cannot calculate the similarity of nodes without meta-path connections, which greatly limits the application scenarios of these methods. Inspired by homogeneous graph representation learning methods, many heterogeneous graph representation learning methods have recently been proposed, and the latent heterogeneous network embeddings have been further applied to various network mining tasks, such as node classification [18], clustering [38, 41], and similarity search [40, 59]. In contrast to conventional meta-path based methods [39], the advantage of latent-

© The Author(s), under exclusive license to Springer Nature Singapore Pte Ltd. 2022
C. Shi et al., *Heterogeneous Graph Representation Learning and Applications*,
Artificial Intelligence: Foundations, Theory, and Algorithms,
https://doi.org/10.1007/978-981-16-6166-2_3

space representation learning lies in its ability to model structural similarities between nodes without connected meta-paths.

In this chapter, we will introduce several representative works which preserve structural information in node representations. In Sects. 3.2 and 3.3, we will first introduce a **H**eterogeneous graph **E**mbedding based **Rec**ommendation model (named **HERec**) [37] and a **Neu**ral network-based **A**spect-level **C**ollaborative **F**iltering model (named **NeuACF**) [11], which combine meta-path with two classic homogeneous graph representation learning methods separately, that is, skip-gram algorithm and attention-based deep neural network. In Sect. 3.4, we will introduce a **R**elation-structure aware **H**eterogeneous **I**nformation **N**etwork **E**mbedding model (named **RHINE**) [23], which divides the relations in heterogeneous graphs into two categories based on mathematical analysis, thereby modeling the structural information in the heterogeneous graphs in a more fine-grained manner. In Sect. 3.5, we will introduce a novel **N**etwork **S**chema preserving **H**eterogeneous graph **E**mbedding method (named **NSHE**) [62], which uses the network schema, that is, the meta template of the heterogeneous graph, to model the structure information more completely, and get rid of the dependence on the hand-designed meta-paths.

3.2 Meta-Path Based Random Walk

3.2.1 Overview

In a graph, there exist many kinds of structures to depict the different relations between nodes, where neighborhood structure, describing the local structure of a node, is an important one. How to appropriately reserve the neighborhood structure in node representations is a fundamental problem. To solve the problem, a typical technique usually used is random walk, which is a process that starts from a target node and randomly chooses a neighbor of the last node as the current one [10, 28]. By exploiting this mechanism, multiple random sequences for nodes are generated, and each describes a local structure over original graph. We can view each sequence as a sentence, and nodes in the sequence as words. In this way, word2vec-based framework proposed in natural language processing (NLP) is adopted to learn node representations, which captures the proximity between neighbor nodes in a window with fixed size.

However, these works focus on representation learning for homogeneous graph, only considering singular type of nodes and relationships. In the field of HG, it is a challenge to effectively combine random walk strategy with complex heterogeneity to capture the local semantic structure. To tackle the challenge, **HG E**mbedding based **Rec**ommendation (named **HERec**) [37] designs a novel mechanism to perform random walk along meta-paths, also called meta-path based random walk, considering that meta-paths are sufficient to depict rich semantics.

Different from random walk in homogeneous graph, HERec constrains generated node sequence aligned with pre-defined meta-path. For example, for a meta-path Author–Paper–Author (APA), if type of the last selected node i is A, then HERec chooses a neighbor of i with type P. The objective of HERec is to simultaneously learn representations of multiple types of nodes and preserve both the structures semantic of a given HG. Due to its effectiveness, HERec, combined with classic matrix factorization framework, is also applied in recommender systems. A basic recommender system usually consists of users and items and aims to help users discover latent items of interest. By the designed framework, HERec effectively extracts semantic and local information from interactions between users and items, and gives a good performance in downstream tasks.

3.2.2 The HERec Model

3.2.2.1 Model Framework

In this section, we provide more details about HERec here. Specifically, HERec is designed for the HG in a recommender system, which consists of two major components. The first part is to learn user/item embeddings from HG, namely HG embedding (Fig. 3.1b). The second part is to involve the learnt embeddings into classic matrix factorization framework, namely recommendation (Fig. 3.1c). Next, we will present detailed illustration of HERec.

3.2.2.2 Heterogeneous Graph Embedding

Given an HG $\mathcal{G} = \{\mathcal{V}, \mathcal{E}\}$, the goal is to learn a low-dimensional embedding $e_v \in \mathbb{R}^d$ for each node $v \in \mathcal{V}$, which is much easier to be used and integrated in subsequent

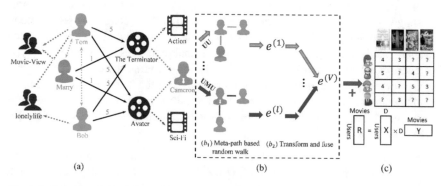

Fig. 3.1 The schematic illustration of the proposed HERec approach. (**a**) An example of HIN (HG). (**b**) HIN (HG) embedding. (**c**) Recommendation

procedures, compared with meta-path based similarity. Inspired by deepwalk [28], a well-known embedding method for homogeneous graph, HERec designs a novel random walk mechanism over HG to learn embeddings.

Meta-Path Based Random Walk　To capture the complex heterogeneity, meta-path is usually adopted to characterize the semantic patterns for HGs [39]. Therefore, HERec also employs a meta-path based random walk method to generate node sequences. Given a heterogeneous graph $\mathcal{G} = \{\mathcal{V}, \mathcal{E}\}$ and a meta-path $\rho : A_1 \xrightarrow{R_1} \cdots A_t \xrightarrow{R_t} A_{t+1} \cdots \xrightarrow{R_l} A_{l+1}$, the walk path is generated according to the following distribution:

$$P(n_{t+1} = x | n_t = v, \rho) \tag{3.1}$$

$$= \begin{cases} \frac{1}{|\mathcal{N}^{A_{t+1}}(v)|}, & (v, x) \in \mathcal{E} \text{ and } \phi(x) = A_{t+1}; \\ 0, & \text{otherwise}, \end{cases}$$

where n_t is the t-th node in the walk, the type of v is A_t, $\phi(\cdot)$ is the object type mapping function, and $\mathcal{N}^{(A_{t+1})}(v)$ is the first-order neighbor set for node v with the type of A_{t+1}. A walk will follow the pattern of a meta-path repetitively until it reaches the pre-defined length.

Example 1　We still take Fig. 3.1a as an example, which represents a movie recommender system. Given a meta-path UMU, we can generate two sample walks (i.e. node sequences) by starting with the user *Tom*: (1) $\text{Tom}_{User} \to$ The $\text{Terminator}_{Movie} \to \text{Mary}_{User}$, and (2) $\text{Tom}_{User} \to \text{Avater}_{Movie} \to \text{Bob}_{User} \to$ The $\text{Terminator}_{Movie} \to \text{Mary}_{User}$. Similarly, given the meta-path $UMDMU$, we can also generate another node sequence: $\text{Tom}_{User} \to$ The $\text{Terminator}_{Movie} \to \text{Cameron}_{Director} \to \text{Avater}_{Movie} \to \text{Mary}_{User}$. It is intuitive to see that these meta-paths can lead to meaningful node sequences corresponding to different semantic relations.

Type Constraint and Filtering　Consider that in recommendation, only user and item should be more emphasized than other types of nodes. So, only meta-paths starting with *user type* or *item type* are selected. To further strengthen user and item, other types of nodes that are different from the starting type will be filtered out from the generated node sequences. In this way, the final sequences only contain one type of nodes (i.e. user of item). Here are two advantages. First, there is only one type, which relaxes the challenging goal of representing all the heterogeneous objects in a unified space. Second, given a fixed-length window, a node is able to utilize more homogeneous neighbors that are more likely to be relevant than others with different types.

Example 2　As shown in Fig. 3.2, only the meta-paths starting from *user type* or *item type* are derived, such as UMU, $UMDMU$, and MUM. Take the meta-path of UMU as an instance. We can generate a sampled sequence "$u_1 \to m_1 \to u_2 \to m_2 \to u_3 \to m_2 \to u_4$" according to Eq. 3.1. And then, we remove the movies and finally obtain a homogeneous node sequence "$u_1 \to u_2 \to u_3 \to u_4$".

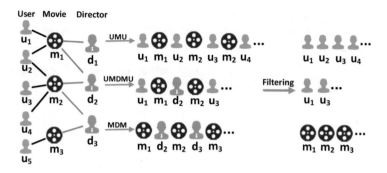

Fig. 3.2 An illustrative example of the proposed meta-path based random walk. First perform random walks guided by some selected meta-paths, and then filter out nodes of types that are different from the starting type

After this step, the next focus will be how to learn effective embeddings for nodes in sequences.

Optimization Objective Given a processed node sequence, HERec defines the neighborhood for node u based on co-occurrence in a fixed-length window in the sequence, denoted as $\mathcal{N}_u$. Following node2vec [10], the final embeddings of nodes are learnt by optimizing the following objective:

$$\max_f \sum_{u \in \mathcal{V}} \log Pr(\mathcal{N}_u | f(u)), \qquad (3.2)$$

where $f : \mathcal{V} \to \mathbb{R}^d$ is a function mapping each node to d-dimensional embedding space. $Pr(\mathcal{N}_u | f(u))$ measures probability of u's neighbors given u's embedding. The embedding mapping function $f(\cdot)$ is trained by utilizing stochastic gradient descent (SGD). A major difference between previous method [7] and HERec lies in the construction of $\mathcal{N}_u$, where HERec selects homogeneous neighbors using meta-path based random walks.

Embedding Fusion For one meta-path l and corresponding node sequence, after optimizing via Eq. 3.2, given a node $v \in \mathcal{V}$, we can obtain its embedding $e_v^{(l)}$ based on meta-path l. In a similar way, we can obtain a set of embeddings $\{e_v^{(l)}\}_{l=1}^{|\mathcal{P}|}$, where $\mathcal{P}$ denotes the set of meta-paths. Then, we propose to use a general function $g(\cdot)$, which aims to fuse the learnt node embeddings to get $e_u^{(U)}$ and $e_i^{(I)}$ for users and items, respectively:

$$e_u^{(U)} \leftarrow g(\{e_u^{(l)}\}), \qquad (3.3)$$
$$e_i^{(I)} \leftarrow g(\{e_i^{(l)}\}).$$

There are three fusion functions that can be used to fuse these embeddings. We discuss the cases for users below. The functions for items are similar and can be derived accordingly.

- *Simple linear fusion.* We simply average the results from different meta-paths after transforming the embeddings with MLP.

$$g\left(\{e_u^{(l)}\}\right) = \frac{1}{|\mathcal{P}|} \sum_{l=1}^{|\mathcal{P}|} \left(\mathbf{M}^{(l)} e_u^{(l)} + b^{(l)}\right), \qquad (3.4)$$

where $\mathbf{M}^{(l)} \in \mathbb{R}^{D \times d}$ and $b^{(l)} \in \mathbb{R}^D$ are the transformation matrix and bias vector w.r.t. the l-th meta-path.

- *Personalized linear fusion.* We further assign each user with a weight vector on the meta-paths, representing the user's personalized preference for each meta-path.

$$g\left(\{e_u^{(l)}\}\right) = \sum_{l=1}^{|\mathcal{P}|} w_u^{(l)} \left(\mathbf{M}^{(l)} e_u^{(l)} + b^{(l)}\right), \qquad (3.5)$$

where $w_u^{(l)}$ is the learnt preference weight of user u over the l-th meta-path.

- *Personalized non-linear fusion.* Moreover, we use non-linear function to enhance the fusion ability.

$$g(\{e_u^{(l)}\}) = \sigma \left(\sum_{l=1}^{|\mathcal{P}|} w_u^{(l)} \sigma \left(\mathbf{M}^{(l)} e_u^{(l)} + b^{(l)}\right) \right), \qquad (3.6)$$

where $\sigma(\cdot)$ is a non-linear function, i.e., sigmoid function or others.

Among the three fusion strategies, HERec demonstrates that personalization and non-linearity are essential for boosting the results. The parameters involved in $g(\cdot)$ are optimized together with the recommendation model, which is introduced in the next section.

3.2.2.3 Integrating Matrix Factorization with Fused HG Embedding for Recommendation

Rating predictor is a component used to predict the ratings of the users on the items. In MF, the rating $r_{u,i}$ of a user u on an item i is simply defined as follows:

$$\widehat{r_{u,i}} = \mathbf{x}_u^\top \cdot \mathbf{y}_i, \qquad (3.7)$$

where $\mathbf{x}_u \in \mathbb{R}^D$ and $\mathbf{y}_i \in \mathbb{R}^D$ denote the latent factors corresponding to user u and item i by factorizing the user–item rating matrix. Since we have also obtained the embeddings for user u and item i, we can further incorporate them into the rating predictor as below:

$$\widehat{r_{u,i}} = \mathbf{x}_u^\top \cdot \mathbf{y}_i + \alpha \cdot e_u^{(U)^\top} \cdot \boldsymbol{\gamma}_i^{(I)} + \beta \cdot \boldsymbol{\gamma}_u^{(U)^\top} \cdot e_i^{(I)}, \qquad (3.8)$$

where $\boldsymbol{\gamma}_u^{(U)}$ and $\boldsymbol{\gamma}_i^{(I)}$ are user-specific and item-specific latent factors to pair with the HG embeddings $e_i^{(I)}$ and $e_u^{(U)}$ respectively, and α and β are the tuning parameters to integrate the three terms.

The overall objective is formulated as follows:

$$\pounds = \sum_{\langle u,i,r_{u,i}\rangle \in \mathcal{R}} (r_{u,i} - \widehat{r_{u,i}})^2 + \lambda \sum_u (\|\mathbf{x}_u\|_2 + \|\mathbf{y}_i\|_2$$
$$+ \|\boldsymbol{\gamma}_u^{(U)}\|_2 + \|\boldsymbol{\gamma}_i^{(I)}\|_2 + \|\boldsymbol{\Theta}^{(U)}\|_2 + \|\boldsymbol{\Theta}^{(I)}\|_2), \qquad (3.9)$$

where $\widehat{r_{u,i}}$ is the predicted rating using Eq. 3.8 by HERec, λ is the regularization parameter, and $\boldsymbol{\Theta}^{(U)}$ and $\boldsymbol{\Theta}^{(I)}$ are the parameters of the function $g(\cdot)$ for users and items, respectively. SGD is utilized to efficiently train HERec.

3.2.3 Experiments

3.2.3.1 Experimental Settings

Datasets Three real datasets are used to perform downstream tasks, consisting of Douban Movie dataset[1] from the movie domain, Douban Book dataset[2] from the book domain, and Yelp dataset[3] from the business domain. The detailed descriptions of the three datasets are shown in Table 3.1. It should be pointed out that these three datasets also have different rating sparsity degrees: The Yelp dataset is very sparse, while the Douban Movie dataset is much denser.

Baselines To demonstrate the effectiveness, HERec is compared with three categories of baselines: classic MF based rating prediction methods { PMF [27], $SoMF$ [24] }, and HG based recommendation methods { FM_{HIN} [30], $HeteMF$ [58], $SemRec$ [35], and DSR [63] }. Moreover, there are two variants of HERec: HERec$_{dw}$ and HERec$_{mp}$, where the difference is the way of random walk during generating node sequences. The former adopts deepwalk [28] to obtain

---

[1] http://movie.douban.com.

[2] http://book.douban.com.

[3] http://www.yelp.com/dataset-challenge.

node embeddings, which ignores the heterogeneity of nodes. The later employs metapath2vec++ [7] to generate node embeddings, which does not filter out different types of nodes.

3.2.3.2 Effectiveness Experiments

In this experiment, we set different training ratios for each dataset, and use MAE and RMSE as metrics to evaluate the performance. We set the embedding dimension as 64 and randomly run experiments 10 times and report the average results shown in Table 3.2. Notice that, HERec adopts the personalized non-linear fusion function here.

From the results, we can draw the following conclusions: (1) HERec is consistently better than other baselines because it provides better information extraction (a new HGs embedding model) and utilization (an extended MF model). (2) HG based methods, especially FM_{HIN}, perform better than traditional MF based methods, which indicates the usefulness of the heterogeneous information. (3) On one hand, compared with $HERec_{mp}$, $HERec_{dw}$ performs much worse, which indicates HG embedding methods are important to HG based recommendation again. On the other hand, HERec outperforms $HERec_{mp}$, which demonstrates that it is more effective to perform task-specific HG embedding for improving the recommendation performance (e.g. only focus on user and item).

3.2.3.3 Cold-Start Prediction

Now, we check the performances of all the methods in cold-start prediction settings, where there are fewer rating records but heterogeneous context information is available. First according to the counts of users' rating records, they are divided into three groups, i.e. (0, 5], (5, 15], and (15, 30]. It is easy to see that the case for the first group is the most difficult, since users from this group have fewest rating records. Here, HERec is only compared with HG based recommendation, including SoMF, HeteMF, SemRec, DSR, and FM_{HIN}, and the results are shown in Fig. 3.3, which reports the improvement ratios w.r.t. PMF. Overall, all the comparison methods are better than PMF (i.e., show positive y-axis values). Moreover, HERec performs the best among all the methods, and the improvement over PMF becomes more significant for users with fewer rating records. The results indicate that HG based information is effective to improve the recommendation performance, and HERec can effectively utilize HG information in a more principled way.

The more detailed method description and experiment validation can be seen in [37].

Table 3.1 Statistics of the three datasets

Dataset (Density)	Relations (A–B)	Number of A	Number of B	Number of (A–B)	Ave. degrees of A	Ave. degrees of B	Meta-paths
Douban Movie (0.63%)	User–Movie	13, 367	12, 677	1, 068, 278	79.9	84.3	UMU, MUM
	User–User	2440	2294	4085	1.7	1.8	UMDMU, MDM
	User–Group	13, 337	2753	570, 047	42.7	207.1	UMAMU, MAM
	Movie–Director	10, 179	2449	11, 276	1.1	4.6	UMTMU, MTM
	Movie–Actor	11, 718	6311	33, 587	2.9	5.3	
	Movie–Type	12, 678	38	27, 668	2.2	728.1	
Douban Book (0.27%)	User–Book	13, 024	22, 347	792, 026	60.8	35.4	UBU, BUB
	User–User	12, 748	12, 748	169, 150	13.3	13.3	UBPBU, BPB
	Book–Author	21, 907	10, 805	21, 905	1.0	2.0	UBYBU, BYB
	Book–Publisher	21, 773	1815	21, 773	1.0	11.9	UBABU
	Book–Year	21, 192	64	21, 192	1.0	331.1	
Yelp (0.08%)	User–Business	16, 239	14, 284	198, 397	12.2	13.9	UBU, BUB
	User–User	10, 580	10, 580	158, 590	15.0	15.0	UBCiBU, BCiB
	User–Compliment	14, 411	11	76, 875	5.3	6988.6	UBCaBU, BCaB
	Business–City	14, 267	47	14, 267	1.0	303.6	
	Business–Category	14, 180	511	40, 009	2.8	78.3	

Table 3.2 Results of effectiveness experiments on three datasets. A smaller MAE or RMSE value indicates a better performance. For easier interpretation, the improvement of each method w.r.t. the PMF model is also reported. A larger improvement ratio indicates a better performance. Bolded values represent the best results

Dataset	Training	Metrics	PMF	SoMF	FM_{HIN}	HeteMF	SemRec	DSR	$HERec_{dw}$	$HERec_{mp}$	HERec
Douban Movie	80%	MAE	0.5741	0.5817	0.5696	0.5750	0.5695	0.5681	0.5703	**0.5515**	0.5519
		Improve		−1.32%	+0.78%	−0.16%	+0.80%	+1.04%	+0.66%	+3.93%	+3.86%
		RMSE	0.7641	0.7680	0.7248	0.7556	0.7399	0.7225	0.7446	0.7121	**0.7053**
		Improve		−0.07%	+5.55%	+1.53%	+3.58%	+5.85%	+2.97%	+7.20%	+8.09%
	60%	MAE	0.5867	0.5991	0.5769	0.5894	0.5738	0.5831	0.5838	0.5611	**0.5587**
		Improve		−2.11%	+1.67%	−0.46%	+2.19%	+0.61%	+0.49%	+4.36%	+4.77%
		RMSE	0.7891	0.7950	0.7842	0.7785	0.7551	0.7408	0.7670	0.7264	**0.7148**
		Improve		−0.75%	+0.62%	+1.34%	+4.30%	+6.12%	+2.80%	+7.94%	+9.41%
	40%	MAE	0.6078	0.6328	0.5871	0.6165	0.5945	0.6170	0.6073	0.5747	**0.5699**
		Improve		−4.11%	+3.40%	−1.43%	+2.18%	−1.51%	+0.08%	+5.44%	+6.23%
		RMSE	0.8321	0.8479	0.7563	0.8221	0.7836	0.7850	0.8057	0.7429	**0.7315**
		Improve		−1.89%	+9.10%	+1.20%	+5.82%	+5.66%	+3.17%	+10.71%	+12.09%
	20%	MAE	0.7247	0.6979	0.6080	0.6896	0.6392	0.6584	0.6699	0.6063	**0.5900**
		Improve		+3.69%	+16.10%	+4.84%	+11.79%	+9.14%	+7.56%	+16.33%	+18.59%
		RMSE	0.9440	0.9852	0.7878	0.9357	0.8599	0.8345	0.9076	0.7877	**0.7660**
		Improve		−4.36%	+16.55%	+0.88%	+8.91%	+11.60%	+3.86%	+16.56%	+18.86%
Douban Book	80%	MAE	0.5774	0.5756	0.5716	0.5740	0.5675	0.5740	0.5875	0.5591	**0.5502**
		Improve		+0.31%	+1.00%	+0.59%	+1.71%	+0.59%	−1.75%	+3.17%	+4.71%
		RMSE	0.7414	0.7302	0.7199	0.7360	0.7283	0.7206	0.7450	0.7081	**0.6811**
		Improve		+1.55%	+2.94%	+0.77%	+1.81%	+2.84%	−0.44%	+4.53%	+8.17%
	60%	MAE	0.6065	0.5903	0.5812	0.5823	0.5833	0.6020	0.6203	0.5666	**0.5600**
		Improve		+2.67%	+4.17%	+3.99%	+3.83%	+0.74%	−2.28%	+6.58%	+7.67%
		RMSE	0.7908	0.7518	0.7319	0.7466	0.7505	0.7552	0.7905	0.7318	**0.7123**
		Improve		+4.93%	+7.45%	+5.59%	+5.10%	+4.50%	+0.04%	+7.46%	+9.93%

Yelp	40%	MAE	0.6800	0.6161	0.6028	0.5982	0.6025	0.6271	0.6976	0.5954	**0.5774**
		Improve		+9.40%	+11.35%	+12.03%	+11.40%	+7.78%	−2.59%	+12.44%	+15.09%
		RMSE	0.9203	0.7936	0.7617	0.7779	0.7751	0.7730	0.9022	0.7703	**0.7400**
		Improve		+13.77%	+17.23%	+15.47%	+15.78%	+16.01%	+1.97%	+16.30%	+19.59%
	20%	MAE	1.0344	0.6327	0.6396	0.6311	0.6481	0.6300	1.0166	0.6785	**0.6450**
		Improve		+38.83%	+38.17%	+38.99%	+37.35%	+39.10%	+1.72%	+34.41%	+37.65%
		RMSE	1.4414	0.8236	0.8188	0.8304	0.8350	0.8200	1.3205	0.8869	**0.8581**
		Improve		+42.86%	+43.19%	+42.39%	+42.07%	+43.11%	+8.39%	+38.47%	+40.47%
	90%	MAE	1.0412	1.0095	0.9013	0.9487	0.9043	0.9054	1.0388	0.8822	**0.8395**
		Improve		+3.04%	+13.44%	+8.88%	+13.15%	+13.04%	+0.23%	+15.27%	+19.37%
		RMSE	1.4268	1.3392	1.1417	1.2549	1.1637	1.1186	1.3581	1.1309	**1.0907**
		Improve		+6.14%	+19.98%	+12.05%	+18.44%	+21.60%	+4.81%	+20.74%	+23.56%
	80%	MAE	1.0791	1.0373	0.9038	0.9654	0.9176	0.9098	1.0750	0.8953	**0.8475**
		Improve		+3.87%	+16.25%	+10.54%	+14.97%	+15.69%	+0.38%	+17.03%	+21.46%
		RMSE	1.4816	1.3782	1.1497	1.2799	1.1771	1.1208	1.4075	1.1516	**1.1117**
		Improve		+6.98%	+22.40%	+13.61%	+20.55%	+24.35%	+5.00%	+22.27%	+24.97%
	70%	MAE	1.1170	1.0694	0.9108	0.9975	0.9407	0.9429	1.1196	0.9043	**0.8580**
		Improve		+4.26%	+18.46%	+10.70%	+15.78%	+15.59%	−0.23%	+19.04%	+23.19%
		RMSE	1.5387	1.4201	1.1651	1.3229	1.2108	1.1582	1.4632	1.1639	**1.1256**
		Improve		+7.71%	+24.28%	+14.02%	+21.31%	+24.73%	+4.91%	+24.36%	+26.85%
	60%	MAE	1.1778	1.1135	0.9435	1.0368	0.9637	1.0043	1.1691	0.9257	**0.8759**
		Improve		+5.46%	+19.89%	+11.97%	+18.18%	+14.73%	+0.74%	+21.40%	+25.63%
		RMSE	1.6167	1.4748	1.2039	1.3713	1.2380	1.2257	1.5182	1.1887	**1.1488**
		Improve		+8.78%	+25.53%	+15.18%	+23.42%	+24.19%	+6.09%	+26.47%	+28.94%

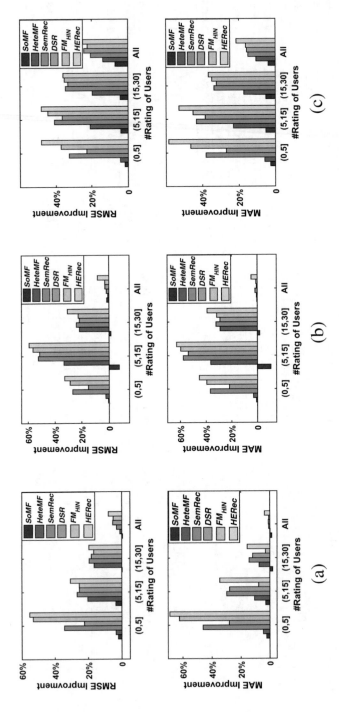

Fig. 3.3 Performance comparison of different methods for cold-start prediction on three datasets. *y*-axis denotes the improvement ratio over PMF. (**a**) Douban movie. (**b**) Douban book. (**c**) Yelp

3.3   Meta-Path Based Decomposition

3.3.1   Overview

In the last section, an HG embedding method exploiting meta-path based random walk is introduced, which generates node sequences that follow the given meta-paths for optimizing the similarity between nodes. However, an observation is that the properties of an object in an HG may stem from different aspects due to the rich type information, which poses a challenge for heterogeneous graph representation. How to effectively extract and fuse different semantic aspect-level information plays an important role in HG representation, which is not considered thoroughly in previous random walk-based methods [7, 9, 37].

To this end, meta-path based decomposition-based techniques aim to decompose HG into several subgraphs according to different meta-paths, each representing a specific semantic aspect, and preserve the proximity of nodes in each subgraph. **NeuACF** (**Neu**ral network-based **A**spect-level **C**ollaborative **F**iltering) [11, 36] is a piece of representative work which exploits decomposition-based HG representation to learn the aspect-level representations and effectively fuse them for recommendation.

Figure 3.4 illustrates an HG composed of 3 types of nodes, which are *User*, *Item*, and *Brand*, and 2 types of relations, which are *User–Item* purchase relation and *Item–Brand* indicating which brand a given item belongs to. Given such an HG, we target to learn the representations of the users and items, and make item recommendation for the users. In this HG, there exist purchase-aspect and brand-aspect information which should be both effectively captured. We can learn the representations of user nodes from the aspect of purchase history with the *User–Item–User* path. Meanwhile, we can also learn the representations from the aspect of brand preference with the *User–Item–Brand–Item–User* path.

Fig. 3.4 A toy example of an HG with different aspect-level information

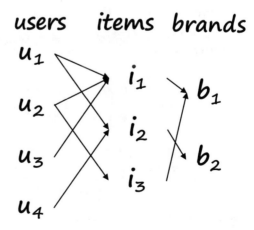

Furthermore, a delicate deep network is designed to learn different aspect-level representations and an attention mechanism is utilized to effectively fuse them for top-N recommendation. Comparing to the above method NeuACF, we further propose NeuACF++ to fuse aspect information with self-attention mechanism which considers different aspect-level representations and learns the attention values simultaneously. More details about NeuACF and NeuACF++ are given in the following subsections.

3.3.2 The NeuACF Model

3.3.2.1 Model Framework

The basic idea of NeuACF is to extract different aspect-level representations for users and items, and then learn and fuse these representations with deep neural network. The model contains three major steps, the architecture of which is illustrated in Fig. 3.5. First, an HG is constructed based on the rich user–item interaction information, and the aspect-level similarity matrices are computed under different meta-paths of HG which reflect different aspect-level features of users and items. Next, a deep neural network is designed to learn the aspect-level representations separately by taking these similarity matrices as inputs. Finally, the aspect-level representations are combined with an attention component to obtain the overall representations for users and items. Moreover, we also employ self-attention mechanism to fuse aspect-level representations more effectively, extending

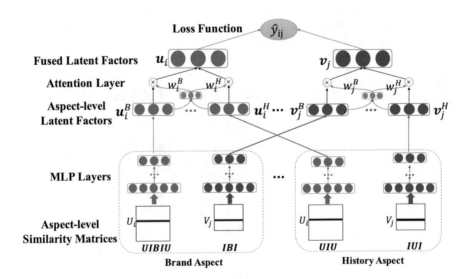

Fig. 3.5 The architecture of the NeuACF model

the model as NeuACF++, which is introduced in Sect. 3.3.2.5. Next we elaborate on the three steps.

3.3.2.2 Aspect-Level Similarity Matrix Extraction

We employ HG to organize objects and relations, due to its power of information fusion and semantics representation [35]. Furthermore, we utilize meta-path to decompose the HG and extract different aspect-level features of users and items.

Given a specific meta-path, similarity matrix is employed to extract the aspect-level features. The popular PathSim [39] is employed to calculate aspect-level similarity matrices under different meta-paths in experiments. For example, the similarity matrices of user–user and item–item are computed based on the meta-path $UIBIU$ and IBI for the brand-aspect features.

The computation of similarity matrix based on meta-path is of great importance in the proposed model, so how to compute similarity matrix quickly is an important problem in the method. In real-world applications, the complexity of similarity matrix computation is not high because the similarity matrix is usually very sparse for most meta-paths. Based on this fact, there are several acceleration computation methods proposed by previous works [34, 39] for similarity matrix computation, for example, PathSim-pruning [39], dynamic programming strategy, and Monte Carlo (MC) strategy [34]. Moreover, there are also many new methods for similarity matrix computation, for example, BLPMC [52], PRSim [50]. In addition, the similarity matrix can be computed offline in advance in our model. The similarity matrix is computed with training data, so we can prepare the similarity matrix before the training processing.

3.3.2.3 Learning Aspect-Level Representations

With the computed user–user and item–item similarity matrices of different aspects, their representations are next learned. A deep neural network is designed to learn their corresponding aspect-level representations separately. Concretely, as shown in Fig. 3.5, for each user in each aspect, the user's similarity vector is extracted from the aspect-specific similarity matrix. Then we take the similarity vector as the input of the MLP, and MLP learns the aspect-level representation as the output. The item representations of each aspect can be learned similarly.

Taking the similarity matrix $S^B \in \mathbb{R}^{N \times N}$ of users under the meta-path $UIBIU$ as an example, user U_i is represented as an N-dimensional vector S_{i*}^B, which means the similarities between U_i and all the other users. Here N means the total number of users in the dataset.

The MLP projects the initial similarity vector S_{i*}^B of user U_i to a low-dimensional aspect-level representation. In each layer of MLP, the input vector is mapped into another vector in a new space. Formally, given the initial input vector S_{i*}^B, and the

l-th hidden layer H_l, the final aspect-level representation u_i^B can be learned through the following multi-layer mapping functions.

From the learning framework in Fig. 3.5, one can see that for each aspect-level similarity matrix of both users and items, there is a corresponding MLP learning component described above to learn the aspect-level representations. Since there are various meta-paths connecting users and items, different aspect-level representations can be learned.

3.3.2.4 Attention-Based Aspect-Level Representations Fusion

After the aspect-level representations are learned separately for users and items, next we need to integrate them together to obtain aggregated representations. A straightforward way is to concatenate all the aspect-level representations to form a higher-dimensional vector. Another intuitive way is to average all the representations. The issue is that both methods do not distinguish their different importance because not all the aspects contribute to the model performance equally, which will be showed in the experiment part.

Therefore, the attention mechanism is chosen to fuse these aspect-level representations. Attention mechanism has shown the effectiveness in various machine learning tasks such as image captioning and machine translation [1, 53, 57]. The advantage of attention mechanism is that it can learn to assign attentive values (normalized by sum to 1) for all the aspect-level representations: higher (lower) values indicate that the corresponding features are more informative (less informative) for recommendation. Specifically, given the user's brand-aspect representations u_i^B, a two-layer network is used to compute the attention score s_i^B by Eq. 3.10,

$$s_i^B = W_2^T f \left(W_1^T \cdot u_i^B + b_1 \right) + b_2, \tag{3.10}$$

where W_* and b_* are the weight matrices and the biases, respectively, and we use the $ReLU$, i.e., $f(x) = max(0, x)$ as the activation function.

The final attention values for the aspect-level representations are obtained by normalizing the above attentive scores with the Softmax function given in Eq. 3.11, which can be interpreted as the contributions of different aspects B to the aggregated latent factor of user U_i,

$$w_i^B = \frac{exp(s_i^B)}{\sum_{A=1}^{L} exp(s_i^A)}, \tag{3.11}$$

where L is the total number of all the aspects.

After obtaining all the attention weights $\boldsymbol{w}_i^B$ of all the aspect-level representations for user U_i, the aggregated representation $\boldsymbol{u}_i$ can be calculated by Eq. 3.12,

$$\boldsymbol{u}_i = \sum_{B=1}^{L} \boldsymbol{w}_i^B \cdot \boldsymbol{u}_i^B. \tag{3.12}$$

This attention method is adopted by NeuACF in the experiments.

3.3.2.5 NeuACF++: Self-Attention-Based Aspect-Level Representations Fusion

Recently, self-attention mechanism has received considerable research interests. For example, Vaswani et al. [46] and Devlin et al. [6] utilize self-attention to learn the relationship between two sequences. Learning dependencies and relationships between aspect-level representations is the most important part in our model, and self-attention has ability to model the relationships between the different aspect-level representations. Therefore, we extend the standard attention mechanism of NeuACF to self-attention, and call the extension version of the model NeuACF++. Next, the self-attention mechanism employed in NeuACF++ will be introduced.

Different from standard attention mechanism, self-attention mainly focuses on the co-learning attentions of two sequences. The vanilla attention mechanism mainly considers computing the attention values based on the user or item representations of one aspect, while self-attention mechanism is able to learn the attention values from different aspects simultaneously. For example, the Brand-level representation of users has strong relationship to the Brand-level representation of items, and the self-attention mechanism can learn this relationship and promote the model performance. So the learned values are able to capture more information on the multi-aspects.

Specifically, we firstly compute the affinity scores between all aspect-level representations. For a user U_i, the affinity score of two different aspect-level representations $\boldsymbol{u}_i^B$ and $\boldsymbol{u}_i^C$ can be calculated by their inner product:

$$M_i^{B,C} = (\boldsymbol{u}_i^B)^T \cdot \boldsymbol{u}_i^C. \tag{3.13}$$

The matrix $\boldsymbol{M}_i = [M_i^{B,C}] \in \mathbb{R}^{L \times L}$ is also called the self-attention matrix, where L is the total number of aspects. There is an affinity matrix $\boldsymbol{M}_i$ for each user. Basically, the matrix $\boldsymbol{M}_i$ characterizes the similarity of aspect-level representations for a specific user U_i, which reflects the correlation between two aspects when recommending for this user. When the aspect B is equal to aspect C, $M_i^{B,C}$ will get a high value due to the inner product operator, so a zero mask is added to avoid a high matching score between identical vectors.

The aspect-level representations learned from self-attention mechanism are not independent. Users can make a trade-off between those aspects. The affinity matrix measures the importance of different aspect-level representations, so the representation of aspect B for the specific user i is computed based on the self-attention matrix as:

$$g_i^B = \sum_{C=1}^{L} \frac{\exp(M_i^{B,C})}{\sum_{A=1}^{L} \exp(M_i^{B,A})} u_i^C.$$ (3.14)

Then for all the aspects, we can obtain the final representation of users or items as:

$$u_i = \sum_{B=1}^{L} g_i^B.$$ (3.15)

The self-attention mechanism can learn self-attentive representations from different aspect-level information effectively. In order to distinguish with the above attention method NeuACF, the self-attention mechanism is implemented as NeuACF++ in the experiments.

3.3.2.6 Objective Function

We model the top-N recommendation as a classification problem which predicts the probability of interaction between users and items in the future. In order to ensure that the output value is a probability, we need to constrain the predicted score $\hat{y}_{ij}$ in the range of [0,1], where we use a Logistic function as the activation function for the output layer. The probability of the interaction between the user U_i and item I_j is calculated according to Eq. 3.16,

$$\hat{y}_{ij} = \text{sigmod}(u_i \cdot v_j) = \frac{1}{1 + e^{-u_i \cdot v_j}},$$ (3.16)

where u_i and v_j are the aggregated representations of user U_i and item I_j, respectively.

Over all the training set, according to the above settings, the likelihood function is:

$$p(\mathcal{Y}, \mathcal{Y}^- | \Theta) = \prod_{i,j \in \mathcal{Y}} \hat{y}_{ij} \prod_{i,k \in \mathcal{Y}^-} (1 - \hat{y}_{ik}),$$ (3.17)

where $\mathcal{Y}$ and $\mathcal{Y}^-$ are the positive and negative instances sets, respectively. The negative instance set $\mathcal{Y}^-$ is sampled from unobserved data for training. Θ is the parameters set.

Since the ground truth y_{ij} is in the set $\{0, 1\}$, Eq. 3.17 can be rewritten as:

$$p(\mathcal{Y}, \mathcal{Y}^- | \Theta) = \prod_{i,j \in \mathcal{Y} \cup \mathcal{Y}^-} (\hat{y}_{ij})^{y_{ij}} (1 - \hat{y}_{ij})^{(1-y_{ij})}. \quad (3.18)$$

Then we can take the negative logarithm of the likelihood function to get the point-wise loss function in Eq. 3.19:

$$\text{Loss} = - \sum_{i,j \in \mathcal{Y} \cup \mathcal{Y}^-} \left(y_{ij} \log \hat{y}_{ij} + (1 - y_{ij}) \log(1 - \hat{y}_{ij}) \right). \quad (3.19)$$

This is the overall objective function of the model, and it can be optimized by stochastic gradient descent or its variants [19].

3.3.3 Experiments

3.3.3.1 Experimental Settings

Datasets The proposed model is evaluated over the publicly available MovieLens dataset [12] and Amazon dataset [13, 25]. We use the origin Movielens dataset for the experiment. For Amazon dataset, we remove the users who bought less than 10 items.

Evaluation Metrics We adopt the leave-one-out method [15] for evaluation. The latest rated item of each user is held out for testing, and the remaining data for training. Following previous work [15], we randomly select 99 items that are not rated by the users as negative samples and rank the 100 sampled items for the users. For a fair comparison with the baseline methods, we use the same negative sample set for each (*user, item*) pair in the test set for all the methods. We evaluate the model performance through the Hit Ratio (HR) and the Normalized Discounted Cumulative Gain (NDCG) defined in Eq. 3.20,

$$\text{HR} = \frac{\#hits}{\#users}, \text{NDCG} = \frac{1}{\#users} \sum_{i=1}^{\#users} \frac{1}{\log_2(p_i + 1)}, \quad (3.20)$$

where $\#hits$ is the number of users whose test item appears in the recommended list and p_i is the position of the test item in the list for the i-th hit. In the experiments, we truncate the ranked list at $K \in [5, 10, 15, 20]$ for both metrics.

Baselines Besides two basic methods (i.e., ItemPop and ItemKNN [32]), the baselines include two MF methods (MF [21] and eALS [14]), one pairwise ranking method (BPR [31]), and two neural network-based methods (DMF [55] and NeuMF [15]). In addition, we use SVD$_{hin}$ to leverage the heterogeneous information for recommendation and also adopt two recent HG based methods (FMG [61] and HeteRS [29]) as baselines.

3.3.3.2 Performance Analysis

The proposed methods are applied to recommendation task. Table 3.3 shows the experiment results of different methods. The proposed methods are marked as NeuACF that implements the attention method in Sect. 3.3.2.4 and NeuACF++ that implements the self-attention mechanism in Sect. 3.3.2.5, respectively. One can draw the following conclusions.

Firstly, NeuACF and NeuACF++ achieve all the best performance over all the datasets and criteria. The improvement of the two models comparing to these baselines is significant. This indicates that the aspect-level information is useful for the downstream task. Besides, NeuACF++ outperforms the NeuACF method in most circumstances. Particularly, the performance of NeuACF++ is significantly improved in Amazon dataset (about +2% at HR and +1% at NDCG). This demonstrates the effectiveness of the self-attention mechanism. Since the affinity matrix evaluates the similarity score of different aspects, we can extract the valuable information from the aspect representations.

Secondly, NeuMF, as one neural network-based method, also performs well on most conditions, while both NeuACF and NeuACF++ outperform NeuMF in almost all the cases. The reason is probably that multiple aspects of representations learned by NeuACF and NeuACF++ provide more features of users and items. Although FMG also utilizes the same features as NeuACF and NeuACF++, the better performance of NeuACF and NeuACF++ implies that the deep neural network and the attention mechanisms in NeuACF and NeuACF++ may have the better ability to learn representations of users and items than the "shallow" model in FMG.

The more detailed method description and experiment validation can be seen in [11] and [36].

3.4 Relation Structure Awareness

3.4.1 Overview

To model the heterogeneity of networks, several attempts have been done on HGs embedding. Representative methods include random walk-based methods [7, 9, 33], decomposition-based methods [37, 43, 54], and neural network-based methods [4, 11, 48, 60]. Although these methods consider the heterogeneity of networks, they usually have an assumption that one single model can handle all relations and nodes,

Table 3.3 HR@K and NDCG@K comparisons of different methods. Bolded values represent the best results

Datasets	Metrics	ItemPop	ItemKNN	MF	eALS	BPR	DMF	NeuMF	SVD$_{hin}$	HeteRS	FMG	NeuACF	NeuACF++
ML100K	HR@5	0.2831	0.4072	0.4634	0.4698	0.4984	0.3483	0.4942	0.4655	0.3747	0.4602	0.5097	**0.5111**
	NDCG@5	0.1892	0.2667	0.3021	0.3201	0.3315	0.2287	0.3357	0.3012	0.2831	0.3014	0.3505	**0.3519**
	HR@10	0.3998	0.5891	0.6437	0.6638	0.6914	0.4994	0.6766	0.6554	0.5337	0.6373	0.6846	**0.6915**
	NDCG@10	0.2264	0.3283	0.3605	0.3819	0.3933	0.2769	0.3945	0.3988	0.3338	0.3588	0.4068	**0.4092**
	HR@15	0.5366	0.7094	0.7338	0.7529	0.7741	0.5873	0.7635	0.7432	0.6524	0.7338	0.7813	**0.7832**
	NDCG@15	0.2624	0.3576	0.3843	0.4056	0.4149	0.3002	0.4175	0.4043	0.3652	0.3844	0.4318	**0.4324**
	HR@20	0.6225	0.7656	0.8144	0.8155	0.8388	0.6519	0.8324	0.8043	0.7224	0.8006	**0.8464**	0.8441
	NDCG@20	0.2826	0.3708	0.4034	0.4204	0.4302	0.3151	0.4338	0.3944	0.3818	0.4002	**0.4469**	0.4469
ML1M	HR@5	0.3088	0.4437	0.5111	0.5353	0.5414	0.4892	0.5485	0.4765	0.3997	0.4732	**0.5630**	0.5584
	NDCG@5	0.2033	0.3012	0.3463	0.3670	0.3756	0.3314	0.3865	0.3098	0.2895	0.3183	**0.3944**	0.3923
	HR@10	0.4553	0.6171	0.6896	0.7055	0.7161	0.6652	0.7177	0.6456	0.5758	0.6528	0.7202	**0.7222**
	NDCG@10	0.2505	0.3572	0.4040	0.4220	0.4321	0.3877	0.4415	0.3665	0.3461	0.3767	0.4453	**0.4454**
	HR@15	0.5568	0.7118	0.7783	0.7914	0.7988	0.7649	0.7982	0.7689	0.6846	0.7536	0.8018	**0.8030**
	NDCG@15	0.2773	0.3822	0.4275	0.4448	0.4541	0.4143	0.4628	0.4003	0.3749	0.4034	**0.4667**	0.4658
	HR@20	0.6409	0.7773	0.8425	0.8409	0.8545	0.8305	0.8586	0.8234	0.7682	0.8169	0.8540	**0.8601**
	NDCG@20	0.2971	0.3977	0.4427	0.4565	0.4673	0.4296	0.4771	0.4456	0.3947	0.4184	0.4789	**0.4790**
Amazon	HR@5	0.2412	0.1897	0.3027	0.3063	0.3296	0.2693	0.3117	0.3055	0.2766	0.3216	0.3268	**0.3429**
	NDCG@5	0.1642	0.1279	0.2068	0.2049	0.2254	0.1848	0.2141	0.1922	0.1800	0.2168	0.2232	**0.2308**
	HR@10	0.3576	0.3126	0.4278	0.4287	0.4657	0.3715	0.4309	0.4123	0.4207	0.4539	0.4686	**0.4933**
	NDCG@10	0.2016	0.1672	0.2471	0.2441	0.2693	0.2179	0.2524	0.2346	0.2267	0.2595	0.2683	**0.2792**
	HR@15	0.4408	0.3901	0.5054	0.5065	0.5467	0.4328	0.5258	0.5056	0.5136	0.5430	0.5591	**0.5948**
	NDCG@15	0.2236	0.1877	0.2676	0.2647	0.2908	0.2332	0.2774	0.2768	0.2513	0.2831	0.2924	**0.3060**
	HR@20	0.4997	0.4431	0.5680	0.5702	0.6141	0.4850	0.5897	0.5607	0.5852	0.6076	0.6257	**0.6702**
	NDCG@20	0.2375	0.2002	0.2824	0.2797	0.3067	0.2458	0.2925	0.2876	0.2683	0.2983	0.3080	**0.3236**

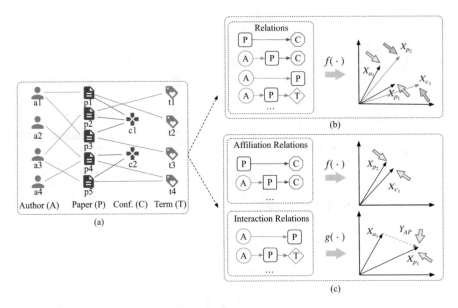

Fig. 3.6 The illustration of HGs and the comparison between conventional methods and our method (non-differentiated relations vs. differentiated relations). (**a**) An example of HG. (**b**) Conventional models. (**c**) Our model

through keeping the representations of two nodes close to each other, as illustrated in Fig. 3.6b.

However, various relations in HGs have significantly different structural characteristics, which should be handled with different models. Let us see a toy example in Fig. 3.6a. The relations in the network include atomic relations (e.g., AP and PC) and composite relations (e.g., APA and APC). Intuitively, AP relation and PC relation reveal rather different structures. That is, some authors write some papers in the AP relation, which shows a peer-to-peer structure. However, many papers are published in one conference in the PC relation, which reveals the structure of one-centered-by-another. Similarly, APA and APC indicate peer-to-peer and one-centered-by-another structures, respectively. The intuitive examples clearly illustrate that relations in HGs indeed have different structural characteristics.

In this section, we present a novel model for HGs embedding, named **R**elation structure-aware **H**eterogeneous **I**nformation **N**etwork **E**mbedding (**RHINE**) [23]. In specific, we first explore the structural characteristics of relations in HGs with thorough mathematical analysis, and we present two structure-related measures which can consistently distinguish the various relations into two categories: Affiliation Relations (ARs) with one-centered-by-another structures and Interaction Relations (IRs) with peer-to-peer structures, as shown in Fig. 3.6c. In order to capture the distinctive structural characteristics of the relations, we then propose two specifically designed models. For ARs where the nodes share similar properties [56], we calculate Euclidean distance as the proximity between nodes, so as to make the nodes directly close in the low-dimensional space. On the other hand, for IRs which bridge two compatible nodes, we model them as translations between the nodes. Since the two models are consistent in terms of mathematical form, they can be optimized in a unified and elegant way.

3.4.2 Preliminary

In this section, we first describe three real-world HGs and analyze the structural characteristics of relations in the HGs. Then we present two structure-related measures which can consistently distinguish various relations quantitatively.

Before analyzing the structural characteristics of relations, we first introduce three datasets used in this section, including DBLP,[4] Yelp,[5] and AMiner[6] [42]. The statistics of these datasets are illustrated in Table 3.4. In order to explore the structural characteristics of relations, we present mathematical analysis on the above datasets.

Since the degree of nodes can well reflect the structures of networks [51], we define a degree-based measure $D(r)$ to explore the distinction of various relations in an HG. Specifically, we compare the average degrees of two types of nodes connected with the relation r, via dividing the larger one by the smaller one $(D(r) \geq 1)$. Formally, given a relation r with nodes u and v (i.e., node-relation triple $\langle u, r, v \rangle$), t_u and t_v are the node types of u and v, we define $D(r)$ as follows:

$$D(r) = \frac{\max [\bar{d}_{t_u}, \bar{d}_{t_v}]}{\min [\bar{d}_{t_u}, \bar{d}_{t_v}]}, \tag{3.21}$$

where $\bar{d}_{t_u}$ and $\bar{d}_{t_v}$ are the average degrees of nodes of the types t_u and t_v respectively.

A large value of $D(r)$ indicates quite inequivalent structural roles of two types of nodes connected via the relation r (one-centered-by-another), while a small value of $D(r)$ means compatible structural roles (peer-to-peer). In other words, relations with a large value of $D(r)$ show much stronger affiliation relationships. Nodes connected via such relations share much more similar properties [8]. While relations with a small value of $D(r)$ implicate much stronger interaction relationships. Therefore, we call the two categories of relations as *Affiliation Relations* (ARs) and *Interaction Relations* (IRs), respectively.

In order to better understand the structural difference between various relations, we take the DBLP network as an example. As shown in Table 3.4, for the relation PC with $D(PC) = 718.8$, the average degree of nodes with type P is 1.0 while that of nodes with type C is 718.8. It shows that papers and conferences are

---

[4] https://dblp.uni-trier.de.

[5] https://www.yelp.com/dataset/.

[6] https://www.aminer.cn/citation.

Table 3.4 Statistics of the three datasets. t_u denotes the type of node u, (u, r, v) is a node-relation triple

Datasets	Nodes	Number of nodes	Relations ($t_u \sim t_v$)	Number of relations	Avg. degree of t_u	Avg. degree of t_v	Measures		Relation category
							$D(r)$	$S(r)$	
DBLP	Term (T)	8811	PC	14,376	1.0	718.8	718.8	0.05	AR
	Paper (P)	14,376	APC	24,495	2.9	2089.7	720.6	0.085	AR
	Author (A)	14,475	AP	41,794	2.8	2.9	1.0	0.0002	IR
	Conference (C)	20	PT	88,683	6.2	10.7	1.7	0.0007	IR
			APT	260,605	18.0	29.6	1.6	0.002	IR
Yelp	User (U)	1286	BR	2614	1.0	1307.0	1307.0	0.5	AR
	Service (S)	2	BS	2614	1.0	1307.0	1307.0	0.5	AR
	Business (B)	2614	BL	2614	1.0	290.4	290.4	0.1	AR
	Star level (L)	9	UB	30,838	23.9	11.8	2.0	0.009	IR
	Reservation (R)	2	BUB	528,332	405.3	405.3	1.0	0.07	IR
AMiner	Paper (P)	127,623	PC	127,623	1.0	1263.6	1264.6	0.01	AR
	Author (A)	164,472	APC	232,659	2.2	3515.6	1598.0	0.01	AR
	Reference (R)	147,251	AP	355,072	2.2	2.8	1.3	0.00002	IR
	Conference (C)	101	PR	392,519	3.1	2.7	1.1	0.00002	IR
			APR	1,084,287	7.1	7.9	1.1	0.00004	IR

structurally inequivalent. Papers are centered by conferences. While $D(AP) = 1.1$ indicates that authors and papers are compatible and peer-to-peer in structure. This is consistent with our common sense. Semantically, the relation PC means that *"papers are published in conferences"*, indicating an affiliation relationship. Differently, AP means that *"authors write papers"*, which explicitly describes an interaction relationship.

In fact, we can also define some other measures to capture the structural difference. For example, we compare the relations in terms of sparsity, which can be defined as:

$$S(r) = \frac{N_r}{N_{t_u} \times N_{t_v}},\tag{3.22}$$

where N_r represents the number of relation instances following r. N_{t_u} and N_{t_v} mean the number of nodes with type t_u and t_v, respectively. The measure can also consistently distinguish the relations into two categories: ARs and IRs. The detailed statistics of all the relations in the three HGs are shown in Table 3.4.

Evidently, Affiliation Relations and Interaction Relations exhibit rather distinct characteristics: (1) ARs indicate one-centered-by-another structures, where the average degrees of the types of end nodes are extremely different. They imply an affiliation relationship between nodes. (2) IRs describe peer-to-peer structures, where the average degrees of the types of end nodes are compatible. They suggest an interaction relationship between nodes.

3.4.3 The RHINE Model

3.4.3.1 Basic Idea

Through our exploration with thorough mathematical analysis, we find that the heterogeneous relations can be typically divided into ARs and IRs with different structural characteristics. In order to exploit their distinct characteristics, we need to specifically design different while appropriate models for the different categories of relations.

For ARs, we propose to take Euclidean distance as a metric to measure the proximity of the connected nodes in the low-dimensional space. There are two motivations behind this: (1) First of all, ARs show affiliation structures between nodes, which indicate that nodes connected via such relations share similar properties. [8, 56]. Hence, nodes connected via ARs could be directly close to each other in the vector space, which is also consistent with the optimization of Euclidean distance [5]. (2) Additionally, one goal of HG embedding is to preserve the high-order proximity. Euclidean distance can ensure that both first-order and second-order proximities are preserved as it meets the condition of the triangle inequality [16].

Different from ARs, IRs indicate strong interaction relationships between compatible nodes, which themselves contain important structural information of two nodes. Thus, we propose to explicitly model an IR as a translation between nodes in the low-dimensional vector space. Additionally, the translation-based distance is consistent with the Euclidean distance in the mathematical form [2]. Therefore, they can be smoothly combined in a unified and elegant manner.

3.4.3.2 Different Models for ARs and IRs

In this subsection, we introduce two different models exploited in RHINE for ARs and IRs, respectively.

Euclidean Distance for Affiliation Relations Nodes connected via ARs share similar properties [8]; therefore, nodes could be directly close to each other in the vector space. We take the Euclidean distance as the proximity measure of two nodes connected by an AR.

Formally, given an affiliation node-relation triple $\langle p, s, q \rangle \in P_{AR}$ where $s \in R_{AR}$ is the relation between p and q with weight w_{pq}, the distance between p and q in the latent vector space is calculated as follows:

$$f(p, q) = w_{pq} \|\mathbf{X}_p - \mathbf{X}_q\|_2^2, \qquad (3.23)$$

in which $\mathbf{X}_p \in \mathbb{R}^d$ and $\mathbf{X}_q \in \mathbb{R}^d$ are the embedding vectors of p and q, respectively. As $f(p, q)$ quantifies the distance between p and q in the low-dimensional vector space, we aim to minimize $f(p, q)$ to ensure that nodes connected by an AR are close to each other. Hence, we define the margin-based loss [2] function as follows:

$$L_{EuAR} = \sum_{s \in R_{AR}} \sum_{\langle p,s,q \rangle \in P_{AR}} \sum_{\langle p',s,q' \rangle \in P'_{AR}} \max[0, \gamma + f(p, q) - f(p', q')], \qquad (3.24)$$

where $\gamma > 0$ is a margin hyperparameter. P_{AR} is the set of positive affiliation node-relation triples, while P'_{AR} is the set of negative affiliation node-relation triples.

Translation-Based Distance for Interaction Relations Interaction Relations demonstrate strong interactions between nodes with compatible structural roles. Thus, different from ARs, we explicitly model IRs as translations between nodes.

Formally, given an interaction node-relation triple $\langle u, r, v \rangle$ where $r \in R_{IR}$ with weight w_{uv}, we define the score function as:

$$g(u, v) = w_{uv} \|\mathbf{X}_u + \mathbf{Y}_r - \mathbf{X}_v\|, \qquad (3.25)$$

where $\mathbf{X}_u$ and $\mathbf{X}_v$ are the node embeddings of u and v, respectively, and $\mathbf{Y}_r$ is the embedding of the relation r. Intuitively, this score function penalizes deviation of $(\mathbf{X}_u + \mathbf{Y}_r)$ from the vector $\mathbf{X}_v$.

For each interaction node-relation triple $\langle u, r, v \rangle \in P_{IR}$, we define the margin-based loss function as follows:

$$L_{TrIR} = \sum_{r \in R_{IR}} \sum_{\langle u,r,v \rangle \in P_{IR}} \sum_{\langle u',r,v' \rangle \in P'_{IR}} \max[0, \gamma + g(u, v) - g(u', v')], \tag{3.26}$$

where P_{IR} is the set of positive interaction node-relation triples, while P'_{IR} is the set of negative interaction node-relation triples.

3.4.3.3 A Unified Model for HG Embedding

Finally, we combine the two models for different categories of relations by minimizing the following loss function:

$$L = L_{EuAR} + L_{TrIR} \tag{3.27}$$

$$= \sum_{s \in R_{AR}} \sum_{\langle p,s,q \rangle \in P_{AR}} \sum_{\langle p',s,q' \rangle \in P'_{AR}} \max[0, \gamma + f(p, q) - f(p', q')]$$

$$+ \sum_{r \in R_{IR}} \sum_{\langle u,r,v \rangle \in P_{IR}} \sum_{\langle u',r,v' \rangle \in P'_{IR}} \max[0, \gamma + g(u, v) - g(u', v')].$$

Sampling Strategy As shown in Table 3.4, the distributions of ARs and IRs are quite unbalanced. What is more, the proportion of relations is unbalanced within ARs and IRs. Traditional edge sampling may suffer from under-sampling for relations with a small amount or over-sampling for relations with a large amount. To address the problems, we draw positive samples according to their probability distributions. As for negative samples, we follow previous work [2] to construct a set of negative node-relation triples $P'_{(u,r,v)} = \{(u', r, v)|u' \in V\} \cup \{(u, r, v')|v' \in V\}$ for the positive node-relation triple (u, r, v), where either the head or tail is replaced by a random node, but not both at the same time.

3.4.4   Experiments

3.4.4.1   Experimental Settings

Datasets As described in Sect. 3.4.2, we conduct experiments on three datasets, including DBLP, Yelp, and AMiner. The statistics of them are summarized in Table 3.4.

Baselines We compare our proposed model RHINE with six state-of-the-art network embedding methods: two classic homogeneous graph embedding methods DeepWalk [28], LINE [44], and four heterogeneous graph embedding methods PTE [43], ESim [33], HIN2Vec [9], and Metapath2vec [7].

Evaluation Metrics We use different evaluation metrics on the following tasks:

- **Node Clustering** The experiments leverage K-means to cluster the nodes and evaluate the results in terms of normalized mutual information (NMI).
- **Link Prediction** We model the link prediction problem as a binary classification problem that aims to predict whether a link exists, and use Area Under Curve (AUC) and F1 score as evaluation metrics.
- **Multi-Class Classification** In this task, we employ the same labeled data used in the node clustering task. We use Micro-F1 and Macro-F1 scores as the metrics for evaluation.

Parameter Settings For a fair comparison, we set the embedding dimension $d = 100$ and the size of negative samples $k = 3$ for all models. For DeepWalk, HIN2Vec, and metapath2vec, we set the number of walks per node $w = 10$, the walk length $l = 100$, and the window size $\tau = 5$. For our model RHINE, the margin γ is set to 1.

3.4.4.2   Node Clustering

As shown in Table 3.5, our model RHINE significantly outperforms all the compared methods. (1) Compared with the best competitors, the clustering performance

Table 3.5 Performance evaluation of node clustering. Bolded values represent the best results

Methods	DBLP	Yelp	AMiner
DeepWalk	0.3884	0.3043	0.5427
LINE-1st	0.2775	0.3103	0.3736
LINE-2nd	0.4675	0.3593	0.3862
PTE	0.3101	0.3527	0.4089
ESim	0.3449	0.2214	0.3409
HIN2Vec	0.4256	0.3657	0.3948
metapath2vec	0.6065	0.3507	0.5586
RHINE	**0.7204**	**0.3882**	**0.6024**

of our model RHINE improves by 18.79%, 6.15%, and 7.84% on DBLP, Yelp, and AMiner, respectively. It demonstrates the effectiveness of our model RHINE by distinguishing the various relations with different structural characteristics in HGs. In addition, it also validates that we utilize appropriate models for different categories of relations. (2) In all baseline methods, homogeneous network embedding models achieve the lowest performance, because they ignore the heterogeneity of relations and nodes. (3) RHINE significantly outperforms existing HGs embedding models (i.e., ESim, HIN2Vec, and metapath2vec) on all datasets. We believe the reason is that our proposed RHINE with appropriate models for different categories of relations can better capture the structural and semantic information of HGs.

3.4.4.3 Link Prediction

The results of link prediction task are reported in Table 3.6 with respect to AUC and F1 score. It is clear that our model performs better than all baseline methods on three datasets. The reason behind the improvement is that our model based on Euclidean distance modeling relations can capture both the first-order and second-order proximities. In addition, our model RHINE distinguishes multiple types of relations into two categories in terms of their structural characteristics, and thus can learn better embeddings of nodes, which are beneficial for predicting complex relationships between two nodes.

3.4.4.4 Multi-Class Classification

We summarize the results of classification in Table 3.7. As we can observe, (1) RHINE achieves better performance than all baseline methods on all datasets except Aminer. It improves the performance of node classification by about 4% on both DBLP and Yelp averagely. In terms of AMiner, the RHINE performs slightly worse than ESim, HIN2vec, and metapath2vec. This may be caused by over-capturing the

Table 3.6 Performance evaluation of link prediction. Bolded values represent the best results

Methods	DBLP (A-A)		DBLP (A-C)		Yelp (U-B)		AMiner (A-A)		AMiner (A-C)	
	AUC	F1	AUC	F1	AUC	F1	AUC	F1	AUC	F1
DeepWalk	0.9131	0.8246	0.7634	0.7047	0.8476	0.6397	0.9122	0.8471	0.7701	0.7112
LINE-1st	0.8264	0.7233	0.5335	0.6436	0.5084	0.4379	0.6665	0.6274	0.7574	0.6983
LINE-2nd	0.7448	0.6741	0.8340	0.7396	0.7509	0.6809	0.5808	0.4682	0.7899	0.7177
PTE	0.8853	0.8331	0.8843	0.7720	0.8061	0.7043	0.8119	0.7319	0.8442	0.7587
ESim	0.9077	0.8129	0.7736	0.6795	0.6160	0.4051	0.8970	0.8245	0.8089	0.7392
HIN2Vec	0.9160	0.8475	0.8966	0.7892	0.8653	0.7709	0.9141	0.8566	0.8099	0.7282
metapath2vec	0.9153	0.8431	0.8987	0.8012	0.7818	0.5391	0.9111	0.8530	0.8902	0.8125
RHINE	**0.9315**	**0.8664**	**0.9148**	**0.8478**	**0.8762**	**0.7912**	**0.9316**	**0.8664**	**0.9173**	**0.8262**

Table 3.7 Performance evaluation of multi-class classification. Bolded values represent the best results

Methods	DBLP		Yelp		AMiner	
	Macro-F1	Micro-F1	Macro-F1	Micro-F1	Macro-F1	Micro-F1
DeepWalk	0.7475	0.7500	0.6723	0.7012	0.9386	0.9512
LINE-1st	0.8091	0.8250	0.4872	0.6639	0.9494	0.9569
LINE-2nd	0.7559	0.7500	0.5304	0.7377	0.9468	0.9491
PTE	0.8852	0.8750	0.5389	0.7342	0.9791	0.9847
ESim	0.8867	0.8750	0.6836	0.7399	0.9910	0.9948
HIN2Vec	0.8631	0.8500	0.6075	0.7361	**0.9962**	**0.9965**
metapath2vec	0.8976	0.9000	0.5337	0.7208	0.9934	0.9936
RHINE	**0.9344**	**0.9250**	**0.7132**	**0.7572**	0.9884	0.9807

information of relations PR and APR (R represents references). Since an author may write a paper referring to various fields, these relations may introduce some noise. (2) Although ESim and HIN2Vec can model multiple types of relations in HGs, they fail to perform well in most cases. Our model RHINE achieves good performance due to the respect of distinct characteristics of various relations.

The more detailed method description and experiment validation can be seen in [23].

3.5 Network Schema Preservation

3.5.1 Overview

Despite the success of meta-path guided HG embedding methods, the selection of meta-paths still remains an open yet challenging problem [39]. The design of meta-path schemes significantly relies on domain knowledge. Manually selecting meta-paths based on prior knowledge may work for a simple HG, while it is difficult to determine meta-paths for a complex HG. Furthermore, different meta-paths will result in different embeddings from different points of view, which leads to another challenging problem, i.e., how to effectively fuse different embeddings to generate uniform embeddings. Some existing works [22, 37, 49] use label information to guide the embedding fusion; unfortunately, this is not applicable in unsupervised scenarios.

To tackle the above challenges, we observe that the network schema [39], as a uniform blueprint of HG, comprehensively retains node types and their relations in an HG. Since network schema is a meta template for HG, guided by it, we can extract subgraphs (i.e., schema instances) from the HG. An example is shown in Fig. 3.7c,d, from which we can see that the schema instance depicts the high-order structure information of these four nodes, besides the first-order structure information of two nodes (i.e., pairwise structure or meta-path based

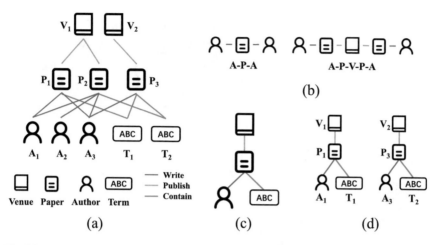

Fig. 3.7 A toy example of an HG on bibliographic data. (**a**) An example of HG. (**b**) Meta-paths. (**c**) Network schema. (**d**) Schema instance

structure as shown in Fig. 3.7b). Moreover, the schema instance also contains rich semantics, i.e., a schema instance (shown in Fig. 3.7d) naturally describes the overall information, such as the author, the term, and the venue of a paper, as well as their relations. More importantly, different from meta-paths, a network schema is a unique structure for an HG, and thus we do not need domain knowledge to make a choice. These benefits of network schema motivate us to study network schema preserving HG embedding. However, it is a non-trivial task. First, *how to effectively preserve the network schema structure?* Moreover, *how to capture the heterogeneity of nodes and links inside network schema?*

In this section, we introduce a novel **N**etwork **S**chema preserving **H**eterogeneous graph **E**mbedding model named **NSHE**. Based on node embedding generated by heterogeneous graph convolutional network, NSHE optimizes the embedding via node pairs and schema instances sampled from the HG. Particularly, in the network schema preserving component, we propose a network schema sampling method, which generates subgraphs (i.e., schema instances) naturally preserving schema structure. Furthermore, for each schema instance, a multi-task learning model is built to predict each node in the instance with other nodes, which tackles the challenge of heterogeneity.

3.5.2 The NSHE Model

3.5.2.1 Model Framework

Consider an HG $\mathcal{G} = \{\mathcal{V}, \mathcal{E}\}$ composed of a node set $\mathcal{V}$ and an edge set $\mathcal{E}$, along with the node type mapping function $\phi : \mathcal{V} \rightarrow \mathcal{A}$, and the edge type mapping function

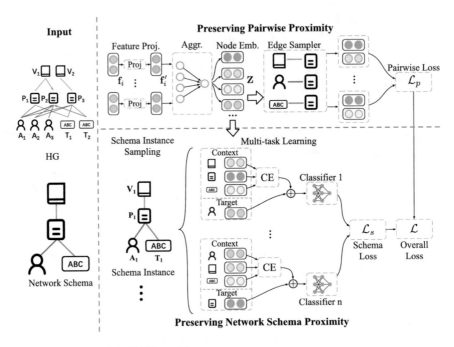

Fig. 3.8 Overview of the NSHE model

$\varphi : \mathcal{E} \to \mathcal{R}$, where $\mathcal{A}$ and $\mathcal{R}$ denote the node and edge types, $|\mathcal{A}| + |\mathcal{R}| > 2$. The task is to learn the representation of nodes $\mathbf{Z} \in \mathbb{R}^{|V| \times d}$, where d is the dimension of representation.

Figure 3.8 illustrates the framework of the proposed NSHE. NSHE preserves the pairwise and schema proximity concurrently. First, to fully exploit complex network structure and heterogeneous node feature together, we propose to learn node embedding via heterogeneous node aggregation. Second, we preserve the pairwise structure and the schema structure simultaneously. While directly performing random walk cannot generate the desired schema structure, we propose to sample schema instances and preserve the proximity inside instances. Moreover, as different types of nodes in the instances carry different context, a multi-task learning model is designed to in turn predict a target node with other context nodes to handle heterogeneity inside schema instances. Finally, NSHE iteratively updates node embeddings via optimizing the aggregation of the pairwise and schema preserving loss.

3.5.2.2 Preserving Pairwise Proximity

Despite that we need to capture the network schema structure in HG embedding, the pairwise proximity between nodes [44], as one of the most direct expressions of an

HG, still needs to be preserved. It demonstrates that two nodes with a link, regardless of their types, should be similar. Specifically, considering the heterogeneity of different node feature, for each node v_i with feature $\mathbf{f}_i$ and type $\phi(v_i)$, we use a type-specific mapping matrix $\mathbf{W}_{\phi(v_i)}$ to map the heterogeneous feature to a common space:

$$\mathbf{f}'_i = \sigma(\mathbf{W}_{\phi(v_i)} \cdot \mathbf{f}_i + \mathbf{b}_{\phi(v_i)}), \tag{3.28}$$

where $\sigma(\cdot)$ denotes an activation function, and $\mathbf{b}_{\phi(v_i)}$ stands for the bias vector of type $\phi(v_i)$. Based on Eq. (3.28), all the nodes with different types are mapped to the common space, and we denote their mapped features as $\mathbf{H} = [\mathbf{f}'_i]$. Then, we use a L-layer graph convolutional network to generate the node embeddings [20] as:

$$\mathbf{H}^{(l+1)} = \sigma\left(\mathbf{D}^{-\frac{1}{2}}(\mathbf{A} + \mathbf{I}_{|V|})\mathbf{D}^{-\frac{1}{2}}\mathbf{H}^{(l)}\mathbf{W}^{(l)}\right), \tag{3.29}$$

where $\mathbf{A}$ is the adjacency matrix, and $\mathbf{A}_{i,j} = 1$ if $(v_i, v_j) \in E$, otherwise $\mathbf{A}_{i,j} = 0$. $\mathbf{D}$ is a diagonal matrix, where $\mathbf{D}_{ii} = \sum_j \mathbf{A}_{ij}$. $\mathbf{I}_{|V|}$ is the identity matrix of $\mathbb{R}^{|V| \times |V|}$. For the first layer, we denote $\mathbf{H}^{(0)} = \mathbf{H}$ and use the output of the L-layer graph convolutional networks as the node embedding, i.e., $\mathbf{Z} = \mathbf{H}^{(L)}$, where the i-th row of $\mathbf{Z}$ is the embedding $\mathbf{z}_{v_i}$ of node v_i.

The objective of preserving the pairwise proximity with parameters Θ can be described as:

$$\mathcal{O}_p = \arg\max_{\Theta} \prod_{v_i \in V} \prod_{v_j \in N_{v_i}} p\left(v_j | v_i; \Theta\right), \tag{3.30}$$

where $N_{v_i} = \{v_j | (v_i, v_j) \in E\}$. The conditional probability $p\left(v_j | v_i; \Theta\right)$ is defined as a softmax function:

$$p\left(v_j | v_i; \Theta\right) = \frac{\exp(\mathbf{z}_{v_j} \cdot \mathbf{z}_{v_i})}{\sum_{v_k \in V} \exp(\mathbf{z}_{v_k} \cdot \mathbf{z}_{v_i})}. \tag{3.31}$$

To calculate $p\left(v_j | v_i; \Theta\right)$ efficiently, we leverage the negative sampling method [26] and optimize Θ with the logarithm of Eq. (3.30), therefore the pairwise loss $\mathcal{L}_p$ can be calculate by:

$$\mathcal{L}_p = \frac{1}{|E|} \sum_{(v_i, v_j) \in E} \left[-\log \delta(\mathbf{z}_{v_j} \cdot \mathbf{z}_{v_i}) \right.$$
$$\left. - \sum_{m=1}^{M_e} \mathbb{E}_{v_{j'} \sim P_n(v)} \log \delta(-\mathbf{z}_{v_{j'}} \cdot \mathbf{z}_{v_i}) \right], \tag{3.32}$$

where $\delta(x) = 1/(1 + \exp(-x))$, $P_n(v)$ is the noisy distribution, and M_e is the negative edge sampling rate. Through minimizing $\mathcal{L}_p$, NSHE preserves the pairwise proximity.

3.5.2.3 Preserving Network Schema Proximity

Network Schema Instance Sampling Network schema is the blueprint of an HG [39]. Given an HG $G = (V, E)$, a network schema $T_G = (\mathcal{A}, \mathcal{R})$ preserves all the node types $\mathcal{A}$ and relation types $\mathcal{R}$ inside G. Network schema proximity implies that all the nodes with different types in a network schema structure should be similar. However, as we mentioned before, the nodes in a network schema structure are usually biased, i.e., the number of nodes of a certain type is larger than those of other types. For example, in Fig. 3.7a, a paper has multiple authors, but only one venue. To alleviate such bias, we propose to sample a network schema instance defined as follows: A *network schema instance* S is the smallest subgraph of an HG, which contains all the node types and edge types defined by the network schema T_G, if existing. By this definition, each network schema instance is composed of all the node types $\mathcal{A}$ and relation types $\mathcal{R}$ defined by the schema, i.e., one node for each type. To illustrate, Fig. 3.7d shows two instances sampled from the given HG. The sampling process is as follows: Starting from a set S with one node, we keep adding a new node to S until $|S| = |\mathcal{A}|$, where the new node satisfies: (1) its type is different from the node types in S; (2) it connects with the node(s) in S.

Schema Preserving with Multi-Task Learning Now, we aim to preserve the network schema proximity by predicting whether a network schema instance exists in an HG. To this end, assume we have a network schema instance $S = \{A_1, P_1, V_1, T_1\}$ as shown in Fig. 3.8, we can predict whether A_1 exists given the set $\{P_1, V_1, T_1\}$, or whether P_1 exists given the set $\{A_1, V_1, T_1\}$, and so on. These two predictions are different, because of the node heterogeneity. Considering this, we are motivated to design a multi-task learning model to handle the heterogeneity within schema.

Without loss of generality, assume we have the schema instance $S = \{v_i, v_j, v_k\}$, if we aim to predict whether v_i exists given $\{v_j, v_k\}$, we call v_i the **target node** and $\{v_j, v_k\}$ the **context nodes**. Therefore, each node will have two roles: one is as the target node and the other is as the context node, as well as two embeddings: target embedding and context embedding. To fully consider the heterogeneity, each node type $\phi(v_i)$ is associated with an encoder $\mathrm{CE}^{\phi(v_i)}$ to learn the context embeddings for the context nodes:

$$\mathbf{c}_{v_j} = \mathrm{CE}^{\phi(v_j)}(\mathbf{z}_{v_j}), \mathbf{c}_{v_k} = \mathrm{CE}^{\phi(v_k)}(\mathbf{z}_{v_k}), \tag{3.33}$$

where each CE stands for a fully connected layer of neural network. Then for the target node v_i, we concatenate its target embedding $\mathbf{z}_{v_i}$ with the context embeddings to obtain the schema instance embedding with target node v_i denoted as $\mathbf{z}_S^{v_i}$ as follows:

$$\mathbf{z}_S^{v_i} = \mathbf{z}_{v_i} \| \mathbf{c}_{v_j} \| \mathbf{c}_{v_k}.$$

(3.34)

After obtaining the embedding $\mathbf{z}_S^{v_i}$, we predict the probability of the existence of S with target node v_i, denoted as $y_S^{v_i}$:

$$y_S^{v_i} = \text{MLP}^{\phi(v_i)}\left(\mathbf{z}_S^{v_i}\right),$$

(3.35)

where $\text{MLP}^{\phi(v_i)}$ is the classifier for schema instances with target node type as $\phi(v_i)$. Similarly, when we treat v_j and v_k as the target nodes, respectively, $y_S^{v_j}$ and $y_S^{v_k}$ can also be obtained following the steps introduced above. Note that, here we take the schema instance with three nodes as an example to explain our method. However, it is easy to extend the model to schema instance with more nodes, since the process is the same.

The schema proximity loss $\mathcal{L}_s$ can be obtained by predicting the multi-tasks of the schema instances $\mathcal{S}$ sampled from HG. Additionally, to avoid trivial solutions, we also draw M_s negative examples of target type for each schema instance via replacing the target node with another node in the same type. The loss of preserving network schema can be described as:

$$\mathcal{L}_s = -\frac{1}{|\mathcal{A}||\mathcal{S}|} \sum_{S \in \mathcal{S}} \sum_{v_i \in S} \left(R_S^{v_i} \log y_S^{v_i} + \left(1 - R_S^{v_i}\right) \log \left(1 - y_S^{v_i}\right)\right),$$

(3.36)

where $R_S^{v_i} = 1$ if S^{v_i} is a positive network schema instance, otherwise $R_S^{v_i} = 0$. By minimizing $\mathcal{L}_s$, the schema structure is preserved.

3.5.2.4 Optimization Objective

To preserve both the pairwise proximity and the network schema proximity of HGs, NSHE optimizes the overall loss $\mathcal{L}$ by aggregating the loss of preserving pairwise proximity $\mathcal{L}_p$ and preserving schema proximity $\mathcal{L}_s$:

$$\mathcal{L} = \mathcal{L}_p + \beta \mathcal{L}_s,$$

(3.37)

where β is a balancing coefficient. At last, we adopt the Adam [19] algorithm to minimize the objective in Eq. (3.37).

3.5.3 Experiments

3.5.3.1 Experimental Settings

Datasets In order to demonstrate the effectiveness of the proposed model, we conduct extensive experiments on three HGs datasets, including **DBLP** [23], **IMDB** [49], and **ACM** [49].

Baselines We compare NSHE with seven state-of-the-art embedding methods including two homogeneous network embedding methods, i.e., DeepWalk [28] and LINE [44] and five heterogeneous networks embedding methods, i.e., **Meta-path2Vec** [7], **HIN2Vec** [9], **HERec** [37], **DHNE** [45], and **HeGAN** [17].

Parameter Settings Here, we briefly introduce the experimental settings. For our proposed model, the feature dimension in common space and the embedding dimension d is set as 128. The negative schema instance sample rate M_s in Sect. 3.5.2.3 is set as 4. We perform neighborhood aggregation via a one-layer-GCN, i.e., $L = 1$, and use two-layer-MLPs for schema instance classification. For models that use meta-paths in modeling, we choose the popular meta-paths adopted in previous methods and report the best result. For models that require node feature, we apply DeepWalk [28] to generate node feature. The code and dataset are publicly available on Github.[7] The more detailed method description and experiment validation can be seen in [62].

3.5.3.2 Node Classification

In this section, we evaluate the performance of node embedding with node classification tasks. After learning the node embeddings, we train a logistic classifier with 80% of the labeled nodes and use the remaining data for testing. We use Micro-F1 and Macro-F1 scores as the metrics for evaluation. The results are shown in Table 3.8, from which we have the following observations: (1) Generally speaking, HG embedding methods perform better than homogeneous network embedding methods, which proves the benefits of considering heterogeneity. (2) Though NSHE does not utilize any prior knowledge, it consistently outperforms the baselines. It demonstrates the effectiveness of our proposed method in classification tasks.

3.5.3.3 Node Clustering

We further conduct clustering tasks to evaluate the embeddings learned by NSHE. Here we utilize the K-Means model to perform node clustering and set the number of clusters for K-Means as the number of classes. The performance in terms of NMI is

---

[7] https://github.com/Andy-Border/NSHE.

Table 3.8 Performance evaluation of multi-class classification. Bolded values represent the best results

	DBLP-P		DBLP-A		IMDB		ACM	
	Micro-F1	Macro-F1	Micro-F1	Macro-F1	Micro-F1	Macro-F1	Micro-F1	Macro-F1
DeepWalk	90.12	89.45	89.44	88.48	56.52	55.24	82.17	81.82
LINE-1st	81.43	80.74	82.32	80.20	43.75	39.87	82.46	82.35
LINE-2nd	84.76	83.45	88.76	87.35	40.54	33.06	82.21	81.32
DHNE	85.71	84.67	73.30	67.61	38.99	30.53	65.27	62.31
Metapath2Vec	92.86	92.44	89.36	87.95	51.90	50.21	83.61	82.77
HIN2Vec	83.81	83.85	90.30	89.46	48.02	46.24	54.30	48.59
HERec	90.47	87.50	86.21	84.55	54.48	53.46	81.89	81.74
HeGAN	88.79	83.81	90.48	89.27	58.56	57.12	83.09	82.94
NSHE	**95.24**	**94.76**	**93.10**	**92.37**	**59.21**	**58.35**	**84.12**	**83.27**

Table 3.9 Performance evaluation of node clustering. Bolded values represent the best results

	DBLP-P	DBLP-A	IMDB	ACM
DeepWalk	46.75	66.25	0.41	**48.81**
LINE-1st	42.18	29.98	0.03	37.75
LINE-2nd	46.83	61.11	0.03	41.80
DHNE	35.33	21.00	0.05	20.25
Metapath2Vec	56.89	68.74	0.09	42.71
HIN2Vec	30.47	65.79	0.04	42.28
HERec	39.46	24.09	0.51	40.70
HeGAN	60.78	68.95	6.56	43.35
NSHE	**65.54**	**69.52**	**7.58**	44.32

shown in Table 3.9. Similarly, the proposed method NSHE significantly outperforms others in most cases, which further demonstrates the effectiveness of NSHE.

The more detailed method description and experiment validation can be seen in [62].

3.6 Conclusions

The structures of heterogeneous graphs are complex and contain rich semantic information. In this chapter, we introduce four heterogeneous graph embedding methods with structure preservation. Specifically, we first introduce a meta-path based random walk method, which generates meaningful node sequences for network embedding. Then, we introduce a meta-path based collaborative filtering framework, which uses an attention mechanism to learn aspect-level network embedding. Besides, we introduce a relation-structure aware method, which distinguishes the various relations into two categories for fine-grained modelling. Finally, we introduce a meta-path independent method, which preserves network schema in network embedding. The experiments not only verify the effectiveness of these methods, but also demonstrate the important role of structural information in the embedding of heterogeneous graphs.

Some interesting future works are to explore other possible methods of distinguishing relations, paths, or network schema to better capture the structural information of heterogeneous graphs, and design more effective graph representation learning methods without the need to manually design meta-paths.

References

1. Bahdanau, D., Cho, K., Bengio, Y.: Neural machine translation by jointly learning to align and translate. Preprint. arXiv:1409.0473 (2014)
2. Bordes, A., Usunier, N., Garcia-Duran, A., Weston, J., Yakhnenko, O.: Translating embeddings for modeling multi-relational data. In: Proceedings of Advances in Neural Information Processing Systems, pp. 2787–2795 (2013)
3. Cao, S., Lu, W., Xu, Q.: GraRep: Learning graph representations with global structural information. In: CIKM '15: Proceedings of the 24th ACM International on Conference on Information and Knowledge Management, pp. 891–900 (2015)
4. Chang, S., Han, W., Tang, J., Qi, G.-J., Aggarwal, C.C., Huang, T.S.: Heterogeneous network embedding via deep architectures. In: Proceedings of the 21th ACM SIGKDD International Conference on Knowledge Discovery and Data Mining, pp. 119–128. ACM, New York (2015)
5. Danielsson, P.-E.: Euclidean distance mapping. Comput. Graphics Image Process. **14**(3), 227–248 (1980)
6. Devlin, J., Chang, M.-W., Lee, K., Toutanova, K.: BERT: pre-training of deep bidirectional transformers for language understanding. Preprint. arXiv:1810.04805 (2018)
7. Dong, Y., Chawla, N.V., Swami, A.: metapath2vec: Scalable representation learning for heterogeneous networks. In: KDD '17: Proceedings of the 23rd ACM SIGKDD International Conference on Knowledge Discovery and Data Mining, pp. 135–144 (2017)
8. Faust, K.: Centrality in affiliation networks. Soc. Netw. **19**(2), 157–191 (1997)
9. Fu, T.-y., Lee, W.-C., Lei, Z.: Hin2vec: explore meta-paths in heterogeneous information networks for representation learning. In: CIKM '17: Proceedings of the 2017 ACM on Conference on Information and Knowledge Management, pp. 1797–1806 (2017)
10. Grover, A., Leskovec, J.: node2vec: scalable feature learning for networks. In: Proceedings of the 22nd ACM SIGKDD International Conference on Knowledge Discovery and Data Mining, pp. 855–864. ACM, New York (2016)
11. Han, X., Shi, C., Wang, S., Philip, S.Y., Song, L.: Aspect-level deep collaborative filtering via heterogeneous information networks. In: Proceedings of the Twenty-Seventh International Joint Conference on Artificial Intelligence, pp. 3393–3399 (2018)
12. Harper, F.M., Konstan, J.A.: The movielens datasets: history and context. ACM Trans. Inter. Intell. Syst. **5**(4), 19 (2016)
13. He, R., McAuley, J.: Ups and downs: modeling the visual evolution of fashion trends with one-class collaborative filtering. In: WWW '16: Proceedings of the 25th International Conference on World Wide Web, pp. 507–517 (2016)
14. He, X., Zhang, H., Kan, M.-Y., Chua, T.-S.: Fast matrix factorization for online recommendation with implicit feedback. In: Proceedings of the 39th International ACM SIGIR conference on Research and Development in Information Retrieval, pp. 549–558 (2016)

15. He, X., Liao, L., Zhang, H., Nie, L., Hu, X., Chua, T.-S.: Neural collaborative filtering. In: WWW '17: Proceedings of the 26th International Conference on World Wide Web, pp. 173–182 (2017)
16. Hsieh, C.-K., Yang, L., Cui, Y., Lin, T.-Y., Belongie, S., Estrin, D.: Collaborative metric learning. In: WWW '17: Proceedings of the 26th International Conference on World Wide Web, pp. 193–201 (2017)
17. Hu, B., Fang, Y., Shi, C.: Adversarial learning on heterogeneous information networks. In: KDD '19: Proceedings of the 25th ACM SIGKDD International Conference on Knowledge Discovery & Data Mining, pp. 120–129 (2019)
18. Ji, M., Han, J., Danilevsky, M.: Ranking-based classification of heterogeneous information networks. In: Proceedings of the 17th ACM SIGKDD International Conference on Knowledge Discovery and Data Mining, pp. 1298–1306 (2011)
19. Kingma, D.P., Ba, J.: Adam: a method for stochastic optimization. In: 3rd International Conference for Learning Representations (2015)
20. Kipf, T.N., Welling, M.: Semi-supervised classification with graph convolutional networks. In: Proceedings of the 5th International Conference on Learning Representations (2017)
21. Koren, Y., Bell, R., and Volinsky, C.: Matrix factorization techniques for recommender systems. Computer **42**, 8 (2009)
22. Linmei, H., Yang, T., Shi, C., Ji, H., Li, X.: Heterogeneous graph attention networks for semi-supervised short text classification. In: Proceedings of the 2019 Conference on Empirical Methods in Natural Language Processing and the 9th International Joint Conference on Natural Language Processing (EMNLP-IJCNLP), pp. 4823–4832 (2019)
23. Lu, Y., Shi, C., Hu, L., Liu, Z.: Relation structure-aware heterogeneous information network embedding. In: Proceedings of the AAAI Conference on Artificial Intelligence, pp. 4456–4463 (2019)
24. Ma, H., Zhou, D., Liu, C., Lyu, M.R., King, I.: Recommender systems with social regularization. In: Proceedings of the Fourth ACM International Conference on Web Search and Data Mining, pp. 287–296 (2011)
25. McAuley, J., Targett, C., Shi, Q., Van Den Hengel, A.: Image-based recommendations on styles and substitutes. In: Proceedings of the 38th International ACM SIGIR Conference on Research and Development in Information Retrieval, pp. 43–52 (2015)
26. Mikolov, T., Sutskever, I., Chen, K., Corrado, G.S., Dean, J.: Distributed representations of words and phrases and their compositionality. In: NIPS'13: Proceedings of the 26th International Conference on Neural Information Processing Systems, pp. 3111–3119 (2013)
27. Mnih, A., Salakhutdinov, R.R.: Probabilistic matrix factorization. Adv. Neural Inf. Proces. Syst. **20**, 1257–1264 (2007)
28. Perozzi, B., Al-Rfou, R., Skiena, S.: Deepwalk: online learning of social representations. In: KDD '14: Proceedings of the 20th ACM SIGKDD International Conference on Knowledge Discovery and Data Mining, pp. 701–710 (2014)
29. Pham, T.-A.N., Li, X., Cong, G., Zhang, Z.: A general recommendation model for heterogeneous networks. IEEE Trans. Knowl. Data Eng. **28**(12), 3140–3153 (2016)
30. Rendle, S.: Factorization machines with libFM. ACM Trans. Intell. Syst. Technol. **3**(3), 1–22 (2012)
31. Rendle, S., Freudenthaler, C., Gantner, Z., Schmidt-Thieme, L.: BPR: bayesian personalized ranking from implicit feedback. In: UAI '09: Proceedings of the Twenty-Fifth Conference on Uncertainty in Artificial Intelligence, pp. 452–461 (2009)
32. Sarwar, B., Karypis, G., Konstan, J., Riedl, J.: Item-based collaborative filtering recommendation algorithms. In: Proceedings of the 10th International Conference on World Wide Web, pp. 285–295 (2001)
33. Shang, J., Qu, M., Liu, J., Kaplan, L.M., Han, J., Peng, J.: Meta-path guided embedding for similarity search in large-scale heterogeneous information networks. Preprint. arXiv:1610.09769 (2016)
34. Shi, C., Kong, X., Huang, Y., Philip, S.Y., Wu, B.: HeteSim: a general framework for relevance measure in heterogeneous networks. IEEE Trans. Knowl. Data Eng. **26**(10), 2479–2492 (2014)

35. Shi, C., Zhang, Z., Luo, P., Yu, P.S., Yue, Y., Wu, B.: Semantic path based personalized recommendation on weighted heterogeneous information networks. In: Proceedings of the 24th ACM International on Conference on Information and Knowledge Management, pp. 453–462. ACM, New York (2015)
36. Shi, C., Han, X., Li, S., Wang, X., Wang, S., Du, J., Yu, P.: Deep collaborative filtering with multi-aspect information in heterogeneous networks. IEEE Trans. Knowl. Data Eng. (2019)
37. Shi, C., Hu, B., Zhao, W.X., Yu, P.S.: Heterogeneous information network embedding for recommendation. IEEE Trans. Knowl. Data Eng. **31**(2), 357–370 (2019)
38. Sun, Y., Yu, Y., Han, J.: Ranking-based clustering of heterogeneous information networks with star network schema. In: Proceedings of the 15th ACM SIGKDD International Conference on Knowledge Discovery and Data Mining, pp. 797–806 (2009)
39. Sun, Y., Han, J., Yan, X., Yu, P.S., Wu, T.: PathSim: meta path-based top-k similarity search in heterogeneous information networks. Proc. VLDB Endow. **4**(11), 992–1003 (2011)
40. Sun, Y., Han, J., Yan, X., Yu, P.S., Wu, T.: PathSim: meta path-based top-k similarity search in heterogeneous information networks. Proc. VLDB Endow. **4**(11), 992–1003 (2011)
41. Sun, Y., Norick, B., Han, J., Yan, X., Philip, S.Y., Yu, X.: Integrating meta-path selection with user-guided object clustering in heterogeneous information networks. In: ACM Transactions on Knowledge Discovery from Data (2012)
42. Tang, J., Zhang, J., Yao, L., Li, J., Zhang, L., Su, Z.: ArnetMiner: extraction and mining of academic social networks. In: Proceedings of the 14th ACM SIGKDD International Conference on Knowledge Discovery and Data Mining, pp. 990–998. ACM, New York (2008)
43. Tang, J., Qu, M., Mei, Q.: PTE: predictive text embedding through large-scale heterogeneous text networks. In: Proceedings of the 21th ACM SIGKDD International Conference on Knowledge Discovery and Data Mining, pp. 1165–1174. ACM, New York (2015)
44. Tang, J., Qu, M., Wang, M., Zhang, M., Yan, J., Mei, Q.: Line: large-scale information network embedding. In: Proceedings of the 24th International Conference on World Wide Web, pp. 1067–1077 (2015)
45. Tu, K., Cui, P., Wang, X., Wang, F., Zhu, W.: Structural deep embedding for hyper-networks. In: Thirty-Second AAAI Conference on Artificial Intelligence, pp. 426–433 (2018)
46. Vaswani, A., Shazeer, N., Parmar, N., Uszkoreit, J., Jones, L., Gomez, A.N., Kaiser, Ł., Polosukhin, I.: Attention is all you need. In: Advances in Neural Information Processing Systems, pp. 5998–6008 (2017)
47. Wang, X., Cui, P., Wang, J., Pei, J., Zhu, W., Yang, S.: Community preserving network embedding. In: Thirty-First AAAI Conference on Artificial Intelligence (2017)
48. Wang, H., Zhang, F., Hou, M., Xie, X., Guo, M., Liu, Q.: SHINE: signed heterogeneous information network embedding for sentiment link prediction. In: Proceedings of the Eleventh ACM International Conference on Web Search and Data Mining, pp. 592–600. ACM, New York (2018)
49. Wang, X., Ji, H., Shi, C., Wang, B., Ye, Y., Cui, P., Yu, P.S.: Heterogeneous graph attention network. In: The World Wide Web Conference, pp. 2022–2032 (2019)
50. Wang, Y., Chen, L., Che, Y., Luo, Q.: Accelerating pairwise SimRank estimation over static and dynamic graphs. VLDB J. Int. J. Very Large Data Bases **28**(1), 99–122 (2019)
51. Wasserman, S., Faust, K.: Social Network Analysis: Methods and Applications, vol. 8. Cambridge University Press, Cambridge (1994)
52. Wei, Z., He, X., Xiao, X., Wang, S., Liu, Y., Du, X., Wen, J.-R.: PRsim: sublinear time SimRank computation on large power-law graphs. Preprint. arXiv:1905.02354 (2019)
53. Xu, K., Ba, J., Kiros, R., Cho, K., Courville, A., Salakhudinov, R., Zemel, R., Bengio, Y.: Show, attend and tell: neural image caption generation with visual attention. In: International Conference on Machine Learning, pp. 2048–2057 (2015)
54. Xu, L., Wei, X., Cao, J., Yu, P.S.: Embedding of embedding (EOE): joint embedding for coupled heterogeneous networks. In: WSDM '17: Proceedings of the Tenth ACM International Conference on Web Search and Data Mining, pp. 741–749. ACM, New York (2017)

55. Xue, H., Dai, X., Zhang, J., Huang, S., Chen, J.: Deep matrix factorization models for recommender systems. In: Proceedings of the Twenty-Sixth International Joint Conference on Artificial Intelligence, pp. 3203–3209 (2017)
56. Yang, J., Leskovec, J.: Community-affiliation graph model for overlapping network community detection. In: 2012 IEEE 12th International Conference on Data Mining, pp. 1170–1175. IEEE, Piscataway (2012)
57. You, Q., Jin, H., Wang, Z., Fang, C., Luo, J.: Image captioning with semantic attention. In: Proceedings of the IEEE Conference on Computer Vision and Pattern Recognition, pp. 4651–4659 (2016)
58. Yu, X., Ren, X., Gu, Q., Sun, Y., Han, J.: Collaborative filtering with entity similarity regularization in heterogeneous information networks. IJCAI HINA **27** (2013)
59. Zhang, J., Tang, J., Ma, C., Tong, H., Jing, Y., Li, J.: Panther: fast top-k similarity search on large networks. In: Proceedings of the 21th ACM SIGKDD International Conference on Knowledge Discovery and Data Mining, pp. 1445–1454 (2015)
60. Zhang, J., Xia, C., Zhang, C., Cui, L., Fu, Y., Philip, S. Y.: BL-MNE: Emerging heterogeneous social network embedding through broad learning with aligned autoencoder. In: 2017 IEEE International Conference on Data Mining (ICDM), pp. 605–614. IEEE, Piscataway (2017)
61. Zhao, H., Yao, Q., Li, J., Song, Y., Lee, D.: Meta-graph based recommendation fusion over heterogeneous information networks. In: Proceedings of the 23rd ACM SIGKDD International Conference on Knowledge Discovery and Data Mining, pp. 635–644 (2017)
62. Zhao, J., Wang, X., Shi, C., Liu, Z., Ye, Y.: Network schema preserving heterogeneous information network embedding. In: Proceedings of the Twenty-Ninth International Joint Conference on Artificial Intelligence, IJCAI 2020, pp. 1366–1372. IJCAI, ijcai.org (2020)
63. Zheng, J., Liu, J., Shi, C., Zhuang, F., Li, J., Wu, B.: Recommendation in heterogeneous information network via dual similarity regularization. Int. J. Data Sci. Anal. **3**(1), 35–48 (2017)

Chapter 4
Attribute-Assisted Heterogeneous Graph Representation

Abstract The previous heterogeneous graph representation methods mainly focus on preserving the complex interactions and rich semantics into node representation. As a matter of fact, diverse types of nodes in heterogeneous graph are assisted with different attributes, providing valuable side information for depicting the characteristics of nodes. Integrating attribution information is also desired for heterogeneous graph representation in a real-world application. Fortunately, heterogeneous graph neural networks naturally provide an alternative way to achieve this, meanwhile, have powerful representation ability. In this chapter, we introduce three attribute-assisted heterogeneous graph representation models including heterogeneous graph attention network (HAN), heterogeneous graph propagation network (HPN), and heterogeneous graph structure learning (HGSL), which simultaneously utilize both complex structural information and rich attribute information to learn node representation.

4.1 Introduction

Besides complex structures and rich semantics, real-world heterogeneous graphs (HGs) are usually associated with diverse types of attributes, called attribute-assisted HGs, providing valuable side information for depicting the characteristics of nodes. For example, in the ACM graph, the user's attribute mainly consists of name, age, and gender, while the paper's attribute usually involves a set of keywords. Ignoring such attributes may lead to sub-optimal heterogeneous graph representation. How to integrate attribution information into node representation is a realistic demand of real-world HG representation.

Heterogeneous graph neural network (HGNN), as a powerful deep learning based technology, naturally provides an elegant way to achieve this. Specifically, HGNN takes node attributes to initial node representation, and then aggregates diverse types of neighbors to update node representation, which simultaneously utilizes both complex structural information and rich attribute information to learn node representations. In this chapter, we introduce three attribute-assisted heterogeneous

© The Author(s), under exclusive license to Springer Nature Singapore Pte Ltd. 2022   71
C. Shi et al., *Heterogeneous Graph Representation Learning and Applications*,
Artificial Intelligence: Foundations, Theory, and Algorithms,
https://doi.org/10.1007/978-981-16-6166-2_4

graph representation models including **H**eterogeneous graph **A**ttention **N**etwork (named **HAN**), **H**eterogeneous graph **P**ropagation **N**etwork (named **HPN**), and Heterogeneous **G**raph **S**tructure **L**earning (named **HGSL**). HAN is a classical HGNN which leverages both node- and semantic-level attention to learn node representation in a hierarchical manner. To alleviate the deep degradation phenomenon (a.k.a, semantic confusion), HPN improves the aggregating process of HGNN via emphasizing the characteristic of each node. After that, to further discover and learn graph structures, HGSL jointly learns the heterogeneous graph and the GNN parameters for downstream tasks.

4.2   Heterogeneous Graph Attention Network

4.2.1   Overview

A recent research trend in deep learning is the attention mechanism, which deals with variable sized data and encourages the model to focus on the most salient parts of data. It has demonstrated the effectiveness in deep neural network framework and is widely applied to various applications, such as text analysis [1], knowledge graph [28], and image processing [37]. Graph Attention Network (GAT) [33], a novel convolution-style graph neural network, leverages attention mechanism for the homogeneous graph which includes only one type of nodes or links.

Despite the success of attention mechanism in deep learning, it has not been considered in the graph neural network framework for heterogeneous graph. As a matter of fact, the real-world graph usually comes with multi-types of nodes and edges, which we uniformly call heterogeneous graph. Because the heterogeneous graph contains more comprehensive information and rich semantics, it has been widely used in many data mining tasks. Meta-path [31] is a widely used structure to capture the semantics. As can be seen in Fig. 4.1, depending on the meta-paths, the relation between nodes in the heterogeneous graph can have different semantics. Due to the complexity of heterogeneous graph, traditional graph neural network cannot be directly applied to heterogeneous graph.

Based on the above analysis, when designing graph neural network architecture with attention mechanism for heterogeneous graph, we need to address the following new requirements. (1) Heterogeneity of graph. The heterogeneity is an intrinsic property of heterogeneous graph, i.e., various types of nodes and edges. How to handle such complex structural information and preserve the diverse feature information simultaneously is an urgent problem that needs to be solved. (2) Semantic-level attention. Different meaningful and complex semantic information are involved in heterogeneous graph, which is usually reflected by meta-paths [31]. Different meta-paths in heterogeneous graph may extract diverse semantic information. How to select the most meaningful meta-paths and fuse the semantic information for the specific task is an open problem [2, 23, 29]. (3) Node-level

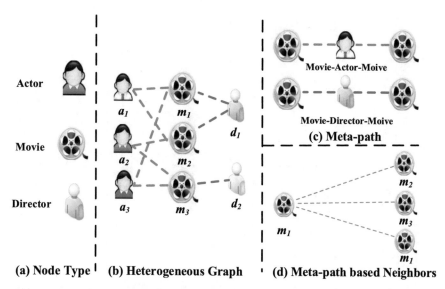

Fig. 4.1 An illustrative example of a heterogenous graph (IMDB). (**a**) Three types of nodes (i.e., actor, movie, director). (**b**) A heterogenous graph IMDB consists of three types of nodes and two types of connections. (**c**) Two meta-paths involved in IMDB (i.e., Movie–Actor–Movie and Movie–Director–Movie). (**d**) Movie m_1 and its meta-path based neighbors (i.e., m_1, m_2, and m_3)

attention. In a heterogeneous graph, nodes can be connected via various types of relation, e.g., meta-path. Given a meta-path, each node has lots of meta-path based neighbors. How to distinguish the subtle difference of their neighbors and select some informative neighbors is required. For each node, node-level attention aims to learn the importance of meta-path based neighbors and assigns different attention values to them. Therefore, how to design a model which can discover the subtle differences of neighbors and learn their weights properly will be desired.

In this section, we introduce a novel **H**eterogeneous graph **A**ttention **N**etwork, named **HAN**, which considers both node-level and semantic-level attentions. In particular, given the node features as input, we use the type-specific transformation matrix to project different types of node features into the same space. Then the node-level attention is able to learn the attention values between the nodes and their meta-path based neighbors, while the semantic-level attention aims to learn the attention values of different meta-paths for the specific task in the heterogeneous graph. Based on the learned attention values in terms of the two levels, our model can get the optimal combination of neighbors and multiple meta-paths in a hierarchical manner, which enables the learned node representations to better capture the complex structure and rich semantic information in a heterogeneous graph. After that, the overall model can be optimized via backpropagation in an end-to-end manner.

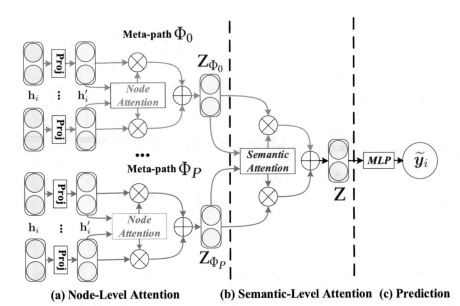

(a) Node-Level Attention   **(b) Semantic-Level Attention (c) Prediction**

Fig. 4.2 The overall framework of the proposed HAN. (**a**) All types of nodes are projected into a unified feature space and the weight of meta-path based node pair can be learned via node-level attention. (**b**) Joint learning the weight of each meta-path and fuse the semantic-specific node representation via semantic-level attention. (**c**) Calculate the loss and end-to-end optimization for the proposed HAN

4.2.2 The HAN Model

In this section, we provide more details about HAN. Specifically, this model follows a hierarchical attention structure: node-level attention $\rightarrow$ semantic-level attention. Figure 4.2 presents the whole framework of HAN. First, we propose a node-level attention to learn the weight of meta-path based neighbors and aggregate them to get the semantic-specific node representation. After that, HAN can tell the difference of meta-paths via semantic-level attention and get the optimal weighted combination of the semantic-specific node representation for the specific task.

4.2.2.1 Node-Level Attention

Before aggregating the information from meta-path neighbors for each node, we should notice that the meta-path based neighbors of each node play a different role and show different importance in learning node representation for the specific task. Here we introduce node-level attention can learn the importance of meta-path based neighbors for each node in a heterogeneous graph and aggregate the representation of these meaningful neighbors to form a node representation.

Due to the heterogeneity of nodes, different types of nodes have different feature spaces. Therefore, for each type of nodes (e.g., node with type ϕ_i), we design the type-specific transformation matrix $\mathbf{M}_{\phi_i}$ to project the features of different types of nodes into the same feature space. Unlike [10], the type-specific transformation matrix is based on node-type rather than edge-type. The projection process can be shown as follows:

$$\mathbf{h}'_i = \mathbf{M}_{\phi_i} \cdot \mathbf{h}_i, \tag{4.1}$$

where $\mathbf{h}_i$ and $\mathbf{h}'_i$ are the original and projected feature of node i, respectively. By type-specific projection operation, the node-level attention can handle arbitrary types of nodes.

After that, we leverage self-attention [32] to learn the weight among various kinds of nodes. Given a node pair (i, j) which are connected via meta-path Φ, the node-level attention e_{ij}^{Φ} can learn the importance e_{ij}^{Φ} which means how important node j will be for node i. The importance of meta-path based node pair (i, j) can be formulated as follows:

$$e_{ij}^{\Phi} = att_{node}(\mathbf{h}'_i, \mathbf{h}'_j; \Phi). \tag{4.2}$$

Here att_{node} denotes the deep neural network which performs the node-level attention. Given meta-path Φ, att_{node} is shared for all meta-path based node pairs. It is because there are some similar connection patterns under one meta-path. The above Eq. (4.2) shows that given meta-path Φ, the weight of meta-path based node pair (i, j) depends on their features. Please note that, e_{ij}^{Φ} is asymmetric, i.e., the importance of node i to node j and the importance of node j to node i can be quite different. It shows node-level attention can preserve the asymmetry which is a critical property of heterogenous graph.

Then we inject the structural information into the model via masked attention which means we only calculate the e_{ij}^{Φ} for nodes $j \in \mathcal{N}_i^{\Phi}$, where $\mathcal{N}_i^{\Phi}$ denotes the meta-path based neighbors of node i (include itself). After obtaining the importance between meta-path based node pairs, we normalize them to get the weight coefficient α_{ij}^{Φ} via softmax function:

$$\alpha_{ij}^{\Phi} = \text{softmax}_j \left(e_{ij}^{\Phi} \right) = \frac{\exp\left(\sigma(\mathbf{a}_{\Phi}^{\mathrm{T}} \cdot [\mathbf{h}'_i \| \mathbf{h}'_j])\right)}{\sum_{k \in \mathcal{N}_i^{\Phi}} \exp\left(\sigma(\mathbf{a}_{\Phi}^{\mathrm{T}} \cdot [\mathbf{h}'_i \| \mathbf{h}'_k])\right)}, \tag{4.3}$$

where σ denotes the activation function, $\|$ denotes the concatenate operation, and $\mathbf{a}_{\Phi}$ is the node-level attention vector for meta-path Φ. As we can see from Eq. (4.3), the weight coefficient of (i, j) depends on their features. Also please note that the weight coefficient α_{ij}^{Φ} is asymmetric which means they make different contribution to each other. This asymmetry is not only because the concatenate order in the numerator, but also because they have different neighbors so the normalize term (denominator) will be quite difference.

Then, the meta-path based representation of node i can be aggregated by the neighbor's projected features with the corresponding coefficients as follows:

$$\mathbf{z}_i^\Phi = \sigma \left(\sum_{j \in \mathcal{N}_i^\Phi} \alpha_{ij}^\Phi \cdot \mathbf{h}'_j \right), \tag{4.4}$$

where $\mathbf{z}_i^\Phi$ is the learned representation of node i for the meta-path Φ. To better understand the aggregating process of node level, we also give a brief explanation in Fig. 4.3a. Every node representation is aggregated by its neighbors. Since the attention weight α_{ij}^Φ is generated for single meta-path, it is semantic specific and able to capture one kind of semantic information.

Since heterogeneous graph present the scale-free property, the variance of graph data is quite high. To tackle the above challenge, we extend node-level attention to multihead attention so that the training process is more stable. Specifically, we repeat the node-level attention for K times and concatenate the learned representations as the semantic-specific representation:

$$\mathbf{z}_i^\Phi = \bigg\Vert_{k=1}^{K} \sigma \left(\sum_{j \in \mathcal{N}_i^\Phi} \alpha_{ij}^\Phi \cdot \mathbf{h}'_j \right). \tag{4.5}$$

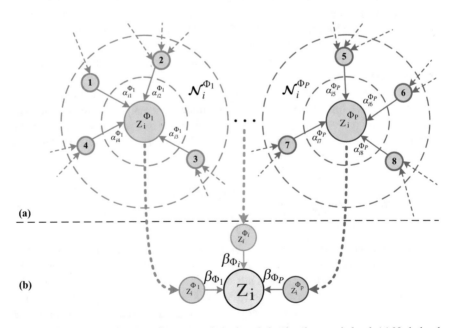

Fig. 4.3 Explanation of aggregating process in both node level and semantic level. (**a**) Node-level aggregating. (**b**) Semantic-level aggregating

Given the meta-path set $\{\Phi_0, \Phi_1, \ldots, \Phi_P\}$, after feeding node features into node-level attention, we can obtain P groups of semantic-specific node representations, denoted as $\{\mathbf{Z}_{\Phi_0}, \mathbf{Z}_{\Phi_1}, \ldots, \mathbf{Z}_{\Phi_P}\}$.

4.2.2.2 Semantic-Level Attention

Generally, every node in a heterogeneous graph contains multiple types of semantic information, and semantic-specific node representation can only reflect node from one aspect. To learn a more comprehensive node representation, we need to fuse multiple semantics which can be revealed by meta-paths. To address the challenge of meta-path selection and semantic fusion in a heterogeneous graph, we propose a novel semantic-level attention to automatically learn the importance of different meta-paths and fuse them for the specific task. Taking P groups of semantic-specific node representations learned from node-level attention as input, the learned weights of each meta-path $(\boldsymbol{\beta}_{\Phi_0}, \boldsymbol{\beta}_{\Phi_1}, \ldots, \boldsymbol{\beta}_{\Phi_P})$ can be shown as follows:

$$(\boldsymbol{\beta}_{\Phi_0}, \boldsymbol{\beta}_{\Phi_1}, \ldots, \boldsymbol{\beta}_{\Phi_P}) = att_{sem}(\mathbf{Z}_{\Phi_0}, \mathbf{Z}_{\Phi_1}, \ldots, \mathbf{Z}_{\Phi_P}). \tag{4.6}$$

Here att_{sem} denotes the deep neural network which performs the semantic-level attention. It shows that the semantic-level attention can capture various types of semantic information behind a heterogeneous graph.

To learn the importance of each meta-path, we first transform semantic-specific representation through a non-linear transformation (e.g., one-layer MLP). Then we measure the importance of the semantic-specific representation as the similarity of transformed representation with a semantic-level attention vector $\mathbf{q}$. Furthermore, we average the importance of all the semantic-specific node representation which can be explained as the importance of each meta-path. The importance of each meta-path, denoted as w_{Φ_i}, is shown as follows:

$$w_{\Phi_i} = \frac{1}{|\mathcal{V}|} \sum_{i \in \mathcal{V}} \mathbf{q}^{\mathrm{T}} \cdot \tanh(\mathbf{W} \cdot \mathbf{z}_i^{\Phi} + \mathbf{b}), \tag{4.7}$$

where $\mathbf{W}$ is the weight matrix, $\mathbf{b}$ is the bias vector, $\mathbf{q}$ is the semantic-level attention vector. Note that for the meaningful comparation, all above parameters are shared for all meta-paths and semantic-specific representation. After obtaining the importance of each meta-path, we normalize them via softmax function. The weight of meta-path Φ_i, denoted as β_{Φ_i}, can be obtained by normalizing the above importance of all meta-paths using softmax function,

$$\beta_{\Phi_i} = \frac{\exp(w_{\Phi_i})}{\sum_{i=1}^{P} \exp(w_{\Phi_i})}, \tag{4.8}$$

which can be interpreted as the contribution of the meta-path Φ_i for specific task. Obviously, the higher β_{Φ_i}, the more important meta-path Φ_i is. Note that for different tasks, meta-path Φ_i may have different weights. With the learned weights as coefficients, we can fuse these semantic-specific representations to obtain the final representation $\mathbf{Z}$ as follows:

$$\mathbf{Z} = \sum_{i=1}^{P} \beta_{\Phi_i} \cdot \mathbf{Z}_{\Phi_i}. \tag{4.9}$$

To better understand the aggregating process of semantic level, we also give a brief explanation in Fig. 4.3b. The final representation is aggregated by all semantic-specific representation. Then we can apply the final representation to specific tasks and design different loss function. For semi-supervised node classification, we can minimize the Cross-Entropy over all labeled node between the ground truth and the prediction:

$$L = -\sum_{l \in \mathcal{Y}_L} \mathbf{Y}^l \ln(\mathbf{C} \cdot \mathbf{Z}^l), \tag{4.10}$$

where $\mathbf{C}$ is the parameter of the classifier, $\mathcal{Y}_L$ is the set of node indices that have labels, $\mathbf{Y}^l$ and $\mathbf{Z}^l$ are the labels and representations of labeled nodes. With the guide of labeled data, we can optimize the proposed model via back propagation and learn the representations of nodes.

4.2.3 Experiments

4.2.3.1 Experimental Settings

Datasets The experiments are conducted on three heterogeneous graphs including DBLP[1], ACM[2], and IMDB[3].

Baselines We compare with some state-of-the-art baselines, including the (heterogeneous) graph representation methods (i.e., DeepWalk [26], ESim [29], metapath2vec [4], and HERec [30]) and graph neural network-based methods (i.e., GCN [20] and GAT [33]), to verify the effectiveness of the proposed HAN. To verify the effectiveness of our node-level attention and semantic-level attention, respectively, we also test two variants of HAN (i.e., HAN_{nd} and HAN_{sem}).

---

[1] https://dblp.uni-trier.de.

[2] http://dl.acm.org/.

[3] https://www.kaggle.com/carolzhangdc/imdb-5000-moviedataset.

Implementation Details For the proposed HAN, we randomly initialize parameters and optimize the model with Adam [19]. For the proposed HAN, we set the learning rate to 0.005, the regularization parameter to 0.001, the dimension of the semantic-level attention vector **q** to 128, the number of attention head K to 8, and the dropout of attention to 0.6. And we use early stopping with a patience of 100, i.e. we stop training if the validation loss does not decrease for 100 consecutive epochs. To make our experiments repeatable, we make our dataset and codes publicly available at website.[4]

4.2.3.2   Classification

Here we employ KNN classifier with $k = 5$ to perform node classification. Since the variance of graph-structured data can be quite high, we repeat the process for 10 times and report the averaged *Macro-F1* and *Micro-F1* in Table 4.1.

Based on Table 4.1, we can see that HAN achieves the best performance. For traditional heterogeneous graph representation method, ESim which can leverage multiple meta-paths performs better than metapath2vec. Generally, graph neural network-based methods which combine the structure and feature information, e.g., GCN and GAT, usually perform better. To go deep into these methods, compared to simply average over node neighbors, e.g., GCN and HAN_{nd}, GAT and HAN can weigh the information properly and improve the performance of the learned representation. Compared to GAT, the proposed HAN, which designs for heterogeneous graph, captures the rich semantics successfully and shows its superiority. Also, without node-level attention (HAN_{nd}) or semantic-level attention (HAN_{sem}), the performance becomes worse than HAN, which indicates the importance of modeling the attention mechanism on both of the nodes and semantics. Note that in ACM and IMDB, HAN improves classification results more significantly than in DBLP. Mainly because *APCPA* is much more important than the rest meta-paths.

Through the above analysis, we can find that the proposed HAN achieves the best performance on all datasets. The results demonstrate that it is quite important to capture the importance of nodes and meta-paths in heterogeneous graph analysis.

4.2.3.3   Analysis of Hierarchical Attention Mechanism

A salient property of HAN is the incorporation of the hierarchical mechanism, which takes the importance of node neighbors and meta-paths into consideration in learning representative representation. Recall that we have learned the node-level attention weight α_{ij}^{Φ} and the semantic-level attention weight β_{Φ_i}. To better

---

[4] https://github.com/Jhy1993/HAN.

Table 4.1 Quantitative results (%) on the node classification task (bold means the best results)

Datasets	Metrics	Training	DeepWalk	ESim	mp2vec	HERec	GCN	GAT	HAN$_{nd}$	HAN$_{sem}$	HAN
ACM	Macro-F1	20%	77.25	77.32	65.09	66.17	86.81	86.23	88.15	89.04	**89.40**
		40%	80.47	80.12	69.93	70.89	87.68	87.04	88.41	89.41	**89.79**
		60%	82.55	82.44	71.47	72.38	88.10	87.56	87.91	**90.00**	89.51
		80%	84.17	83.00	73.81	73.92	88.29	87.33	88.48	90.17	**90.63**
	Micro-F1	20%	76.92	76.89	65.00	66.03	86.77	86.01	87.99	88.85	**89.22**
		40%	79.99	79.70	69.75	70.73	87.64	86.79	88.31	89.27	**89.64**
		60%	82.11	82.02	71.29	72.24	88.12	87.40	87.68	**89.85**	89.33
		80%	83.88	82.89	73.69	73.84	88.35	87.11	88.26	89.95	**90.54**
DBLP	Macro-F1	20%	77.43	91.64	90.16	91.68	90.79	90.97	91.17	92.03	**92.24**
		40%	81.02	92.04	90.82	92.16	91.48	91.20	91.46	92.08	**92.40**
		60%	83.67	92.44	91.32	92.80	91.89	90.80	91.78	92.38	**92.80**
		80%	84.81	92.53	91.89	92.34	92.38	91.73	91.80	92.53	**93.08**
	Micro-F1	20%	79.37	92.73	91.53	92.69	91.71	91.96	92.05	92.99	**93.11**
		40%	82.73	93.07	92.03	93.18	92.31	92.16	92.38	93.00	**93.30**
		60%	85.27	93.39	92.48	93.70	92.62	91.84	92.69	93.31	**93.70**
		80%	86.26	93.44	92.80	93.27	93.09	92.55	92.69	93.29	**93.99**
IMDB	Macro-F1	20%	40.72	32.10	41.16	41.65	45.73	49.44	49.78	**50.87**	50.00
		40%	45.19	31.94	44.22	43.86	48.01	50.64	52.11	50.85	**52.71**
		60%	48.13	31.68	45.11	46.27	49.15	51.90	51.73	52.09	**54.24**
		80%	50.35	32.06	45.15	47.64	51.81	52.99	52.66	51.60	**54.38**
	Micro-F1	20%	46.38	35.28	45.65	45.81	49.78	55.28	54.17	55.01	**55.73**
		40%	49.99	35.47	48.24	47.59	51.71	55.91	56.39	55.15	**57.97**
		60%	52.21	35.64	49.09	49.88	52.29	56.44	56.09	56.66	**58.32**
		80%	54.33	35.59	48.81	50.99	54.61	56.97	56.38	56.49	**58.51**

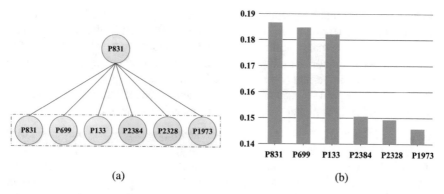

(a)   (b)

Fig. 4.4 Meta-path based neighbors of node P831 and corresponding attention values (Different colors mean different classes, e.g., *green* means Data Mining, *blue* means Database, *orange* means Wireless Communication). (**a**) Meta-path based neighbors of P831. (**b**) Attention values of P831's neighbors

understand the importance of the neighbors and meta-paths, we provide a detailed analysis on the hierarchical attention mechanism.

Analysis of Node-Level Attention   As mentioned before, given a specific task, our model can learn the attention values between nodes and its neighbors in a meta-path. Some important neighbors which are useful for the specific task tend to have larger attention values. Here we take the paper P831[5] in ACM dataset as an illustrative example. Given a meta-path Paper–Author–Paper which describes the co-author of different papers, we enumerate the meta-path based neighbors of paper P831 and their attention values are shown in Fig. 4.4. From Fig. 4.4a, we can see that P831 connects to P699[6] and P133,[7] which all belong to *Data Mining*; connects to P2384[8] and P2328[9] while P2384 and P2328 both belong to *Database*; connects to P1973[10] while P1973 belongs to *Wireless Communication*. From Fig. 4.4b, we can see that paper P831 gets the highest attention value from node-level attention which means the node itself plays the most important role in learning its representation. It is reasonable because all information supported by neighbors are usually viewed as

[5] Xintao Wu, Daniel Barbara, Yong Ye. Screening and Interpreting Multi-item Associations Based on Log-linear Modeling, KDD'03.

[6] Xintao Wu, Jianpin Fan, Kalpathi Subramanian. B-EM: a classifier incorporating bootstrap with EM approach for data mining, KDD'02.

[7] Daniel Barbara, Carlotta Domeniconi, James P. Rogers. Detecting outliers using transduction and statistical testing, KDD'06.

[8] Walid G. Aref, Daniel Barbara, Padmavathi Vallabhaneni. The Handwritten Trie: Indexing Electronic Ink, SIGMOD'95.

[9] Daniel Barbara, Tomasz Imielinski. Sleepers and Workaholics: Caching Strategies in Mobile Environments, VLDB'95.

[10] Hector Garcia-Holina, Daniel Barbara. The cost of data replication, SIGCOMM'81.

a kind of supplementary information. Beyond itself, P699 and P133 get the second and third largest attention values. This is because P699 and P133 also belong to *Data Mining* and they can make significant contribution to identify the class of P831. The rest neighbors get minor attention values because they do not belong to *Data Mining* and cannot make important contribution to identify the P831's class. Based on the above analysis, we can see that the node-level attention can tell the difference among neighbors and assigns higher weights to some meaningful neighbors.

Analysis of Semantic-Level Attention As mentioned before, the proposed HAN can learn the importance of meta-paths for the specific task. To verify the ability of semantic-level attention, taking DBLP and ACM as examples, we report the clustering results (*NMI*) of single meta-path and corresponding attention values in Fig. 4.5. Obviously, there is a positive correlation between the performance of a single meta-path and its attention value. For DBLP, HAN gives *APCPA* the largest weight, which means that HAN considers the *APCPA* as the most critical meta-path in identifying the author's research area. It makes sense because the author's research area and the conferences they submitted are highly correlated. For example, some natural language processing researchers mainly submit their papers to ACL or EMNLP, whereas some data mining researchers may submit their papers to KDD or WWW. Meanwhile, it is difficult for *APA* to identify the author's research area well. If we treat these meta-paths equally, e.g., HAN$_{sem}$, the performance will drop significantly. Based on the attention values of each meta-path, we can find that the meta-path *APCPA* is much more useful than *APA* and *APTPA*. So even the proposed HAN can fuse them, *APCPA* still plays a leading role in identifying the author's research area while *APA* and *APTPA* do not. It also explains why the performance of HAN in DBLP may not be as significant as in ACM and IMDB. We get the similar conclusions on ACM. For ACM, the results show that HAN gives the most considerable weight to *PAP*. Since the performance of *PAP* is slightly better than

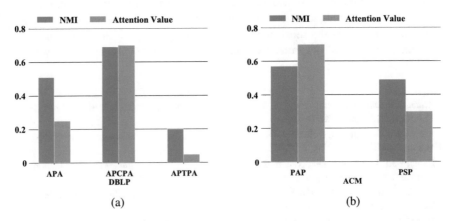

Fig. 4.5 Performance of single meta-path and corresponding attention value. (**a**) NMI values on DBLP. (**b**) NMI values on ACM

PSP, so HAN$_{sem}$ can achieve good performance by simple average operation. We can see that semantic-level attention can reveal the difference between these meta-paths and weights them adequately.

The detailed method description and validation experiments can been seen in [35].

4.3 Heterogeneous Graph Propagation Network

4.3.1 Overview

Recently, several HGNNs [13, 13, 35, 38] have been proposed to better analyze heterogeneous graphs, which usually follow two step aggregating process in a hierarchical manner: aggregate neighbors via single meta-path in node level and then aggregate rich semantics via multiple meta-paths in semantic level. When applying HGNNs in practice, we find an important phenomenon, called *semantic confusion*. Similar to over-smoothing in homogeneous GNNs [18], semantic confusion means HGNNs inject confused semantics extracted via multiple meta-paths into node representation, which makes the learned node representation indistinguishable and leads to worse performance with more hidden layers. Figure 4.6 shows the clustering performance of HAN on ACM academic graph [35]. It clearly displays that with the growth of model depth, the performance of HGNNs is getting worse and worse.

Semantic confusion makes HGNNs hard to become a really deep model, which severely limits their representation capabilities and hurts the performance of downstream tasks. Alleviating the semantic confusion phenomenon to build a more powerful deeper HGNNs is an urgent problem.

In this section, we theoretically analyze the semantic confusion in HGNNs and prove that HGNNs and multiple meta-paths based random walk [22] are essentially equivalent. Then we propose a novel **H**eterogeneous Graph **P**ropagation **N**etwork (HPN) to alleviate semantic confusion from the perspective of multiple meta-paths based random walk, which mainly consists of two parts: semantic propagation mechanism and semantic fusion mechanism. Besides aggregating information from meta-path based neighbors, the semantic propagation mechanism also absorbs node's local semantics with a proper weight. So even with more hidden layers, semantic propagation mechanism can capture the characteristics of each node rather than inject confused semantics into node representation and thus alleviates semantic confusion. The semantic fusion mechanism aims to learn the importance of meta-paths and fuses them for comprehensive node representation.

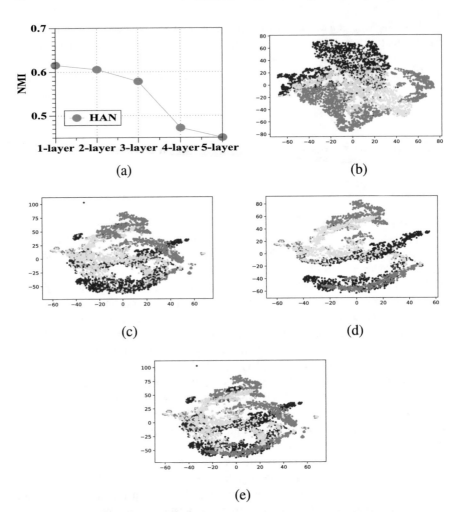

Fig. 4.6 The clustering results and visualization of paper representations via HAN with different layers. Each point denotes one paper and corresponding color indicates the label (i.e., research areas). With the growth of model depth, semantic confusion happens which means the learned node representations become indistinguishable. For example, the paper representations belonging to different research areas which are learned via 1-layer HAN located in different positions, while the paper representations learned via 4-layer HAN mixed together. (**a**) Cluster results of HAN with 1, 2, 3, 4, 5 layers. (**b**) Paper representation via 1-layer HAN. (**c**) Paper representation via 2-layer HAN. (**d**) Paper representation via 3-layer HAN. (**e**) Paper representation via 4-layer HAN

4.3.2 Semantic Confusion Analysis

We first give a brief review of HGNNs, and then prove that HGNNs and multiple meta-paths based random walk are essentially equivalent. Lastly, we explain why semantic confusion happens from the perspective of limit distribution of multiple meta-paths based random walk.

4.3.2.1 Heterogeneous Graph Neural Network

As shown in Fig. 4.3, HGNNs (e.g., HAN) aggregate information from multiple meta-paths and update node representation in both node level and semantic level. Specially, as shown in Fig. 4.3a, given a meta-path Φ_1 and node i, node level attention in HAN aggregates the meta-path Φ_1 based neighbors $\{1, 2, 3, 4\}$ with attentions $\{\alpha_{i1}^{\Phi_1}, \alpha_{i2}^{\Phi_1}, \alpha_{i3}^{\Phi_1}, \alpha_{i4}^{\Phi_1}\}$ to learn the semantic-specific node representation $\mathbf{z}_i^{\Phi_1}$ for node i. Given one meta-path Φ, the node-level aggregating is defined as:

$$
\begin{aligned}
\mathbf{Z}^{\Phi,0} &= \mathbf{X}, \\
\mathbf{Z}^{\Phi,1} &= \sigma\left(\boldsymbol{\alpha}^{\Phi,0} \cdot \mathbf{Z}^{\Phi,0}\right), \\
&\cdots, \\
\mathbf{Z}^{\Phi} = \mathbf{Z}^{\Phi,k} &= \sigma\left(\boldsymbol{\alpha}^{\Phi,k-1} \cdot \mathbf{Z}^{\Phi,k-1}\right),
\end{aligned}
\tag{4.11}
$$

where X denotes node feature matrix, where the i-th row corresponds to the i-th node. And σ is an activate function, the element $\alpha_{ij}^{\Phi,k}$ of $\boldsymbol{\alpha}^{\Phi,k}$ denotes the learned attention weight between meta-path based node pair (i, j) via node-level attention by the k-th layer. Note that $\boldsymbol{\alpha}^{\Phi,k}$ is a (row-normalized) probability matrix and $Z^{\Phi,k}$ denotes the learned representation matrix by the k-th layer, where the i-th row corresponds to the i-th node. As shown in Fig. 4.3b, given a node i and a set of meta-paths $\{\Phi_1, \Phi_2, \cdots, \Phi_P\}$, semantic-level aggregating in HAN fuses P semantic-specific node representations $\left\{\mathbf{z}_i^{\Phi_1}, \cdots, \mathbf{z}_i^{\Phi_P}\right\}$ with attentions $\left\{\beta_{\Phi_1}, \cdots, \beta_{\Phi_P}\right\}$ to get the final representation $\mathbf{z}_i$ for node i. The semantic-level aggregating is shown as follows:

$$
\mathbf{Z} = \sum_{p=1}^{P} \beta_{\Phi_p} \cdot \mathbf{Z}^{\Phi_p},
\tag{4.12}
$$

where $\mathbf{Z}$ denotes the final node representation.

4.3.2.2   Relationship Between HGNNs and Multiple Meta-Paths Based Random Walk

As a classical heterogeneous graph algorithm, multiple meta-paths based random walk [22] mainly contains: single meta-path based random walk and multiple meta-path combinations. Given a meta-path Φ, we have the meta-path based probability matrix $\mathbf{M}^{\Phi}$ whose element $\mathbf{M}_{ij}^{\Phi}$ denotes the transition probability from node i to j via meta-path Φ. Then, the k-step single meta-path based random walk is defined as:

$$\pi^{\Phi,k} = \mathbf{M}^{\Phi} \cdot \pi^{\Phi,k-1}, \tag{4.13}$$

where $\pi^{\Phi,k}$ denotes the distribution of k-step single meta-path based random walk. Considering a set of meta-paths $\{\Phi_1, \Phi_2, \cdots, \Phi_P\}$ and their weights $\{w_{\Phi_1}, w_{\Phi_2}, \cdots, w_{\Phi_P}\}$, the k-step multiple meta-paths based random walk is defined as:

$$\pi^k = \sum_{p=1}^{P} w_{\Phi_p} \cdot \pi^{\Phi_p,k}, \tag{4.14}$$

where π^k denotes the distribution of k-step multiple meta-paths based random walk. For k-step single meta-path based random walk:

Theorem 1 *Assuming a heterogeneous graph is aperiodic and irreducible, if we take the limit $k \to \infty$, then k-step meta-path based random walk will converge to a meta-path specific limit distribution $\pi^{\Phi,\lim}$ which is independent of nodes:*

$$\pi^{\Phi,\lim} = \mathbf{M}^{\Phi} \cdot \pi^{\Phi,\lim}. \tag{4.15}$$

Different nodes connected via some relationships will influence each other and [18] demonstrates the influence distribution between two nodes is proportional to random walk distribution, shown as the following theorem:

Theorem 2 ([18]) *For the aggregation models (e.g., graph neural networks) on homogeneous graph, if the graph is aperiodic and irreducible, then the influence distribution I_i of node i is equivalent, in expectation, to the k-step random walk distribution.*

By Theorems 1 and 2, we conclude the influence distribution revealed by single meta-path based random walk is independent of nodes. Comparing Eq. 4.11 with Eq. 4.13, we find they both propagate and aggregate information via meta-path Φ. The difference is that $\alpha^{\Phi,k}$ is a parameter matrix learned via node-level attention, while $\mathbf{M}^{\Phi}$ is a pre-defined matrix. Since $\mathbf{M}^{\Phi}$ and $\alpha^{\Phi,k}$ are both probability matrix, they are actually meta-path related Markov Chain. So we find that node-level aggregation in HGNNs is essentially equivalent to meta-path based random walk if activate function is a linear function. So, if we stack infinite layers in node-level aggregating, the learned node representations $\mathbf{Z}^\Phi$ will only be influenced by the meta-path Φ and therefore are independent of nodes. So the learned node representations cannot capture the characteristics of each node and therefore are indistinguishable. For k-step multiple meta-paths based random walk, we have:

Theorem 3 *Assuming k-step single meta-path based random walk is independent of each other, if we take the limit $k \to \infty$, then the limit distribution of k-step multiple meta-paths based random walk is a weighted combination of single meta-path based random walk limit distribution, shown as follows:*

$$\boldsymbol{\pi}^{\lim} = \sum_{p=1}^{P} w_{\Phi_p} \cdot \boldsymbol{\pi}^{\Phi_p, \lim}. \tag{4.16}$$

By Theorems 2 and 3, we conclude the influence distribution revealed by multiple meta-paths based random walk is independent of nodes. Comparing Eq. 4.12 with Eq. 4.14, we can see that they both combine multiple meta-paths according to their weights. The difference is that semantic-level aggregating in HAN leverages neural network to learn the weight of meta-path β_{Φ_p}, while multiple meta-paths based random walk assigns pre-defined weight w_{Φ_p} to meta-path Φ_p by hand. Recall that in node-level aggregation, the node representations learned via single meta-path cannot capture the characteristics of each node and therefore are indistinguishable. In semantic-level aggregation, HGNNs fuse multiple node representations learned via multiple node-level aggregations with semantic-wise weights. In summary, the final node representations learned via node and semantic level were only influenced by a set of meta-paths and still remain indistinguishable, which is the critical limitation of HGNNs and leads to semantic confusion.

4.3.3 The HPN Model

4.3.3.1 Semantic Propagation Mechanism

Given one meta-path Φ, the semantic propagation mechanism $\mathcal{P}_\Phi$ first projects node into semantic space via semantic projection function f_Φ. Then, it aggregates information from meta-path based neighbors via semantic aggregation function g_Φ to learn semantic-specific node representation, shown as follows:

$$\mathbf{Z}^\Phi = \mathcal{P}_\Phi(\mathbf{X}) = g_\Phi(f_\Phi(\mathbf{X})), \tag{4.17}$$

where X denotes initial feature matrix and Z^Φ denotes semantic-specific node representation. To handle heterogeneity graph, the semantic projection function f_Φ projects node into semantic space, shown as follows:

$$\mathbf{H}^\Phi = f_\Phi(\mathbf{X}) = \sigma(\mathbf{X} \cdot \mathbf{W}^\Phi + \mathbf{b}^\Phi), \tag{4.18}$$

where $\mathbf{H}^\Phi$ is the projected node feature matrix, $\mathbf{W}^\Phi$ and $\mathbf{b}^\Phi$ denote weight matrix and bias vector for meta-path Φ, respectively. To alleviate semantic confusion, we design semantic aggregation function g_Φ, shown as follows:

$$\mathbf{Z}^{\Phi,k} = g_\Phi(\mathbf{Z}^{\Phi,k-1}) = (1 - \gamma) \cdot \mathbf{M}^\Phi \cdot \mathbf{Z}^{\Phi,k-1} + \gamma \cdot \mathbf{H}^\Phi, \tag{4.19}$$

where $\mathbf{Z}^{\Phi,k}$ denotes node representation learned by k-th layer semantic propagation mechanism and we take it as the semantic-specific node representation $\mathbf{Z}^\Phi$. Note that $\mathbf{H}^\Phi$ reflects the characteristics of each node in meta-path Φ (also can be viewed as $\mathbf{Z}^{\Phi,0}$) and $\mathbf{M}^\Phi \cdot \mathbf{Z}^{\Phi,k-1}$ means aggregating information from meta-path based neighbors. Here γ is a weight scalar which indicates the importance of characteristic of node in aggregating process.

Why Semantic Aggregation Function g_Φ Works   Here we establish the relationship between semantic aggregation function g_Φ and k-step meta-path based random walk with restart. k-step meta-path based random walk with restart for node i is defined as:

$$\pi^{\Phi,k}(i) = (1 - \gamma) \cdot \mathbf{M}^\Phi \cdot \pi^{\Phi,k-1}(i) + \gamma \cdot i, \tag{4.20}$$

where i is a one-hot vector of node i, γ means the restart probability. For k-step meta-path based random walk with restart:

Theorem 4 *Assuming a heterogeneous graph is aperiodic and irreducible, if we take the limit $k \to \infty$, then* k*-step meta-path based random walk with restart will converge to $\pi^{\Phi,\mathrm{lim}}(i)$ which is related to the start node i:*

$$\pi^{\Phi,\mathrm{lim}}(i) = \gamma \cdot (\mathbf{I} - (1 - \gamma) \cdot \mathbf{M}^\Phi)^{-1} \cdot i. \tag{4.21}$$

By Theorems 2 and 4, we conclude that the influence distribution revealed by meta-path based random walk with restart is related to nodes. Comparing Eq. 4.19 to Eq. 4.20, we find they both emphasis node's local semantics with a proper weight γ. By Theorem 4, we can see that the semantic aggregation function g_Φ absorbs node's local semantics and makes semantic-specific node representation $\mathbf{Z}^{\Phi,k}$ distinguish from each other even if we take the limit $k \to \infty$. So semantic propagation mechanism can alleviate the semantic confusion.

4.3.3.2 Semantic Fusion Mechanism

Generally, every node in a heterogeneous graph contains multiple types of semantic information and semantic-specific node representation can only reflect node from one aspect. To describe node more comprehensively, we leverage multiple meta-paths to capture rich semantics and describe node from different aspects.

Given a set of meta-paths $\{\Phi_1, \Phi_2, \cdots, \Phi_P\}$, we have P group semantic-specific node representations $\{\mathbf{Z}^{\Phi_1}, \mathbf{Z}^{\Phi_2}, \cdots, \mathbf{Z}^{\Phi_P}\}$. Then, we propose the semantic fusion mechanism $\mathcal{F}$ to fuse them for the specific task. Taking P groups of semantic-specific node representations learned from semantic propagation mechanism as input, the final node representation $\mathbf{Z}$ learned via semantic fusion mechanism $\mathcal{F}$ is shown as follows:

$$\mathbf{Z} = \mathcal{F}(\mathbf{Z}^{\Phi_1}, \mathbf{Z}^{\Phi_2}, \cdots, \mathbf{Z}^{\Phi_P}). \tag{4.22}$$

Intuitively, not all meta-paths should be treated equally. So semantic fusion mechanism should be able to tell the difference of meta-paths and assign different weights to them. To learn the importance of meta-paths, we project each semantic-specific node representation into the same latent space and adopt semantic fusion vector $\mathbf{q}$ to learn the importance of meta-paths. The importance of meta-path Φ_p, denoted as w_{Φ_p}, is defined as:

$$w_{\Phi_p} = \frac{1}{|\mathcal{V}|} \sum_{i \in \mathcal{V}} \mathbf{q}^{\mathrm{T}} \cdot \tanh(\mathbf{W} \cdot \mathbf{z}_i^{\Phi_p} + \mathbf{b}), \tag{4.23}$$

where $\mathbf{W}$ and $\mathbf{b}$ denote weight matrix and bias vector, respectively, which are shared for all meta-paths. Note that all parameters in semantic fusion mechanism are shared for all nodes and semantics. After obtaining the importance of meta-paths, we normalize them via softmax function to get the weight of each meta-path. The weight of meta-path Φ_p, denoted as β_{Φ_p}, is defined as:

$$\beta_{\Phi_p} = \frac{\exp(w_{\Phi_p})}{\sum_{p=1}^{P} \exp(w_{\Phi_p})}. \tag{4.24}$$

Obviously, the higher β_{Φ_p}, the more important meta-path Φ_p is. With the learned weights as coefficients, we can fuse P semantic-specific representations to obtain the final representation $\mathbf{Z}$ as follows:

$$\mathbf{Z} = \sum_{p=1}^{P} \beta_{\Phi_p} \cdot \mathbf{Z}^{\Phi_p}. \tag{4.25}$$

Then we can optimize the whole model for the specific task and learn the final node representation. Note that semantic fusion mechanism is quite flexible and be optimized for various types of tasks. For different tasks, each semantic may make different contribution which means β_{Φ_p} may change a lot. For semi-supervised node classification, we calculate Cross-Entropy and update parameters in HPN:

$$\mathcal{L} = - \sum_{l \in \mathcal{Y}_L} \mathbf{Y}_l \cdot \ln(\mathbf{Z}_l \cdot \mathbf{C}), \tag{4.26}$$

where $\mathbf{C}$ is a projection matrix which projects the node representation as a node label vector, $\mathcal{Y}_L$ is the set of labeled nodes, $\mathbf{Y}_l$ and $\mathbf{Z}_l$ are the label vector and representation of the labeled node l, respectively.

For unsupervised node recommendation, we leverage BPR loss with negative sampling [34] to update parameters in HPN:

$$\mathcal{L} = - \sum_{(u,v) \in \Omega} \log \sigma \left(\mathbf{z}_u^\top \mathbf{z}_v \right) - \sum_{(u,v') \in \Omega^-} \log \sigma \left(-\mathbf{z}_u^\top \mathbf{z}_{v'} \right), \tag{4.27}$$

where $(u, v) \in \Omega$ and $(u, v') \in \Omega^-$ denote the set of observed (positive) node pairs and the set of negative node pairs sampled from all unobserved node pairs, respectively.

4.3.4 Experiments

4.3.4.1 Experimental Settings

Datasets Three datasets including Yelp[11], ACM[12], and IMDB[13], and MovieLens[14] (ML for short) are used to evaluate the proposed model.

Baselines HPN is compared with some state-of-the-art baselines, including the (heterogeneous) graph representation including metapath2vec [4] and HERec [30], and (heterogeneous) GNNs including GCN [20], GAT [33], PPNP [21], HAN [35], MEIRec [6], MAGNN [8], and HGT [14] to verify the effectiveness of the proposed HPN. Meanwhile, we also test two variants of HPN (i.e., HPN_{pro} and HPN_{fus}) to verify the effectiveness of different parts in our model.

---

[11] https://www.yelp.com.

[12] http://dl.acm.org/.

[13] https://www.kaggle.com/carolzhangdc/imdb-5000-moviedataset.

[14] https://grouplens.org/datasets/movielens/.

4.3.4.2 Clustering

Following the previous work [35], we get the learned node representations of all models via feed forward, and then leverage classical node clustering to test their effectivenesses. Here we utilize the K-Means to perform node clustering and the number of clusters K is set to the number of classes, then select *NMI* and *ARI* to evaluate the clustering task and report the averaged results of 10 runs in Table 4.2.

As can be seen, the proposed HPN performs significantly better than all baselines. It shows the importance of alleviating semantic confusion in HGNNs. We also find that graph neural networks always perform better than graph representation methods. Moreover, heterogeneous graph neural networks including HAN, MEIRec, HGT, MAGNN, and HPN outperform homogeneous GNNs because they can capture rich semantics and describe the characteristic of node more comprehensively. Note that the performance of HPN_{pro} and HPN_{fus} both show different degradations, which imply the importance of semantic propagation mechanism and semantic fusion mechanism. Based on the above analysis, we can find that the proposed HPN can propagate and fuse semantic information effectively and shows significant improvements.

4.3.4.3 Robustness to Model Depth

A salient property of HPN is the incorporation of the semantic propagation mechanism which is able to alleviate the semantic confusion and build a deeper and more powerful HGNN. Comparing to the previous HGNNs (e.g., HAN), the proposed HPN can stack more layers and learn more representative node representation. To show the superiority of semantic propagation in HPN, we test HAN and HPN with 1, 2, 3, 4, 5 layers, shown in Fig. 4.7.

As can be seen, with the growth of model depth, the performance of HAN performs worse and worse on both ACM and IMDB and we believe this phenomenon is the semantic confusion, leading to the degradation of previous heterogeneous GNNs (e.g., HAN). Obviously, semantic confusion makes HGNNs hard to become a really deep model, which severely limits their representation capabilities and hurts the performance of downstream tasks (e.g., node clustering). On the other hand, with the growth of model depth, the performance of the proposed HPN is getting better and better, indicating that semantic propagation mechanism is able to effectively alleviate the semantic confusion. So even stacking for more layers, the node representations learned via the proposed HPN are still distinguishable. In summary, the proposed HPN is able to capture high-order semantics and learns more representative node representation with deeper architecture, rather than learning indistinguishable node representation.

The detailed method description and validation experiments can been seen in [15].

Table 4.2 Quantitative results (%) on the node clustering task. The larger values, the better performance (bold means the best results)

Datasets	Metrics	mp2vec	HERec	GCN	GAT	PPNP	MEIRec	HAN	HGT	MG	HPN$_{pro}$	HPN$_{fus}$	HPN
Yelp	NMI	42.04	0.30	32.58	42.30	40.60	30.09	45.46	47.82	47.56	44.36	12.86	**48.90**
	ARI	38.27	0.41	23.30	41.52	37.72	27.88	41.39	42.91	43.24	42.57	10.54	**44.89**
ACM	NMI	21.22	40.70	51.40	57.29	61.68	61.56	61.56	60.89	64.12	65.60	67.55	**68.21**
	ARI	21.00	37.13	53.01	60.43	65.15	61.46	64.39	59.85	66.29	69.30	71.53	**72.33**
IMDB	NMI	1.20	1.20	5.45	8.45	10.20	11.32	10.87	11.59	11.79	9.45	12.01	**12.31**
	ARI	1.70	1.65	4.40	7.46	8.20	10.40	10.01	9.92	10.32	8.02	12.32	**12.55**

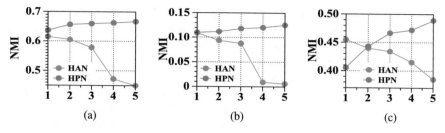

Fig. 4.7 Clustering results of HAN/HPN with 1,2,3,4,5 layers. (**a**) ACM. (**b**) IMDB. (**c**) Yelp

4.4   Heterogeneous Graph Structure Learning

4.4.1   Overview

Most HGNNs follow a message passing scheme where the node representation is learned by aggregating and transforming the representations of its original neighbors [11, 39, 42] or meta-path based neighbors [8, 14, 35, 38], which rely on one fundamental assumption, i.e., the raw heterogeneous graph structure is good. However, as heterogeneous graphs are usually extracted from complex interaction systems by some pre-defined rules, such assumption cannot be always satisfied. One reason is that, these interaction systems inevitably contain some uncertain information or mistakes. Taking a user–item graph built from recommendation as an example, it is well accepted that users may misclick some unwanted items, bringing noisy information to the graph. The other reason is that, the heterogeneous graphs are often extracted with data cleaning, feature extraction, and feature transformation by some pre-defined rules, which are usually independent to the downstream tasks and lead to the gap between the extracted graph and the optimal graph structure for the downstream tasks. Therefore, learning an optimal heterogeneous graph for GNN is a fundamental problem.

Recently, to alleviate the limitation of GNNs and adaptively learn graph structures for GNNs, graph structure learning (GSL) methods [3, 7, 16, 17] are proposed to jointly learn the GNN parameters and graph structure simultaneously. However, these methods cannot be directly applied to HGs with the following challenges: (1) The heterogeneity of graphs. When learning a homogeneous graph with only one type of relation, we usually only need to parameterize one adjacency matrix. However, a heterogeneous graph consists of multiple relations, each of which reflects one aspect of the heterogeneous graph. Treating the heterogeneous graph as a homogeneous graph and failure to distinguish between these heterogeneous relations will inevitably restrict the capability of graph structure learning, but how to deal with this heterogeneity is a challenging problem. (2) The complex interactions in heterogeneous graphs. Different relations and node features have complex interactions, which drives the formation of different kinds of underlying graph structure [40]. Moreover, the combination of different relations further forms a large number of high-order relationships with diverse semantics, which also implies distinct ways of graph generation. The heterogeneous graph structure will be affected by all these factors; therefore, these complex interactions must be thoroughly considered in heterogeneous graph structure learning.

In this section, we introduce an interesting work that makes the first attempt to investigate **Heterogeneous Graph Structure Learning** for graph neural networks, called **HGSL**. In HGSL, the heterogeneous graph and the GNN parameters are jointly learned towards better node classification performance. Particularly, in the graph learning part, aiming to capture the heterogeneous metric of different relation generation, each relation subgraph is separately learned. Specifically, for each relation, three types of candidate graphs, i.e. the feature similarity graph, feature propagation graphs, and semantic graphs, are generated by mining the complex correlations from heterogeneous node features and graph structures. The learned graphs are further fused to a heterogeneous graph and fed to a GNN. The graph learning parameters and the GNN parameters are jointly optimized towards classification objective.

4.4.2 The HGSL Model

We firstly introduce some basic concepts and formalize the problem of heterogeneous graph structure learning as follows:

Definition 1 Node-Relation Triple A node-relation triple $\langle v_i, r, v_j \rangle$ describes that two nodes v_i (head node) and v_j (tail node) are connected by relation $r \in \mathcal{R}$. We further define the type mapping functions $\phi_h, \phi_t : \mathcal{R} \to \mathcal{T}$ that map the relation to its head node type and tail node type, respectively.

Example 1 In a user–item heterogeneous graph, say $r = ``UI''$ (a user buys an item), then we have $\phi_h(r) = ``User''$ and $\phi_t(r) = ``Item''$.

Definition 2 Relation Subgraph Given a heterogeneous graph $G = (V, E, \mathcal{F})$, a relation subgraph G_r is a subgraph of G that contains all node-relation triples with relation r. The adjacency matrix of G_r is $\mathbf{A}_r \in \mathbb{R}^{|V_{\phi_h(r)}| \times |V_{\phi_t(r)}|}$, where $\mathbf{A}_r[i, j] = 1$ if $\langle v_i, r, v_j \rangle$ exists in G_r, otherwise $\mathbf{A}_r[i, j] = 0$. $\mathcal{A}$ denotes the relation subgraph set of all the relation subgraphs in G, i.e. $\mathcal{A} = \{\mathbf{A}_r, r \in \mathcal{R}\}$.

Definition 3 Heterogeneous Graph Structure Learning (HGSL) Given a heterogeneous graph G, the task of heterogeneous graph structure learning is to jointly learn a heterogeneous graph structure, i.e. a new relational subgraph set $\mathcal{A}'$, and the parameters of GNN for downstream tasks.

4.4.2.1 Model Framework

Figure 4.8a illustrates the framework of the proposed HGSL. As we can see, given a heterogeneous graph, HGSL firstly constructs the semantic representation matrices

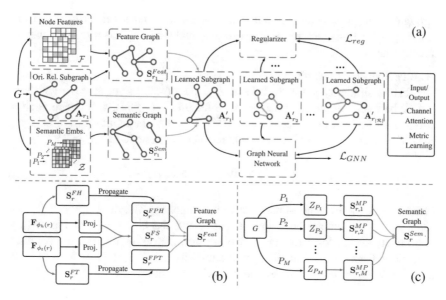

Fig. 4.8 Overview of the HGSL framework. (**a**) Model framework. (**b**) Feature graph generator. (**c**) Semantic graph generator

$\mathcal{Z}$ by meta-path based node representations from M meta-paths. Afterwards, the heterogeneous graph structure and GNN parameters are trained jointly. For the graph learning part, HGSL takes the information from the original relation subgraph, the node features, and the semantic representations as input and generates relation subgraphs separately. Specifically, taking relation r_1 as an example, HGSL learns a feature graph $\mathbf{S}_{r_1}^{Feat}$ and a semantic graph $\mathbf{S}_{r_1}^{Sem}$ and fuse them with the original graph $\mathbf{A}_{r_1}$ to obtain the learned relation subgraph $\mathbf{A}'_{r_1}$. Then, the learned subgraphs are fed into a GNN and a regularizer to perform node classification with regularization. By minimizing the regularized classification loss, HGSL optimizes the graph structure and the GNN parameters jointly.

4.4.2.2 Feature Graph Generator

Since the original graph may not be optimal for downstream task, a natural idea would be to augment the original graph structure via fully utilizing the rich information inside heterogeneous node features. Usually, there are two factors that affect the formation of graph structure based on features. One is the similarity between node features, and the other is the relationship between node feature and relation in HG [36]. As shown in Fig. 4.8b, we first propose to generate a feature similarity graph that captures the potential relationship generated by node features via heterogeneous feature projection and metric learning. Then we propose to generate the feature propagation graph, by propagating feature similarity matrices through topology structure. Finally, the generated feature similarity graph and feature propagation graph are aggregated to a final feature graph through a channel attention layer.

Feature Similarity Graph

The feature similarity graph $\mathbf{S}_r^{FS}$ determines the possibility of an edge with type $r \in \mathcal{R}$ between two nodes based on node features. Specifically, for each node v_i of type $\phi(v_i)$ with feature vector $\mathbf{f}_i \in \mathbb{R}^{1 \times d_{\phi(v_i)}}$, we adopt a type-specific mapping layer to project the feature $\mathbf{f}_i$ to a d_c-dimensional common feature $\mathbf{f}_i' \in \mathbb{R}^{1 \times d_c}$:

$$\mathbf{f}_i' = \sigma \left(\mathbf{f}_i \cdot \mathbf{W}_{\phi(v_i)} + \mathbf{b}_{\phi(v_i)} \right), \tag{4.28}$$

where $\sigma(\cdot)$ denotes a non-linear activation function, $\mathbf{W}_{\phi(v_i)} \in \mathbb{R}^{d_{\phi(v_i)} \times d_c}$ and $\mathbf{b}_{\phi(v_i)} \in \mathbb{R}^{1 \times d_c}$ denote the mapping matrix and the bias vector of type $\phi(v_i)$, respectively. Then, for a relation r, we perform metric learning on the common features and obtain the learned feature similarity graph $\mathbf{S}_r^{FS} \in \mathbb{R}^{|V_{\phi_h(r)}| \times |V_{\phi_t(r)}|}$, where the edge between nodes v_i and v_j is obtained by:

$$\mathbf{S}_r^{FS}[i, j] = \begin{cases} \Gamma_r^{FS}(\mathbf{f}_i', \mathbf{f}_j') & \Gamma_r^{FS}(\mathbf{f}_i', \mathbf{f}_j') \geq \epsilon^{FS} \\ 0 & \text{otherwise}, \end{cases} \tag{4.29}$$

where $\epsilon^{FS} \in [0, 1]$ is the threshold that controls the sparsity of feature similarity graph, and larger ϵ^{FS} implies a more sparse feature similarity graph. Γ_r^{FS} is a K-head weighted cosine similarity function defined as:

$$\Gamma_r^{FS}(\mathbf{f}_i', \mathbf{f}_j') = \frac{1}{K} \sum_k^K \cos \left(\mathbf{w}_{k,r}^{FS} \odot \mathbf{f}_i', \mathbf{w}_{k,r}^{FS} \odot \mathbf{f}_j' \right), \tag{4.30}$$

where $\odot$ denotes the Hadamard product, and $\mathbf{W}_r^{FS} = [\mathbf{w}_{k,r}^{FS}]$ is the learnable parameter matrix of Γ_r^{FS} that weights the importance of different dimensions of the feature vectors. By performing metric learning as in Eq. 4.30 and ruling out edges with little feature similarity by threshold ϵ^{FS}, HGSL learns the candidate feature similarity graph $\mathbf{S}_r^{FS}$.

Feature Propagation Graph

The feature propagation graph is the underlying graph structure generated by the interaction between node features and topology structure. The key insight is that two nodes with similar features may have similar neighbors. Therefore, the process of generating feature propagation graph is two-fold: Firstly, generate the feature similarity graphs, i.e. find the similar nodes; secondly, propagate the feature similarity graph by topological structure to generate new edges, i.e. find the neighbors of the nodes with similar features.

Specifically, for each relation r, assume that we have two types of nodes $V_{\phi_h(r)}$ and $V_{\phi_t(r)}$, and the topology structure between them is $\mathbf{A}_r \in \mathbb{R}^{|V_{\phi_h(r)}| \times |V_{\phi_t(r)}|}$. For the nodes $v_i, v_j \in V_{\phi_h(r)}$ with the same type $\phi_h(r)$, we can obtain the feature similarity:

$$\mathbf{S}_r^{FH}[i, j] = \begin{cases} \Gamma_r^{FH}(\mathbf{f}_i, \mathbf{f}_j) & \Gamma_r^{FH}(\mathbf{f}_i, \mathbf{f}_j) \geq \epsilon^{FP} \\ 0 & \text{otherwise,} \end{cases} \tag{4.31}$$

where the threshold ϵ^{FP} controls the sparsity of feature similarity graph $\mathbf{S}_r^{FH}$. Γ_r^{FH} is the metric learning function in the framework of Eq. 4.30 with different parameters $\mathbf{W}_r^{FH}$. Then we can model the head feature propagation graph $\mathbf{S}_r^{FPH} \in \mathbb{R}^{|V_{\phi_h(r)}| \times |V_{\phi_t(r)}|}$ using $\mathbf{S}_r^{FH}$ and $\mathbf{A}_r$ as follows:

$$\mathbf{S}_r^{FPH} = \mathbf{S}_r^{FH} \mathbf{A}_r. \tag{4.32}$$

As we can see, the feature similarity is propagated through the original graph topological structure and further generates the potential feature propagation graph structure. As for the nodes $V_{\phi_t(r)}$ with the same type $\phi_t(r)$, similar to Eq. 4.31, we can obtain the corresponding feature similarity graph $\mathbf{S}_r^{FT}$ with parameters $\mathbf{W}_r^{FT}$. Therefore, the corresponding feature propagation graph $\mathbf{S}_r^{FPT}$ can be obtained as follows:

$$\mathbf{S}_r^{FPT} = \mathbf{A}_r \mathbf{S}_r^{FT}. \tag{4.33}$$

Now, we have generated one feature similarity graph $\mathbf{S}_r^{FS}$ and two feature propagation graphs $\mathbf{S}_r^{FPH}$ and $\mathbf{S}_r^{FPT}$. The overall feature graph for relation r, denoted as $\mathbf{S}_r^{Feat} \in \mathbb{R}^{|V_{\phi_h(r)}| \times |V_{\phi_t(r)}|}$, can be obtained by fusing these graphs through a channel attention layer [38]:

$$\mathbf{S}_r^{Feat} = \Psi_r^{Feat}([\mathbf{S}_r^{FS}, \mathbf{S}_r^{FPH}, \mathbf{S}_r^{FPT}]), \tag{4.34}$$

where $[\mathbf{S}_r^{FS}, \mathbf{S}_r^{FPH}, \mathbf{S}_r^{FPT}] \in \mathbb{R}^{|V_{\phi_h(r)}| \times |V_{\phi_t(r)}| \times 3}$ is the stacked matrix of the feature candidate graphs, and Ψ_r^{Feat} denotes a channel attention layer with parameters $\mathbf{W}_{\Psi,r}^{Feat} \in \mathbb{R}^{1 \times 1 \times 3}$ which performs 1×1 convolution on the input using softmax($\mathbf{W}_{\Psi,r}^{Feat}$). In this way, HGSL balances the importance of each candidate feature graph for each relation r by learning different weights, respectively.

4.4.2.3 Semantic Graph Generator

The semantic graph is generated depending on the high-order topology structure in HG, describing the multi-hop structural interactions between two nodes. Notably, in heterogeneous graphs, these high-order relationships differ from each other with different semantics determined by meta-paths. In light of this, we propose to learn semantic graph structures from different semantics.

Given a meta-path P with the corresponding relations $r_1 \circ r_2 \circ \cdots \circ r_l$, a straightforward way to generate semantic graph would be fusing the adjacency matrices, i.e. $\mathbf{A}_{r_1} \cdot \mathbf{A}_{r_2} \cdot \cdots \cdot \mathbf{A}_{r_l}$ [38]. However, this method not only costs large memory with the computation of stacking multiple layers of adjacency matrices, but also discards the intermediate nodes which lead to information loss [8].

Alternatively, we propose a semantic graph generator shown in Fig. 4.8c. The semantic graph generator generates the potential semantic graph structure by metric learning on trained meta-path based node representations. Specifically, for an interested meta-path set $\mathcal{P} = \{P_1, P_2, \ldots, P_M\}$ with M meta-paths, HGSL uses trained MP2Vec [5] representations, denoted as $\mathcal{Z} = \{\mathbf{Z}_{P_1}, \mathbf{Z}_{P_2}, \cdots, \mathbf{Z}_{P_M} \in \mathbb{R}^{|V| \times d}\}$, to generate semantic graphs. Since the training process of semantic representations is off-line, the computation cost and model complexity is largely reduced. Moreover, thanks to the mechanism of heterogeneous skip-gram, the information of intermediate nodes is well preserved.

After obtaining the semantic representations $\mathcal{Z}$, for each meta-path P_m, we generate a candidate semantic subgraph adjacency matrix $\mathbf{S}_{r,m}^{MP} \in \mathbb{R}^{|V_{\phi_h(r)}| \times |V_{\phi_t(r)}|}$, where each edge is calculated by:

$$\mathbf{S}_{r,m}^{MP}[i, j] = \begin{cases} \Gamma_{r,m}^{MP}(\mathbf{z}_i^m, \mathbf{z}_j^m) & \Gamma_{r,m}^{MP}(\mathbf{z}_i^m, \mathbf{z}_j^m) \geq \epsilon^{MP} \\ 0 & \text{otherwise,} \end{cases} \tag{4.35}$$

where $\mathbf{z}_i^m$ stands for the ith row of $\mathbf{Z}_{P_m}$, and $\Gamma_{r,m}^{MP}$ is the metric learning function with parameters $\mathbf{W}_{r,m}^{MP}$. We can see that a relation r will generate M candidate semantic subgraphs, so the overall semantic subgraph for relation r, denoted as $\mathbf{S}_r^{Sem}$, can be obtained by aggregating them:

$$\mathbf{S}_r^{Sem} = \Psi_r^{MP}\left(\left[\mathbf{S}_{r,1}^{MP}, \mathbf{S}_{r,2}^{MP}, \cdots, \mathbf{S}_{r,M}^{MP}\right]\right), \tag{4.36}$$

where $[\mathbf{S}_{r,1}^{MP}, \mathbf{S}_{r,2}^{MP}, \ldots, \mathbf{S}_{r,M}^{MP}]$ is the stacked matrix of M candidate semantic graphs. Ψ_r^{MP} denotes a channel attention layer whose weight matrix $\mathbf{W}_{\Psi,r}^{MP} \in \mathbb{R}^{1 \times 1 \times M}$ represents the importance of different meta-path based candidate graphs. After we obtain the aggregated semantic graph $\mathbf{S}_r^{Sem}$, the overall generated graph structure $\mathbf{A}_r'$ for relation r can be obtained by aggregating the learned feature graph and semantic graph along with the original graph structure:

$$\mathbf{A}_r' = \Psi_r\left(\left[\mathbf{S}_r^{Feat}, \mathbf{S}_r^{Sem}, \mathbf{A}_r\right]\right), \tag{4.37}$$

where $[\mathbf{S}_r^{Feat}, \mathbf{S}_r^{Sem}, \mathbf{A}_r] \in \mathbb{R}^{|V_{\phi_h(r)}| \times |V_{\phi_t(r)}| \times 3}$ is the stacked matrix of the candidate graphs. Ψ_r is the channel attention layer whose weight matrix $\mathbf{W}_{\Psi,r} \in \mathbb{R}^{1 \times 1 \times 3}$ denotes the importance of candidate graphs in fusing the overall relation subgraph $\mathbf{A}_r'$. With a new relation adjacency matrix $\mathbf{A}_r'$ for each relation r, a new heterogeneous graph structure is generated, i.e. $\mathcal{A}' = \{\mathbf{A}_r', r \in \mathcal{R}\}$.

4.4.2.4 Optimization

In this section, we show how HGSL jointly optimizes the graph structure $\mathcal{A}'$ and the GNN parameters for downstream task. Here we focus on GCN [20] and node classification. Please note that, with the learned graph structure $\mathcal{A}'$, our model can be applied to other homogeneous or heterogeneous GNN methods and other tasks. A two layer GCN with parameters $\theta = (\mathbf{W}_1, \mathbf{W}_2)$ on the learned graph structure $\mathcal{A}'$, can be described as:

$$f_\theta(\mathbf{X}, \mathbf{A}') = \text{softmax}\left(\hat{\mathbf{A}}\sigma\left(\hat{\mathbf{A}}\mathbf{X}\mathbf{W}_1\right)\mathbf{W}_2\right), \tag{4.38}$$

where $\mathbf{X}$ is the original node feature matrix, i.e. $\mathbf{X}[i, :] = \mathbf{f}_i^T$ if the dimensions of all features are identical; otherwise, we use the common feature to construct $\mathbf{X}$, i.e. $\mathbf{X}[i, :] = \mathbf{f}_i'^T$. The adjacency matrix $\mathbf{A}'$ is constructed from the learned heterogeneous graph $\mathcal{A}'$ by considering all nodes as one type. $\hat{\mathbf{A}} = \tilde{\mathbf{D}}^{-1/2}(\mathbf{A}' + \mathbf{I})\tilde{\mathbf{D}}^{-1/2}$, where $\tilde{\mathbf{D}}_{ii} = 1 + \sum_j \mathbf{A}'_{ij}$. Thus, the classification loss of GNN, i.e. $\mathcal{L}_{GNN}$, on the learned graph can be obtained by:

$$\mathcal{L}_{GNN} = \sum_{v_i \in V_L} \ell\left(f_\theta(\mathbf{X}, \mathbf{A}')_i, y_i\right), \tag{4.39}$$

where $f_\theta(\mathbf{X}, \mathbf{A}')_i$ is the predicted label of node $v_i \in V_L$ and $\ell(\cdot, \cdot)$ measures the difference between prediction and the true label y_i such as cross-entropy.

Since graph structure learning methods enable the original GNN with stronger ability to fit the downstream task, it would be easier for them to over-fit. Thus, we apply regularization term $\mathcal{L}_{reg}$ to the learned graph as follows:

$$\mathcal{L}_{reg} = \alpha\|\mathbf{A}'\|_1. \tag{4.40}$$

This term encourages the learned graph to be sparse. The overall loss $\mathcal{L}$ can be obtained by:

$$\mathcal{L} = \mathcal{L}_{GNN} + \mathcal{L}_{reg}. \tag{4.41}$$

By minimizing $\mathcal{L}$, HGSL optimizes heterogeneous graph structure and the GNN parameters θ jointly towards better downstream task performance.

4.4.3 Experiments

4.4.3.1 Experimental Settings

Datasets Three datasets including Yelp,[15] ACM,[16] and DBLP[17] are used to evaluate the proposed model.

Baselines HGSL are compared with eleven state-of-the-art representation methods including four homogeneous graph representation methods, i.e., DeepWalk [27], GCN [20], GAT [33], and GraphSAGE [9], four heterogeneous graph representation methods, i.e., MP2Vec [5], HAN [35], HeGAN [12], and GTN [38], and three graph structure learning related methods, i.e. LDS [7], Pro-GNN [17], and Geom-GCN [25].

Experimental Settings For all GNN-related models, the number of layers is set as 2 for a fair comparison. The feature dimension in common space d_c and the representation dimension d for all methods are set as 16 and 64, respectively. We choose the popular meta-paths adopted in previous methods [5, 24, 35] for meta-path based models and report the best result. For our proposed model, we use 2-head cosine similarity function defined in Eq. 4.30, i.e. $K = 2$. We set learning rate and weight decay as 0.01 and 0.0005, respectively. Other hyper-parameters, namely ε^{FS}, ε^{FP}, ε^{MP}, and α, are tuned by grid search. The code and datasets are publicly available on Github.[18]

4.4.3.2 Node Classification

In this section, the performance of HGSL on node classification task is shown. Macro-F1 and Micro-F1 are selected as the metrics for evaluation. The mean and standard deviation of percentage of the metric values are shown in Table 4.3, from which we have following observations: (1) With the capability to adaptively learn the heterogeneous graph structure, HGSL consistently outperforms all the baselines. It demonstrates the effectiveness of our proposed model. (2) Graph structure learning methods generally outperform the original GCN since it enables GCN to aggregate feature from the learned structure. (3) HGNN methods, i.e. HAN, GTN, and HGSL achieve better performance compared to GNNs since the heterogeneity is addressed. (4) GNN-based methods mostly outperform random walk-based graph representation methods since the node features are utilized. This phenomenon becomes more obvious when it comes to Yelp dataset, since the node features, i.e. keywords, are helpful in classifying business categories.

---

[15] https://www.yelp.com.

[16] http://dl.acm.org/.

[17] https://dblp.uni-trier.de.

[18] https://github.com/Andy-Border/HGSL.

Table 4.3 Performance evaluation of node classification (mean in percentage ± standard deviation) (bold means the best results)

	DBLP		ACM		Yelp	
	Macro-F1	Micro-F1	Macro-F1	Micro-F1	Macro-F1	Micro-F1
DeepWalk	88.00 ± 0.47	89.13 ± 0.41	80.65 ± 0.60	80.32 ± 0.61	68.68 ± 0.83	73.16 ± 0.96
GCN	83.38 ± 0.67	84.40 ± 0.64	91.32 ± 0.61	91.22 ± 0.64	82.95 ± 0.43	85.22 ± 0.55
GAT	77.59 ± 0.72	78.63 ± 0.72	92.96 ± 0.28	92.86 ± 0.29	84.35 ± 0.74	86.22 ± 0.56
GraphSage	78.37 ± 1.17	79.39 ± 1.17	91.19 ± 0.36	91.12 ± 0.36	93.06 ± 0.35	92.08 ± 0.31
MP2Vec	88.86 ± 0.19	89.98 ± 0.17	78.63 ± 1.11	78.27 ± 1.14	59.47 ± 0.57	65.11 ± 0.53
HAN	90.53 ± 0.24	91.47 ± 0.22	91.67 ± 0.39	91.57 ± 0.38	88.49 ± 1.73	88.78 ± 1.40
HeGAN	87.02 ± 0.37	88.34 ± 0.38	82.04 ± 0.77	81.80 ± 0.79	62.41 ± 0.76	68.17 ± 0.79
GTN	90.42 ± 1.29	91.41 ± 1.09	91.91 ± 0.58	91.78 ± 0.59	92.84 ± 0.28	92.19 ± 0.29
LDS	75.65 ± 0.20	76.63 ± 0.18	92.14 ± 0.16	92.07 ± 0.15	85.05 ± 0.16	86.05 ± 0.50
Pro-GNN	89.20 ± 0.15	90.28 ± 0.16	91.62 ± 1.28	91.55 ± 1.31	74.12 ± 2.03	77.45 ± 2.12
Geom-GCN	79.43 ± 1.01	80.94 ± 1.06	70.20 ± 1.23	70.00 ± 1.06	84.28 ± 0.70	85.36 ± 0.60
HGSL	**91.92 ± 0.11**	**92.77 ± 0.11**	**93.48 ± 0.59**	**93.37 ± 0.59**	**93.55 ± 0.52**	**92.76 ± 0.60**

4.4.3.3 Importance Analysis of Candidate Graphs

In order to investigate whether HGSL can distinguish the importance of candidate graphs, we analyze the weight distribution of the channel attention layer for fusing each relation subgraphs, i.e. the weights of Ψ_r in Eq. 4.37, on three datasets. We train HGSL 20 times and set all the thresholds of HGSL as 0.2. The attention distributions are shown in Fig. 4.9. As we can observe, for relation subgraphs in ACM and DBLP, the original graph structure is the most important structure for GNN-based classification. However, as for Yelp, the channel attention values of different relation subgraphs differ from each other. Specifically, for B-U (business-user) and B-L (business-rating level) relation subgraphs, the feature graphs are assigned with large channel attention value in graph structure learning. This phenomenon implies that the information in node features plays a more important role than that of semantic representations which agrees with the previously discussed experiments and further demonstrates the capability of HGSL in adaptively learning a larger channel attention value for more important information.

The detailed method description and validation experiments can been seen in [41].

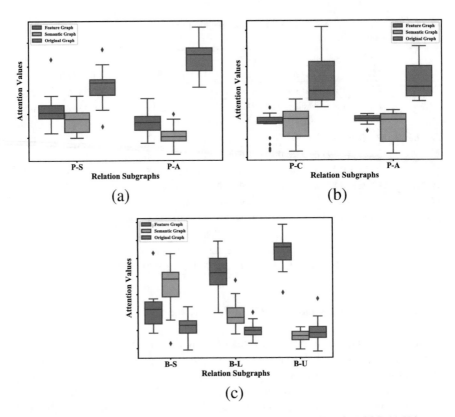

Fig. 4.9 Channel attention distributions of relation subgraphs. (**a**) ACM. (**b**) DBLP. (**c**) Yelp

4.5   Conclusions

In attribute-assisted HG, diverse types of nodes are assisted with different attributes, reflecting the characteristics of nodes. In this chapter, we introduce three attribute-assisted HG representation models including HAN, HPN, and HGSL, which simultaneously integrate both structural information and attribute information into the node representations. Taking node attributes and graph structures as inputs, HAN learns the importance of neighbor and meta-path and then aggregates them to learn node representation in a hierarchical manner. Further, HPN improves the node-level aggregating process in HAN via emphasizing the local semantic of each node, significantly alleviating the deep degradation phenomenon (a.k.a, semantic confusion). After that, HGSL mines the complex correlations from heterogeneous node features and graph structures, and then jointly learns the GNN parameters and graph structure simultaneously.

References

1. Bahdanau, D., Cho, K., Bengio, Y.: Neural machine translation by jointly learning to align and translate. In: International Conference on Learning Representations (2015)
2. Chen, T., Sun, Y.: Task-guided and path-augmented heterogeneous network embedding for author identification. In: Proceedings of the Tenth ACM International Conference on Web Search and Data Mining, pp. 295–304 (2017)
3. Chen, Y., Wu, L., Zaki, M.J.: Deep iterative and adaptive learning for graph neural networks. Preprint. arXiv:1912.07832 (2019)
4. Dong, Y., Chawla, N.V., Swami, A.: metapath2vec: Scalable representation learning for heterogeneous networks. In: Proceedings of the 23rd ACM SIGKDD International Conference on Knowledge Discovery and Data Mining, pp. 135–144 (2017)
5. Dong, Y., Chawla, N.V., Swami, A.: metapath2vec: Scalable representation learning for heterogeneous networks. In: Proceedings of the 23rd ACM SIGKDD International Conference on Knowledge Discovery and Data Mining, pp. 135–144 (2017)
6. Fan, S., Zhu, J., Han, X., Shi, C., Hu, L., Ma, B., Li, Y.: Metapath-guided heterogeneous graph neural network for intent recommendation. In: Proceedings of the 25th ACM SIGKDD International Conference on Knowledge Discovery & Data Mining, pp. 2478–2486. ACM, New York (2019)
7. Franceschi, L., Niepert, M., Pontil, M., He, X.: Learning discrete structures for graph neural networks. In: International Conference on Machine Learning, pp. 1972–1982 (2019)
8. Fu, X., Zhang, J., Meng, Z., King, I.: MAGNN: metapath aggregated graph neural network for heterogeneous graph embedding. In: Proceedings of The Web Conference, pp. 2331–2341 (2020)
9. Hamilton, W.L., Ying, R., Leskovec, J.: Inductive representation learning on large graphs. In: Proceedings of the 31st International Conference on Neural Information Processing Systems, pp. 1024–1034 (2017)
10. Hamilton, W., Bajaj, P., Zitnik, M., Jurafsky, D., Leskovec, J.: Embedding logical queries on knowledge graphs. In: Advances in Neural Information Processing Systems, pp. 2030–2041 (2018)
11. Hong, H., Guo, H., Lin, Y., Yang, X., Li, Z., Ye, J.: An attention-based graph neural network for heterogeneous structural learning. In: Proceedings of the AAAI Conference on Artificial Intelligence, pp. 4132–4139 (2020)
12. Hu, B., Fang, Y., Shi, C.: Adversarial learning on heterogeneous information networks. In: Proceedings of the 25th ACM SIGKDD International Conference on Knowledge Discovery & Data Mining, pp. 120–129 (2019)
13. Hu, L., Yang, T., Shi, C., Ji, H., Li, X.: Heterogeneous graph attention networks for semi-supervised short text classification. In: Proceedings of the 2019 Conference on Empirical Methods in Natural Language Processing and the 9th International Joint Conference on Natural Language Processing, pp. 4820–4829 (2019)
14. Hu, Z., Dong, Y., Wang, K., Sun, Y.: Heterogeneous graph transformer. In: Proceedings of The Web Conference 2020, pp. 2704–2710 (2020)
15. Ji, H., Wang, X., Shi, C., Wang, B., Yu, P.: Heterogeneous graph propagation network. IEEE Trans. Knowl. Data Eng. (2021)
16. Jiang, B., Zhang, Z., Lin, D., Tang, J., Luo, B.: Semi-supervised learning with graph learning-convolutional networks. In: Proceedings of the IEEE/CVF Conference on Computer Vision and Pattern Recognition, pp. 11313–11320 (2019)
17. Jin, W., Ma, Y., Liu, X., Tang, X., Wang, S., Tang, J.: Graph structure learning for robust graph neural networks. In: Proceedings of the 26th ACM SIGKDD International Conference on Knowledge Discovery & Data Mining, pp. 66–74 (2020)
18. Keyulu, X., Chengtao, L., Yonglong, T., Tomohiro, S., Ken-ichi, K., Stefanie, J.: Representation learning on graphs with jumping knowledge networks. In: International Conference on Machine Learning, pp. 5453–5462 (2018)

19. Kingma, D.P., Ba, J.: Adam: a method for stochastic optimization. In: International Conference on Learning Representations (2015)
20. Kipf, T. N., Welling, M.: Semi-supervised classification with graph convolutional networks. In: International Conference on Learning Representations (2017)
21. Klicpera, J., Bojchevski, A., Gunnemann, S.: Predict then propagate: graph neural networks meet personalized pagerank. In: International Conference on Learning Representations (2019)
22. Lee, S., Park, S., Kahng, M., Lee, S.-G.: PathRank: ranking nodes on a heterogeneous graph for flexible hybrid recommender systems. Expert Syst. Appl. **40**(2), 684–697 (2013)
23. Li, X., Wu, Y., Ester, M., Kao, B., Wang, X., Zheng, Y.: Semi-supervised clustering in attributed heterogeneous information networks. In: Proceedings of the 26th International Conference on World Wide Web, pp. 1621–1629 (2017)
24. Lu, Y., Shi, C., Hu, L., Liu, Z.: Relation structure-aware heterogeneous information network embedding. In: Proceedings of the AAAI Conference on Artificial Intelligence, pp. 4456–4463 (2019)
25. Pei, H., Wei, B., Chang, K.C., Lei, Y., Yang, B.: Geom-GCN: geometric graph convolutional networks. In: International Conference on Learning Representations (2020)
26. Perozzi, B., Al-Rfou, R., Skiena, S.: DeepWalk: online learning of social representations. In: Proceedings of the 20th ACM SIGKDD International Conference on Knowledge Discovery and Data Mining, pp. 701–710 (2014)
27. Perozzi, B., Al-Rfou, R., Skiena, S.: DeepWalk: online learning of social representations. In: Proceedings of the 20th ACM SIGKDD International Conference on Knowledge Discovery and Data Mining, pp. 701–710 (2014)
28. Schlichtkrull, M., Kipf, T.N., Bloem, P., van den Berg, R., Titov, I., Welling, M.: Modeling relational data with graph convolutional networks. In: European Semantic Web Conference, pp. 593–607. Springer, Berlin (2018)
29. Shang, J., Qu, M., Liu, J., Kaplan, L.M., Han, J., Peng, J.: Meta-path guided embedding for similarity search in large-scale heterogeneous information networks. CoRR abs/1610.09769 (2016)
30. Shi, C., Hu, B., Zhao, W.X., Philip, S.Y.: Heterogeneous information network embedding for recommendation. IEEE Trans. Knowl. Data Eng. **31**(2), 357–370 (2018)
31. Sun, Y., Han, J., Yan, X., Yu, P.S., Wu, T.: PathSim: meta path-based top-k similarity search in heterogeneous information networks. VLDB **4**(11), 992–1003 (2011)
32. Vaswani, A., Shazeer, N., Parmar, N., Uszkoreit, J., Jones, L., Gomez, A.N., Kaiser, L., Polosukhin, I.: Attention is all you need. In: Neural Information Processing Systems (NIPS 2017), pp. 5998–6008 (2017)
33. Veličković, P., Cucurull, G., Casanova, A., Romero, A., Liò, P., Bengio, Y.: Graph attention networks. In: International Conference on Learning Representations (2018)
34. Wang, X., He, X., Wang, M., Feng, F., Chua, T.-S.: Neural graph collaborative filtering. In: SIGIR'19: Proceedings of the 42nd International ACM SIGIR Conference on Research and Development in Information Retrieval, pp. 165–174 (2019)
35. Wang, X., Ji, H., Shi, C., Wang, B., Ye, Y., Cui, P., Yu, P.S.: Heterogeneous graph attention network. In: The World Wide Web Conference, pp. 2022–2032 (2019)
36. Wang, X., Zhu, M., Bo, D., Cui, P., Shi, C., Pei, J.: AM-GCN: adaptive multi-channel graph convolutional networks. In: Gupta, R., Liu, Y., Tang, J., Prakash, B.A. (eds.) Proceedings of the 26th ACM SIGKDD International Conference on Knowledge Discovery & Data Mining (2020)
37. Xu, K., Ba, J., Kiros, R., Cho, K., Courville, A., Salakhudinov, R., Zemel, R., Bengio, Y.: Show, attend and tell: neural image caption generation with visual attention. In: International Conference on Machine Learning, pp. 2048–2057 (2015)
38. Yun, S., Jeong, M., Kim, R., Kang, J., Kim, H.J.: Graph transformer networks. In: Advances in Neural Information Processing Systems, pp. 11960–11970 (2019)
39. Zhang, C., Song, D., Huang, C., Swami, A., Chawla, N.V.: Heterogeneous graph neural network. In: Proceedings of the 25th ACM SIGKDD International Conference on Knowledge Discovery & Data Mining, pp. 793–803 (2019)

40. Zhang, C., Swami, A., Chawla, N.V.: SHNE: representation learning for semantic-associated heterogeneous networks. In: Proceedings of the Twelfth ACM International Conference on Web Search and Data Mining, pp. 690–698 (2019)
41. Zhao, J., Wang, X., Shi, C., Hu, B., Song, G., Ye, Y.: Heterogeneous graph structure learning for graph neural networks. In: 35th AAAI Conference on Artificial Intelligence (AAAI) (2021)
42. Zhao, J., Wang, X., Shi, C., Liu, Z., Ye, Y.: Network schema preserving heterogeneous information network embedding. In: 29th International Joint Conference on Artificial Intelligence, pp. 1366–1372 (2020)

Chapter 5
Dynamic Heterogeneous Graph Representation

Abstract Graphs are gradually generated with multiple temporal heterogeneous interactions in real-world scenarios, containing both abundant structures and complex dynamics. Compared to static heterogeneous graphs, the dynamics express not only the changing graph topology but also the sequential evolution as well as multiple temporal preferences, indicating the necessity of dynamic heterogeneous graph modeling. This chapter focuses on simultaneously modeling both the evolving dynamics and heterogeneous semantics, and introduces three representative approaches, including DyHNE to handle structure changes via matrix perturbation theory based incremental learning, SHCF to tackle evolving sequences via heterogeneous sequential neural collaborative filtering, and THIGE to model multiple long-and short-term preferences via temporal heterogeneous GNNs.

5.1 Introduction

Heterogeneous graphs in real-world scenarios usually exhibit high dynamics with the evolution of various types of nodes and edges, e.g., the newly added (deleted) nodes or edges, forming the dynamic heterogeneous graph or temporal heterogeneous graph. The complex but valuable temporal heterogeneity introduces essential challenges for representation. Specifically, heterogeneous interactions are accumulated over time, leading to the continuous changes of graphic topologies. Besides, interactions are often sequentially generated, which indicate the corresponding evolution of interests. Furthermore, there are multiple temporal interactions, expressing both heterogeneous sequential evolution of current demands and multi-view historical habits.

However, the current heterogeneous graph embedding methods [3, 6, 16, 36, 42] cannot model the dynamics within structural semantics, which have to be retrained repeatedly at each time step. Focusing on modeling the evolving dynamics, some researchers attempt to integrate with Recurrent Neural Networks (RNNs) [13, 14, 22, 31] and Transformers [18, 37, 46], and some recent approaches [25, 26, 39] pay attention to combine both short-term and long-term interests to generate fine-

© The Author(s), under exclusive license to Springer Nature Singapore Pte Ltd. 2022   107
C. Shi et al., *Heterogeneous Graph Representation Learning and Applications*,
Artificial Intelligence: Foundations, Theory, and Algorithms,
https://doi.org/10.1007/978-981-16-6166-2_5

grained node representations. Unfortunately, almost all of these sequential methods construct user embeddings based on only their own sequential item interactions while ignoring the abundant heterogeneous information.

In this chapter, we introduce three dynamic heterogeneous graph representation learning methods to solve the challenges in incremental learning, sequential information, and temporal interactions. First, we introduce the **Dy**namic **H**eterogeneous **N**etwork **E**mbedding (named **DyHNE**) [26] to handle the incremental learning of temporal semantics by utilizing matrix perturbation theory. Second, we introduce the **S**equence-aware **H**eterogeneous graph neural **C**ollaborative **F**iltering (called **SHCF**) [24] to model the heterogeneous evolution of sequential information for recommendation. Third, we introduce the **T**emporal **H**eterogeneous **I**nteraction **G**raph **E**mbedding (named **THIGE**) [17] to model both long-term habits and short-term demands of temporal interactions via temporal heterogeneous GNNs.

5.2   Incremental Learning

5.2.1   Overview

Heterogeneous graph are often gradually formatted with dynamic edges. The current HG embedding methods can hardly handle such complex evolution effectively in a dynamic heterogeneous graph. Basically, there are two fundamental problems which need to be carefully considered for dynamic heterogeneous graph embedding. One is how to effectively preserve the structure and semantics in a dynamic heterogeneous graph. As the heterogeneous graph evolves with a newly added node, the local structure centered on this node will be changed, and such changes will be gradually propagated across all the nodes via different meta-paths, leading to changes in the global structure. The other problem is how to efficiently update the node embeddings without retraining on the whole heterogeneous graph, when the heterogeneous graph evolves over time. For each time step, retraining a heterogeneous graph embedding method is the most straightforward way to get the optimal embeddings. However, apparently, this strategy is very time-consuming, especially when the change of network structure is very slight. In the era of big data, retraining manner becomes unrealistic. These problems motivate us to seek an effective and efficient method to preserve the structure and semantics for dynamic heterogeneous graph embedding.

This section designs the DyHNE with meta-path based proximity to effectively and efficiently learn the node embeddings. Inspired by the perturbation theory [10] widely used for capturing changes of a system, we learn the node embeddings by solving the generalized eigenvalue problem and model the evolution of the heterogeneous graph with the eigenvalue perturbation. We firstly adopt meta-path augmented adjacency matrices to capture the structure and semantics in dynamic heterogeneous graphs. For capturing the evolution of the heterogeneous graph, we then utilize the perturbations of multiple meta-path augmented adjacency matrices to model the changes of the structure and semantics of the heterogeneous graph in a natural manner. Finally, we employ the eigenvalue perturbation theory to incorporate the changes and derive the node embeddings efficiently.

5.2.2 The DyHNE Model

The core idea of DyHNE is to build an effective and efficient architecture that can capture the changes of structure and semantics in a dynamic heterogeneous graph and derive the node embeddings efficiently.

To achieve this, we first introduce the meta-path based first-order and second-order proximity to preserve structure and semantics in heterogeneous graphs. As shown in Fig. 5.1, three augmented adjacency matrices based on meta-path *APA*, *APCPA*, and *APTPA* are defined and fused with weights, which gives rise to the fused matrix $\mathbf{W}^{(t)}$ at time t. Then, we learn node embeddings $\mathbf{U}^{(t)}$ by solving the generalized eigenvalue problem in terms of the fused matrix $\mathbf{W}^{(t)}$. As the heterogeneous graph evolves from time t to $t + 1$, new nodes and edges are added into the network (i.e. nodes a_3, p_4, and t_3; edges (a_3, p_4), (a_1, p_4), (p_4, c_2), (p_4, t_2), and (p_4, t_3)), leading to the changes of meta-path augmented adjacency matrices. Since these matrices are actually the realization of structure and semantics in the heterogeneous graph, we naturally capture changes of structure and semantics with the perturbation of the fused matrix (i.e. $\Delta\mathbf{W}$). Further, we tailor the embeddings update formulas for dynamic heterogeneous graph with matrix perturbation theory, so that our DyHNE can efficiently derive the changed embedding $\Delta\mathbf{U}$ and update network embedding from $\mathbf{U}^{(t)}$ to $\mathbf{U}^{(t+1)}$ with $\mathbf{U}^{(t+1)} = \Delta\mathbf{U} + \mathbf{U}^{(t)}$.

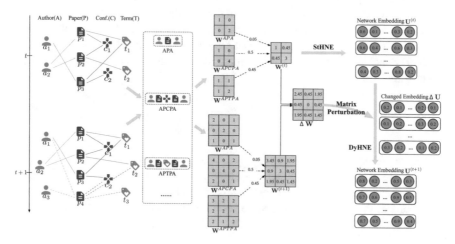

Fig. 5.1 The overall architecture of the proposed StHNE and DyHNE

In a nutshell, the proposed StHNE is capable of capturing the structures and semantics in a heterogeneous graph with meta-path based first-order and second-order proximity, and DyHNE achieves the efficient update of network embeddings with the perturbation of meta-path augmented adjacency matrices.

5.2.2.1 Static Modeling

Before achieving effective update node embeddings when the heterogeneous graph evolves over time, a proper static heterogeneous graph embedding for capturing structural and semantic information is a must. Hence, we next propose a **Static Heterogeneous Network Embedding** model (named **StHNE**), which preserves the meta-path based first- and second-order proximity. The meta-path based first-order proximity models the local proximity in heterogeneous graphs, which means that the nodes connected via path instances are similar. Given a node pair (v_i, v_j) connected via path instances following m, we model the meta-path based first-order proximity as:

$$p_1^m(v_i, v_j) = w_{ij}^m \|\mathbf{u}_i - \mathbf{u}_j\|_2^2, \tag{5.1}$$

where $\mathbf{u}_i \in \mathbb{R}^d$ is the d-dimension representation vector of node v_i. To preserve the meta-path based first-order proximity in heterogeneous graphs, we minimize the following objective function:

$$\mathcal{L}_1^m = \sum_{v_i, v_j \in \mathcal{V}} w_{ij}^m \|\mathbf{u}_i - \mathbf{u}_j\|_2^2. \tag{5.2}$$

As larger w_{ij}^m indicates that v_i and v_j have more connections via the meta-path m, which makes nodes v_i and v_j closer in the low-dimensional space.

The meta-path based second-order proximity is determined through the shared neighborhood structure of nodes. Given the neighbors of node v_p under the meta-path m, denoted as $\mathcal{N}(v_p)^m$, we can model the second-order proximity based on meta-path as follows:

$$p_2^m(v_p, \mathcal{N}(v_p)^m) = \|\mathbf{u}_p - \sum_{v_q \in \mathcal{N}(v_p)^m} w_{pq}^m \mathbf{u}_q\|_2^2. \tag{5.3}$$

Here, we normalize w_{pq}^m so that $\sum_{v_q \in \mathcal{N}(v_p)^m} w_{pq} = 1$.

With Eq. (5.3), we keep the node p close to its neighbors under a specific meta-path. Equation (5.3) guarantees that unconnected nodes are close to each other if they contain the similar neighbors. To preserve the meta-path based second-order proximity in heterogeneous graphs, we minimize the following object function, namely

$$\mathcal{L}_2^m = \sum_{v_p \in \mathcal{V}} ||\mathbf{u}_p - \sum_{v_q \in \mathcal{N}(v_p)^m} w_{pq}^m \mathbf{u}_q||_2^2. \tag{5.4}$$

Intuitively, minimizing Eq. (5.4) will cause the small distance between node v_p and its neighbors in the low-dimensional space. Thus, nodes that share the same neighbors with node v_p will also be close to v_p. In this way, the meta-path based second-order proximity can be preserved. Considering multiple semantic relations in a heterogeneous graph, we define a set of meta-paths $\mathcal{M}$ and assign weights $\{\theta_1, \theta_2, \ldots, \theta_{|\mathcal{M}|}\}$ to each meta-path, where $\forall \theta_i > 0$ and $\sum_{i=1}^{|\mathcal{M}|} \theta_i = 1$. Thus, our unified model combines multiple meta-paths while preserving both of the meta-path based first- and second-order proximity, namely

$$\mathcal{L} = \sum_{m \in \mathcal{M}} \theta_m (\mathcal{L}_1^m + \gamma \mathcal{L}_2^m), \tag{5.5}$$

where γ is the trade-off factor. Now, the static heterogeneous graph embedding problem is turned to:

$$\arg \min_{\mathbf{U}^\top \mathbf{D} \mathbf{U} = \mathbf{I}} \sum_{m \in \mathcal{M}} \theta_m (\mathcal{L}_1^m + \gamma \mathcal{L}_2^m), \tag{5.6}$$

where $\mathbf{D}$ is the degree matrix that will be described later. The constraint $\mathbf{U}^\top \mathbf{D} \mathbf{U} = \mathbf{I}$ removes an arbitrary scaling factor in the embedding and avoids the degenerate case where all node embeddings are equal.

Inspired by spectral theory [2, 27], we transform the problem of Eq. (5.6) as the generalized eigenvalue problem, so that we can get a closed-form solution and dynamically update embeddings with the eigenvalue perturbation theory [10]. Hence, we reformulate Eq. (5.2) as follows:

$$\mathcal{L}_1^m = \sum_{v_i, v_j \in \mathcal{V}} w_{ij}^m ||\mathbf{u}_i - \mathbf{u}_j||_2^2 = 2tr(\mathbf{U}^\top \mathbf{L}^m \mathbf{U}), \tag{5.7}$$

where $tr(\cdot)$ is the trace of the matrix, $\mathbf{U}$ is the embedding matrix, $\mathbf{L}^m = \mathbf{D}^m - \mathbf{W}^m$ is the Laplacian matrix under the meta-path m, and $\mathbf{D}^m$ is a diagonal matrix with $\mathbf{D}_{ii}^m = \sum_j w_{ij}^m$. Similarly, Eq. (5.4) can be rewritten as follows:

$$\mathcal{L}_2^m = \sum_{v_p \in \mathcal{V}} ||\mathbf{u}_p - \sum_{v_q \in \mathcal{N}(v_p)^m} w_{pq}^m \mathbf{u}_q||_2^2 = 2tr(\mathbf{U}^\top \mathbf{H}^m \mathbf{U}), \tag{5.8}$$

where $\mathbf{H}^m = (\mathbf{I} - \mathbf{W}^m)^\top (\mathbf{I} - \mathbf{W}^m)$ is symmetric. As discussed earlier, we fuse all meta-paths in $\mathcal{M}$, which gives rise to:

$$\mathbf{W} = \sum_{m \in \mathcal{M}} \theta_m \mathbf{W}^m, \qquad \mathbf{D} = \sum_{m \in \mathcal{M}} \theta_m \mathbf{D}^m. \qquad (5.9)$$

Hence, the StHNE can be reformulated as:

$$\mathcal{L} = tr(\mathbf{U}^\top (\mathbf{L} + \gamma \mathbf{H})\mathbf{U})), \qquad (5.10)$$

where $\mathbf{L} = \mathbf{D} - \mathbf{W}$ and $\mathbf{H} = (\mathbf{I} - \mathbf{W})^\top (\mathbf{I} - \mathbf{W})$. Now, the problem of static heterogeneous graph embedding reduces to:

$$\arg \min_{\mathbf{U}^\top \mathbf{D} \mathbf{U} = \mathbf{I}} tr(\mathbf{U}^\top (\mathbf{L} + \gamma \mathbf{H})\mathbf{U}), \qquad (5.11)$$

where $\mathbf{L} + \gamma \mathbf{H}$ is symmetric. The problem of Eq. (5.11) boils down to the generalized eigenvalue problem [40] as follows:

$$(\mathbf{L} + \gamma \mathbf{H})\mathbf{U} = \mathbf{D}\lambda\mathbf{U}, \qquad (5.12)$$

where $\lambda = diag(\lambda_1, \lambda_2, \ldots, \lambda_{N_\mathcal{M}})$ is the eigenvector matrix, $N_\mathcal{M}$ is the number of nodes in the meta-path set $\mathcal{M}$.

Having transformed the StHNE as the generalized eigenvalue problem, the embedding matrix $\mathbf{U}$ is given by the top-d eigenvectors with the smallest non-zero eigenvalues. As the heterogeneous graph evolves from time t to $t + 1$, the dynamic heterogeneous graph embedding model focuses on efficiently updating $\mathbf{U}^{(t)}$ to $\mathbf{U}^{(t+1)}$. That is, update the eigenvectors and eigenvalues.

5.2.2.2 Dynamic Modeling

The core idea of a dynamic heterogeneous graph embedding model is to learn node embeddings efficiently in a dynamic manner, thus we next to effectively update the eigenvectors and eigenvalues based on matrix perturbation.

Following the previous works [21, 50], we assume that the network evolves on a common node set of cardinality N. A nonexistent node is treated as an isolated node with zero degree, and thereby the evolution of a network can be regarded as the change of edges [1]. Besides, the addition (deletion) of edges may vary by types. It is naturally appealing to capture the evolution of a dynamic heterogeneous graph with the perturbation of meta-path augmented adjacency matrix $\Delta \mathbf{W} = \sum_{m \in \mathcal{M}} \theta_m \Delta \mathbf{W}^m$. Thus, the changes of $\mathbf{L}$ and $\mathbf{H}$ can be calculated as follows:

$$\Delta \mathbf{L} = \Delta \mathbf{D} - \Delta \mathbf{W}, \tag{5.13}$$

$$\Delta \mathbf{H} = \Delta \mathbf{W}^\top \Delta \mathbf{W} - (\mathbf{I} - \mathbf{W})^\top \Delta \mathbf{W} - \Delta \mathbf{W}^\top (\mathbf{I} - \mathbf{W}). \tag{5.14}$$

Since perturbation theory can give approximate solution to a problem by adding a perturbation term [10], we can update eigenvalues and eigenvectors from the eigenvalues and eigenvectors at the previous time with the eigenvalue perturbation. Hence, at new time step, we have the following equation based on Eq. (5.12):

$$(\mathbf{L} + \Delta \mathbf{L} + \gamma \mathbf{H} + \gamma \Delta \mathbf{H})(\mathbf{U} + \Delta \mathbf{U}) = (\mathbf{D} + \Delta \mathbf{D})(\lambda + \Delta \lambda)(\mathbf{U} + \Delta \mathbf{U}), \tag{5.15}$$

where $\Delta \mathbf{U}$ and $\Delta \lambda$ are the changes of the eigenvectors and eigenvalues. Here, we omit the (t) superscript for brevity since the perturbation process for any time step t is the same. Let us focus on a specific eigen-pair $(\mathbf{u}_i, \lambda_i)$, Eq. (5.15) is rewritten as

$$(\mathbf{L} + \Delta \mathbf{L} + \gamma \mathbf{H} + \gamma \Delta \mathbf{H})(\mathbf{u}_i + \Delta \mathbf{u}_i) = (\lambda_i + \Delta \lambda_i)(\mathbf{D} + \Delta \mathbf{D})(\mathbf{u}_i + \Delta \mathbf{u}_i). \tag{5.16}$$

Hence, the dynamic heterogeneous graph embedding problem is how to calculate the changes of the i-th eigen-pair $(\Delta \mathbf{u}_i, \Delta \lambda_i)$, because if we have $\Delta \mathbf{U}$ and $\Delta \lambda$ between t and $t + 1$, we can efficiently update the embedding matrix with $\mathbf{U}^{(t+1)} = \mathbf{U}^{(t)} + \Delta \mathbf{U}$.

We first introduce how to calculate $\Delta \lambda_i$. By expanding Eq. (5.16) and removing the higher-order terms that have limited effects on the accuracy of the solution [10], such as $\Delta \mathbf{L} \Delta \mathbf{u}_i$ and $\Delta \lambda_i \Delta \mathbf{D} \Delta \mathbf{u}_i$, then based on the fact $(\mathbf{L} + \gamma \mathbf{H})\mathbf{u}_i = \lambda_i \mathbf{D} \mathbf{u}_i$, we have

$$(\mathbf{L} + \gamma \mathbf{H})\Delta \mathbf{u}_i + (\Delta \mathbf{L} + \gamma \Delta \mathbf{H})\mathbf{u}_i = \lambda_i \mathbf{D} \Delta \mathbf{u}_i + \lambda_i \Delta \mathbf{D} \mathbf{u}_i + \Delta \lambda_i \mathbf{D} \mathbf{u}_i. \tag{5.17}$$

Furthermore, left multiplying both sides by $\mathbf{u}_i^\top$, we have

$$\mathbf{u}_i^\top (\mathbf{L} + \gamma \mathbf{H})\Delta \mathbf{u}_i + \mathbf{u}_i^\top (\Delta \mathbf{L} + \gamma \Delta \mathbf{H})\mathbf{u}_i = \lambda_i \mathbf{u}_i^\top \mathbf{D} \Delta \mathbf{u}_i + \lambda_i \mathbf{u}_i^\top \Delta \mathbf{D} \mathbf{u}_i + \Delta \lambda_i \mathbf{u}_i^\top \mathbf{D} \mathbf{u}_i. \tag{5.18}$$

As $\mathbf{L} + \gamma \mathbf{H}$ and $\mathbf{D}$ are symmetric, then based on the fact $(\mathbf{L} + \gamma \mathbf{H})\mathbf{u}_i = \lambda_i \mathbf{D} \mathbf{u}_i$ and right multiplying both sides by $\Delta \mathbf{u}_i$, we have $\mathbf{u}_i^\top (\mathbf{L} + \gamma \mathbf{H})\Delta \mathbf{u}_i = \lambda_i \mathbf{u}_i^\top \mathbf{D} \Delta \mathbf{u}_i$. Thus, we can rewrite Eq. (5.18) as follows:

$$\mathbf{u}_i^\top (\Delta \mathbf{L} + \gamma \Delta \mathbf{H})\mathbf{u}_i = \lambda_i \mathbf{u}_i^\top \Delta \mathbf{D} \mathbf{u}_i + \Delta \lambda_i \mathbf{u}_i^\top \mathbf{D} \mathbf{u}_i. \tag{5.19}$$

Based on Eq. (5.19), we get the changes of the eigenvalue λ_i:

$$\Delta\lambda_i = \frac{\mathbf{u}_i^\top \Delta\mathbf{L}\mathbf{u}_i + \gamma\mathbf{u}_i^\top \Delta\mathbf{H}\mathbf{u}_i - \lambda_i\mathbf{u}_i^\top \Delta\mathbf{D}\mathbf{u}_i}{\mathbf{u}_i^\top \mathbf{D}\mathbf{u}_i}. \tag{5.20}$$

It is easy to see that $\mathbf{D}$ is a positive-semidefinite matrix, so we have $\mathbf{u}_i^\top \mathbf{D}\mathbf{u}_i = 1$ and $\mathbf{u}_i^\top \mathbf{D}\mathbf{u}_j = 0 (i \neq j)$ [10, 29]. Thus,

$$\Delta\lambda_i = \mathbf{u}_i^\top \Delta\mathbf{L}\mathbf{u}_i + \gamma\mathbf{u}_i^\top \Delta\mathbf{H}\mathbf{u}_i - \lambda\mathbf{u}_i^\top \Delta\mathbf{D}\mathbf{u}_i. \tag{5.21}$$

Having got the change of eigenvalue $\Delta\lambda_i$ between two continuous time steps, our next goal is to calculate the changes of eigenvectors $\Delta\mathbf{u}_i$.

As a heterogeneous graph usually evolves smoothly [1], the network changes based on meta-paths (i.e., $\Delta\mathbf{W}$) are subtle. We assume the perturbation of the eigenvectors $\Delta\mathbf{u}_i$ is linearly weighted by the top-d eigenvectors with the smallest non-zero eigenvalues [10]:

$$\Delta\mathbf{u}_i = \sum_{j=2, j\neq i}^{d+1} \alpha_{ij}\mathbf{u}_j, \tag{5.22}$$

where α_{ij} indicates the weight of $\mathbf{u}_j$ on $\Delta\mathbf{u}_i$. Thus, the problem of calculating $\Delta\mathbf{u}_i$ now is transformed into how to determine these weights. Considering Eq. (5.16), by replacing $\Delta\mathbf{u}_i$ with Eq. (5.22) and removing the higher-order terms that have limited effects on the accuracy of the solution [19], we obtain the following:

$$(\mathbf{L} + \gamma\mathbf{H}) \sum_{j=2, j\neq i}^{d+1} \alpha_{ij}\mathbf{u}_j + (\Delta\mathbf{L} + \gamma\Delta\mathbf{H})\mathbf{u}_i$$
$$= \lambda_i\mathbf{D} \sum_{j=2, j\neq i}^{d+1} \alpha_{ij}\mathbf{u}_j + \lambda_i\Delta\mathbf{D}\mathbf{u}_i + \Delta\lambda_i\mathbf{D}\mathbf{u}_i. \tag{5.23}$$

With the fact that $(\mathbf{L} + \gamma\mathbf{H}) \sum_{j=2}^{d+1} \alpha_{ij}\mathbf{u}_j = \mathbf{D} \sum_{j=2}^{d+1} \alpha_{ij}\lambda_j\mathbf{u}_j$, and by multiplying $\mathbf{u}_p^\top (2 \leq p \leq d+1, p \neq i)$ on both sides of Eq. (5.23), we get

$$\mathbf{u}_p^\top\mathbf{D} \sum_{j=2, j\neq i}^{d+1} \alpha_{ij}\lambda_j\mathbf{u}_j + \mathbf{u}_p^\top(\Delta\mathbf{L} + \gamma\Delta\mathbf{H})\mathbf{u}_i$$
$$= \lambda_i\mathbf{u}_p^\top\mathbf{D} \sum_{j=2, j\neq i}^{d+1} \alpha_{ij}\mathbf{u}_j + \lambda_i\mathbf{u}_p^\top\Delta\mathbf{D}\mathbf{u}_i + \Delta\lambda_i\mathbf{u}_p^\top\mathbf{D}\mathbf{u}_i. \tag{5.24}$$

Based on $\mathbf{u}_i^\top \mathbf{Du}_i = 1$ and $\mathbf{u}_i^\top \mathbf{Du}_j = 0(i \neq j)$, we can simplify the above formula and get

$$\lambda_p \alpha_{ip} + \mathbf{u}_p^\top (\Delta \mathbf{L} + \gamma \Delta \mathbf{H})\mathbf{u}_i = \lambda_i \alpha_{ip} + \lambda_i \mathbf{u}_p^\top \Delta \mathbf{Du}_i. \tag{5.25}$$

Finally, we obtain the weight α_{ip} as follows:

$$\alpha_{ip} = \frac{\mathbf{u}_p^\top \Delta \mathbf{Lu}_i + \gamma \mathbf{u}_p^\top \Delta \mathbf{Hu}_i - \lambda_i \mathbf{u}_p^\top \Delta \mathbf{Du}_i}{\lambda_i - \lambda_p}, i \neq p. \tag{5.26}$$

To sum up, we now have the changes of eigenvalues and eigenvectors based on Eqs. (5.21), (5.22), and (5.26). The new eigenvalues and eigenvectors at $t + 1$ can be updated as follows:

$$\boldsymbol{\lambda}^{(t+1)} = \boldsymbol{\lambda}^{(t)} + \Delta \lambda, \qquad \mathbf{U}^{(t+1)} = \mathbf{U}^{(t)} + \Delta \mathbf{U}. \tag{5.27}$$

5.2.2.3 Acceleration

A straightforward idea to update the embeddings is to calculate Eqs. (5.21), (5.22), and (5.26) for Eq. (5.27). However, the calculation of Eq. (5.21) is time-consuming due to the definition of $\Delta \mathbf{H}$ (i.e., Eq. (5.14)). Thus, we propose an acceleration solution tailored for dynamic heterogeneous graph embedding.

Let us focus on $\Delta \lambda_i$ and α_{ij} in a more detailed way. We replace $\Delta \mathbf{H}$ with Eq. (5.14) and remove the higher-order terms as earlier, Eqs. (5.21) and (5.26) can be reformulated as follows:

$$\Delta \lambda_i = \mathbf{u}_i^\top \Delta \mathbf{Lu}_i - \lambda_i \mathbf{u}_i^\top \Delta \mathbf{Du}_i + \gamma \{[(\mathbf{W} - \mathbf{I})\mathbf{u}_i]^\top \Delta \mathbf{Wu}_i + (\Delta \mathbf{Wu}_i)^\top (\mathbf{W} - \mathbf{I})\mathbf{u}_i\}, \tag{5.28}$$

$$\alpha_{ij} = \frac{\mathbf{u}_j^\top \Delta \mathbf{Lu}_i - \lambda_i \mathbf{u}_j^\top \Delta \mathbf{Du}_i}{\lambda_i - \lambda_j} \tag{5.29}$$

$$+ \frac{\gamma \{[(\mathbf{W} - \mathbf{I})\mathbf{u}_j]^\top \Delta \mathbf{Wu}_i + (\Delta \mathbf{Wu}_j)^\top (\mathbf{W} - \mathbf{I})\mathbf{u}_i\}}{\lambda_i - \lambda_j}.$$

For the sake of convenience, we rewrite $\Delta \lambda_i$ and α_{ij} as follows:

$$\Delta \lambda_i = \mathbf{C}(i, i) + \gamma [\mathbf{A}(:, i)^\top \mathbf{B}(:, i) + \mathbf{B}(:, i)^\top \mathbf{A}(:, i)], \tag{5.30}$$

$$\alpha_{ij} = \frac{\mathbf{C}(j, i) + \gamma [\mathbf{A}(:, j)^\top \mathbf{B}(:, i) + \mathbf{B}(:, j)^\top \mathbf{A}(:, i)]}{\lambda_i - \lambda_j}, \tag{5.31}$$

where $\mathbf{A}(:, i) = (\mathbf{W} - \mathbf{I})\mathbf{u}_i$, $\mathbf{B}(:, i) = \Delta \mathbf{Wu}_i$, and $\mathbf{C}(i, j) = \mathbf{u}_i^\top \Delta \mathbf{Lu}_j - \lambda_i \mathbf{u}_i^\top \Delta \mathbf{Du}_j$.

Obviously, the calculation of $\mathbf{A}$ is time-consuming. Hence, we define $\mathbf{A}^{(t+1)}(:, i)$ at time step $t + 1$ as follows:

$$\mathbf{A}^{(t+1)}(:, i) = (\mathbf{W} - \mathbf{I} + \Delta \mathbf{W})(\mathbf{u}_i + \Delta \mathbf{u}_i). \tag{5.32}$$

Replacing $\Delta \mathbf{u}_i$ with Eq. (5.22), we have

$$\mathbf{A}^{(t+1)}(:, i) = (\mathbf{W} - \mathbf{I} + \Delta \mathbf{W}) \left(\mathbf{u}_i + \sum_{j=2, j \neq i}^{d+1} \alpha_{ij} \mathbf{u}_j \right) = \sum_{j=2}^{d+1} \beta_{ij} (\mathbf{W} - \mathbf{I} + \Delta \mathbf{W}) \mathbf{u}_j, \tag{5.33}$$

where $\beta_{ij} = \alpha_{ij}$ if $i \neq j$, otherwise, $\beta_{ij} = 1$. Furthermore, we can obtain the following:

$$\mathbf{A}^{(t+1)}(:, i) = \sum_{j=2}^{d+1} \beta_{ij} (\mathbf{A}^t(:, j) + \mathbf{B}^t(:, j)). \tag{5.34}$$

Now, we reduce the time complexity of updating $\mathbf{A}^{(t+1)}$ from $O(ed)$ to $O(d^2)$, which guarantees the efficiency of DyHNE.

5.2.3 Experiments

5.2.3.1 Experimental Settings

Datasets We evaluate models on three datasets, including two academic networks (i.e., DBLP and AMiner) and a social Yelp. Yelp dataset extracts information related to restaurants of three sub-categories: "American (New) Food", "Fast Food," and "Sushi Bars" [23]. The meta-paths that we are interested in are *BRURB* (i.e., the user reviewed on two businesses) and *BSB* (i.e., the same star level businesses). DBLP is an academic network in computer science where the authors are labeled with their research areas. We consider meta-paths including *APA* (i.e., the co-author relationship), *APCPA* (i.e., authors sharing conferences), and *APTPA* (i.e., authors sharing terms). AMiner is also an academic network, which evolved from 1990 to 2005 in five research domains and the meta-paths are *APA*, *APCPA*, and *APTPA*.

Baselines We compare our proposed StHNE and DyHNE with comprehensive state-of-the-art alternatives, including two homogeneous network embedding methods (i.e., DeepWalk [30] and LINE [38]); two heterogeneous information network embedding methods (i.e., ESim [33] and metapath2vec [8]); and two dynamic homogeneous network embedding methods (i.e., DANE [21], DHPE [50], and DHNE [47]). Additionally, in order to verify the effectiveness of the meta-path based first-order and second-order proximity, we test the performance of StHNE-1st and StHNE-2nd. We use codes of the baseline methods provided by their authors.

5.2.3.2 Effectiveness of StHNE

To evaluate the effectiveness of StHNE, here we learn the node embeddings with the static embedding methods on the whole heterogeneous graph without considering the evolution of the network. In other words, given a dynamic network with 10 time steps $\{\mathcal{G}^1, \cdots, \mathcal{G}^{10}\}$, we conduct all static network embedding methods, including StHNE, on the union network, i.e., $\mathcal{G}^1 \cup \mathcal{G}^1 \cup \cdots \mathcal{G}^{10}$.

Node Classification Node classification is a common task to evaluate the performance of representation learning on networks. In this task, after learning the node embeddings on the fully evolved network, we train a logistic regression classifier with node embeddings as input features. The ratio of training set is set as 40%, 60%, and 80%. We set the weights of *BSB* and *BRURB* in Yelp to 0.4 and 0.6. In DBLP, we assign weights $\{0.05, 0.5, 0.45\}$ to $\{APA, APCPA, APTPA\}$. In AMiner, we assign weights $\{0.25, 0.5, 0.25\}$ to $\{APA, APCPA, APTPA\}$. We report the results in terms of Macro-F1 and Micro-F1 in Table 5.1.

As we can observe, the StHNE outperforms all baselines on three datasets. It improves classification performance by about 8.7% in terms of Macro-F1 averagely with 80% training ratio, which is due to the weighted integration of meta-paths and the preservation of network structure. Both our model StHNE, ESim and metapath2vec fuse multiple meta-paths with weights, but the performances of ESim and metapath2vec are slight worse on three datasets. This may be caused by the separation of meta-paths fusion and model optimization, which loses some information between multiple relationships for heterogeneous graph embedding. We also notice that StHNE-1st and StHNE-2nd both outperform LINE-1st and LINE-2nd in most cases, which shows the superiority of the meta-path based first- and second-order proximity in heterogeneous graphs. From a vertical comparison, our StHNE continues to perform best against different sizes of training data, which implies the stability and robustness of our model.

Relationship Prediction For DBLP and AMiner, we are interested in the co-author relationships (*APA*). Hence, we generate training networks by randomly hiding 20% *AP* in DBLP and 40% *AP* in AMiner as AMiner is much larger. For Yelp, we want to find two businesses that one person has reviewed (*BRURB*), which can be used to recommend businesses for users. Thus, we randomly hide 20% *BR* to generate the training network. We set the weights of *BSB* and *BRURB* in Yelp to 0.4 and 0.6. In DBLP, we assign weights $\{0.9, 0.05, 0.05\}$ to $\{APA, APCPA, APTPA\}$. In AMiner, we assign weights $\{0.4, 0.3, 0.3\}$ to $\{APA, APCPA, APTPA\}$. We evaluate the prediction performance on testing networks with AUC and Accuracy.

Table 5.2 shows the comparison results of different methods. Overall, we can see that StHNE achieves better relation prediction performance than other methods on

Table 5.1 Performance evaluation of node classification on static heterogeneous graphs (Tr.Ratio means the training ratio). Bold values denote the best performance of all baselines and the proposed model

Datasets	Metric	Tr.Ratio	DeepWalk	LINE-1st	LINE-1st	ESim	metapath2vec	StHNE-1st	StHNE-2nd	StHNE
Yelp	Macro-F1	40%	0.6021	0.5389	0.5438	0.6387	0.5872	0.6193	0.5377	**0.6421**
		60%	0.5954	0.5865	0.5558	0.6464	0.6081	0.6639	0.5691	**0.6644**
		80%	0.6101	0.6012	0.6068	0.6793	0.6374	0.6909	0.5783	**0.6922**
	Micro-F1	40%	0.6520	0.6054	0.6105	0.6896	0.6427	0.6838	0.6118	**0.6902**
		60%	0.6472	0.6510	0.6233	0.7011	0.6681	0.7103	0.6309	**0.7017**
		80%	0.6673	0.6615	0.6367	0.7186	0.6875	0.7232	0.6367	**0.7326**
DBLP	Macro-F1	40%	0.9295	0.9271	0.9172	0.9354	0.9213	0.9392	0.9283	**0.9473**
		60%	0.9355	0.9298	0.9252	0.9362	0.9311	0.9436	0.9374	**0.9503**
		80%	0.9368	0.9273	0.9301	0.9451	0.9432	0.9511	0.9443	**0.9611**
	Micro-F1	40%	0.9331	0.9310	0.9219	0.9394	0.9228	0.9421	0.9312	**0.9503**
		60%	0.9383	0.9328	0.9291	0.9406	0.9305	0.9487	0.9389	**0.9519**
		80%	0.9392	0.9323	0.9347	0.9502	0.9484	0.9543	0.9496	**0.9643**
AMiner	Macro-F1	40%	0.8838	0.8929	0.8972	0.9449	**0.9487**	0.9389	0.9309	0.9452
		60%	0.8846	0.8909	0.8967	0.9482	0.9490	0.9401	0.9354	**0.9499**
		80%	0.8853	0.8947	0.8962	0.9491	0.9493	0.9412	0.9381	**0.9521**
	Micro-F1	40%	0.8879	0.8925	0.8958	0.9465	**0.9469**	0.9407	0.9412	0.9467
		60%	0.8881	0.8936	0.8960	0.9482	0.9497	0.9423	0.9431	**0.9509**
		80%	0.8882	0.8960	0.8962	0.9500	0.9511	0.9448	0.9423	**0.9529**

Table 5.2 Performance evaluation of relationship prediction on static heterogeneous graphs. Bold values denote the best performance of all baselines and the proposed model

Datasets	Metric	DeepWalk	LINE-1st	LINE-1st	ESim	metapath2vec	StHNE-1st	StHNE-2nd	StHNE
Yelp	AUC	0.7404	0.6553	0.7896	0.6651	0.8187	0.8046	0.8233	**0.8364**
	F1	0.6864	0.6269	0.7370	0.6361	0.7355	0.7348	0.7397	**0.7512**
	ACC	0.6819	0.6115	0.7326	0.6386	0.7436	0.7286	0.7526	**0.7661**
DBLP	AUC	0.9235	0.8368	0.7672	0.9074	0.9291	0.9002	0.9246	**0.9385**
	F1	0.8424	0.7680	0.7054	0.8321	0.8645	0.8359	0.8631	**0.8850**
	ACC	0.8531	0.7680	0.6805	0.8416	0.8596	0.8266	0.8577	**0.8751**
AMiner	AUC	0.7366	0.5163	0.5835	0.8691	0.8783	0.8935	**0.9180**	0.8939
	F1	0.5209	0.5012	0.5276	0.6636	0.6697	0.7037	**0.8021**	0.7085
	ACC	0.6686	0.6475	0.6344	0.7425	0.7506	0.7622	**0.8251**	0.7701

two metrics. The improvement indicates the effectiveness of our model to preserve structural information in heterogeneous graphs. Benefiting from the second-order proximity preserved based on meta-path, StHNE-2nd outperforms than StHNE-1st significantly. The reason is that the higher-order proximity is more conducive for preserving complex relationships in heterogeneous graphs.

5.2.3.3 Effectiveness of DyHNE

In this section, our goal is to verify the effectiveness of DyHNE compared with these baselines designed for dynamic networks (i.e., DANE and DHPE). Since some baselines (e.g., DeepWalk, LINE, and StHNE) cannot handle dynamic networks and we have reported the performance of these methods in Sect. 5.2, here we only apply these methods to initial networks as in [21, 50]. Specifically, given a dynamic network with 10 time steps $\{\mathcal{G}^1, \cdots, \mathcal{G}^{10}\}$, for the static network embedding methods, including StHNE, we only conduct them on $\mathcal{G}^1$ and report the results, while for the dynamic network embedding methods, i.e., DANE, DHPE, and DyHNE, we conduct them from $\mathcal{G}^1$ to $\mathcal{G}^{10}$ to update the embedding incrementally, and report the final results to evaluate their performance in a dynamic environment.

Node Classification For each dataset, we generate the initial and growing heterogeneous graph from the original network. Each growing heterogeneous graph contains ten time steps. In Yelp, reviews are time-stamped, we randomly add 0.1% new *UR* and *BR* to the initial network at each time step. For DBLP, we randomly add 0.1% new *PA*, *PC*, and *PT* to the initial network at each time step. Since AMiner itself contains the published year of each paper, we divide the edges appearing in 2005 into 10 time steps uniformly.

We vary the size of the training set from 40% to 80% with the step size of 20% and the remaining nodes as testing. We repeat each classification experiment ten times and report the average performance in terms of both Macro-F1 and Micro-F1 scores, as shown in Table 5.3. We can see that DyHNE consistently performs better than other baselines on all datasets with all varying sizes of training data, which demonstrates the effectiveness and robustness of our learned node embeddings when served as features for node classification. (a) Especially, our DyHNE significantly outperforms the two dynamic homogeneous network embedding methods, DANE and DHPE. The reason is that our model considers the different types of nodes and relations and can capture the structure and semantic information in heterogeneous graphs. (b) We also notice that our DyHNE achieves better performance than DHNE which is also designed for dynamic heterogeneous graphs. We believe that the improvement is due to the preserved meta-path based first-order and second-order proximity in node embeddings learned by our DyHNE. (c) Compared with the baselines designed for static heterogeneous graphs (i.e., DeepWalk, LINE, ESim, and metapath2vec), our method also achieves the best performance, which proves the effectiveness of the update algorithm without losing important structure and semantic information in heterogeneous graphs.

Table 5.3 Performance evaluation of node classification on dynamic heterogeneous graphs (Tr.Ratio means the training ratio). Bold values denote the best performance of all baselines and the proposed model

Datasets	Metric	Tr.Ratio	DeepWalk	LINE-1st	LINE-1st	ESim	metapath2vec	StHNE	DANE	DHPE	DHNE	DyHNE
Yelp	Macro-F1	40%	0.5840	0.5623	0.5248	0.6463	0.5765	0.6118	0.6102	0.5412	0.6293	**0.6459**
		60%	0.5962	0.5863	0.5392	0.6642	0.6192	0.6644	0.6342	0.5546	0.6342	**0.6641**
		80%	0.6044	0.6001	0.6030	0.6744	0.6285	0.6882	0.6471	0.5616	0.6529	**0.6893**
	Micro-F1	40%	0.6443	0.6214	0.5901	0.6932	0.6457	0.6826	0.6894	0.5823	0.6689	**0.6933**
		60%	0.6558	0.6338	0.5435	0.6941	0.6656	0.7074	0.6921	0.5981	0.6794	**0.6998**
		80%	0.6634	0.6424	0.6297	0.7104	0.6722	0.7281	0.6959	0.6034	0.6931	**0.7298**
DBLP	Macro-F1	40%	0.9269	0.9266	0.9147	0.9372	0.9162	0.9395	0.8862	0.8893	0.9302	**0.9434**
		60%	0.9297	0.9283	0.9141	0.9369	0.9253	0.9461	0.8956	0.8946	0.9351	**0.9476**
		80%	0.9322	0.9291	0.9217	0.9376	0.9302	0.9502	0.9051	0.9087	0.9423	**0.9581**
	Micro-F1	40%	0.9375	0.9310	0.9198	0.9383	0.9254	0.9438	0.8883	0.8847	0.9352	**0.9467**
		60%	0.9346	0.9245	0.9192	0.9404	0.9281	0.9496	0.8879	0.8931	0.9404	**0.9505**
		80%	0.9371	0.9297	0.9261	0.9415	0.9354	0.9543	0.9071	0.9041	0.9489	**0.9617**
AMiner	Macro-F1	40%	0.8197	0.8219	0.8282	0.8797	0.8673	0.8628	0.7642	0.7694	0.8903	**0.9014**
		60%	0.8221	0.8218	0.8323	0.8807	0.8734	0.8651	0.7704	0.7735	0.9011	**0.9131**
		80%	0.8235	0.8238	0.8351	0.8821	0.8754	0.8778	0.7793	0.7851	0.9183	**0.9212**
	Micro-F1	40%	0.8157	0.8189	0.8323	0.8729	0.8652	0.8563	0.7698	0.7633	0.8992	**0.9117**
		60%	0.8175	0.8182	0.8361	0.8734	0.8693	0.8574	0.7723	0.7698	0.9045	**0.9178**
		80%	0.8191	0.8201	0.8298	0.8751	0.8725	0.8728	0.7857	0.7704	0.9132	**0.9203**

Relationship Prediction For each dataset, we generate the initial, growing, and testing heterogeneous graph from the original heterogeneous graph. For Yelp, we first build the testing network containing 20% *BR*. The remaining constitutes the initial and growing network, where the growing network is divided into 10 time steps, and 0.1% new *UR* and *BR* are added to the initial network at each time step. For DBLP, we use the similar approach as described above. For AMiner, we take the data involved in 1990–2003 as the initial network, 2004 as the growing network, and 2005 as the testing network.

We report the prediction performance in Table 5.4 and have some findings: (a) Our method consistently improves the relationships prediction accuracy on the three datasets, which is attributed to the structural information preserved by the meta-path based first-order and second-order proximity. (b) DANE and DHPE obtain poor performance due to the neglect of multiple types of nodes and relations in heterogeneous graphs. (c) Compared to DHNE, our DyHNE consistently performances better on three datasets, which is benefit from the effectiveness of update algorithm. Additionally, the meta-path based second-order proximity ensures that our DyHNE captures the high-order structures of heterogeneous graph, which is also preserved with the updated node embeddings.

The more detailed method description and experiment validation can be seen in [45]. While DyHNE mainly focuses on incremental learning of dynamic semantics, the sequential interactions of entities indeed express the evolution of heterogeneous interests, and we introduce the following model SHCF to extract such valuable information for recommender systems.

5.3 Sequence Information

5.3.1 Overview

Heterogeneous graph modeling, as an effective information fusion method containing different types of nodes and links, can be used to integrate multiple types of objects and their complex interactions in the recommendation system that may produce more accurate recommendation results [35]. These methods mainly model a user's static preference [34], while ignoring the interactions' sequential pattern that a small set of the most recent interactions can better reflect a user's dynamic interests over time. Considering the sequential pattern to model a user's latest interests, there is another line of work called sequential recommendation. The sequential recommendation system is to predict which item a user most probably would like to interact with next time given her sequential interaction data as context. However, almost all of the sequential recommendation methods [13, 14, 22, 31] model user embeddings based on only their own sequential item interactions while ignoring the heterogeneous information widely existing in recommendation system, such like

Table 5.4 Performance evaluation of relationship prediction on dynamic heterogeneous graphs. Bold values denote the best performance of all baselines and the proposed model

Datasets	Metric	DeepWalk	LINE-1st	LINE-1st	ESim	metapath2vec	StHNE	DANE	DHPE	DHNE	DyHNE
Yelp	AUC	0.7316	0.6549	0.7895	0.6521	0.8164	0.8341	0.7928	0.7629	0.8023	**0.8346**
	F1	0.6771	0.6125	0.7350	0.6168	0.7293	0.7506	0.7221	0.6809	0.7194	**0.7504**
	ACC	0.6751	0.6059	0.7300	0.6185	0.7395	0.7616	0.7211	0.7023	0.7024	**0.7639**
DBLP	AUC	0.9125	0.8261	0.7432	0.9053	0.9196	0.9216	0.5413	0.6411	0.8945	**0.9278**
	F1	0.8421	0.7840	0.7014	0.8215	0.8497	0.8621	0.7141	0.6223	0.8348	**0.8744**
	ACC	0.8221	0.7227	0.6754	0.8306	0.8405	0.8436	0.5511	0.5734	0.8195	**0.8635**
AMiner	AUC	0.8660	0.6271	0.5648	0.8459	0.8694	0.8659	0.8405	0.8412	0.8289	**0.8823**
	F1	0.7658	0.5651	0.6071	0.7172	0.7761	0.7567	0.7167	0.7158	0.7386	**0.7792**
	ACC	0.7856	0.5328	0.5828	0.7594	0.7793	0.7733	0.7527	0.7545	0.7498	**0.7889**

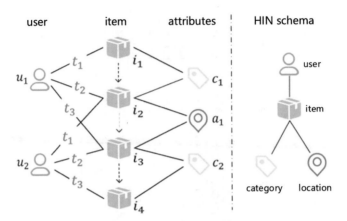

Fig. 5.2 A toy example of user–item interactions with heterogeneous information

item attributes. When the data is sparse and there are few user interaction behaviors, these methods also suffer from cold-start problem.

In this section, we construct a heterogeneous graph with user–item interactions and item attributes in Fig. 5.2 and propose the SHCF to fully consider both the sequential patterns and the high-order heterogeneous collaborative signals. For user embedding, we aggregate the static representation with a novel element-wise attention mechanism, and the dynamic interests by aggregating her interacted item sequence with a sequence-aware self-attention mechanism. For item embedding, we aggregate the heterogeneous information of its neighboring nodes including users and item attributes with dual-level attention. We can not only learn the importance of different nodes but also pay attention to important types of nodes. By stacking multiple message passing layers, we can enforce the embeddings to capture the high-order collaborative relationships.

5.3.2 The SHCF Model

Figure 5.3 shows the framework of our SHCF, which contains three major steps. First, we construct a heterogeneous graph with user–item interactions and item attributes as shown in Fig. 5.2. Notice that here we only consider item attributes to focus on clearly illustrating how to handle sequence patterns and heterogeneous information. In fact, user attributes and other heterogeneous information can be easily added into our SHCF through concatenating embedding learned from these heterogeneous attributes as item embedding does. Then we apply an embedding layer to initialize the representations of users, items, and item attributes (e.g., item categories). Second, we design multiple message passing layers over the heterogeneous graph to learn the user and item embeddings. For user embedding,

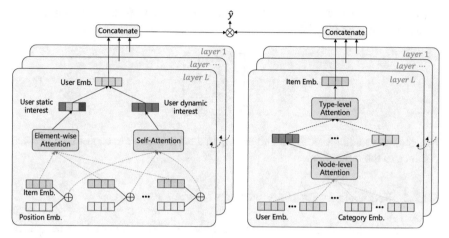

Fig. 5.3 The framework of SHCF

we capture a user's fine-grained static interests on different aspects of an item with an element-wise attention mechanism. We also consider a user's dynamic interests by aggregating her interacted item sequence with a sequence-aware self-attention mechanism. For item embedding, we aggregate the heterogeneous information of its neighborhoods including users and item attributes with dual-level attention which considers the importance of different neighboring nodes with different types. Finally, the prediction layer aggregates the learned embeddings from different message passing layers for both user and item representations, and outputs the prediction of target user–item pairs.

5.3.2.1   Embedding Layer

The users, items, and item attributes in real datasets are usually identified by some unique IDs, whereas these original IDs have a very limited representation capacity. Therefore, we create a user embedding matrix $\mathbf{U} \in \mathbb{R}^{|\mathcal{U}| \times d}$, where d is dimensionality of the latent embedding spaces. The jth row of the embedding matrix $\mathbf{U}$ encodes the user u_j to the real-valued embedding $\mathbf{u}_j$, which is more informative. In the same way, we respectively create an item embedding matrix $\mathbf{I} \in \mathbb{R}^{|\mathcal{I}| \times d}$ and item attribute embedding matrices, e.g. item category embedding matrix $\mathbf{C} \in \mathbb{R}^{|\mathcal{C}| \times d}$.

Positional Embedding   Motivated by the recent works of transformer [7, 41], as for the interacted item sequence $S_u = \{i_1, \cdots, i_{t-1}, i_t\}$ sorted by the time t, we correlate each item with a learnable position embedding $\mathbf{p} \in \mathbb{R}^d$ to capture the sequential pattern of the items.

$$\hat{\mathbf{I}}_u = \begin{bmatrix} \mathbf{i}_t + \mathbf{p}_1 \\ \mathbf{i}_{t-1} + \mathbf{p}_2 \\ \dots \\ \mathbf{i}_1 + \mathbf{p}_t \end{bmatrix}. \tag{5.35}$$

Notice that we add the position embedding in a reverse order to capture the relevant distance to the target item.

5.3.2.2   Sequence-Aware Heterogeneous Message Passing

To capture high-order heterogeneous collaborative information and the sequential information, we first construct a heterogeneous graph enriching user–item interactions with added item attributes as shown in Fig. 5.2. We also model a user's interacted item sequence with a sequence-aware attention mechanism, in order to capture the user's dynamic interests. In this way, we can not only better model user preferences but also alleviate the sparsity of the interactions. In the following, we will first present a single graph convolution layer to model item embedding and user embedding with considering heterogeneous information and sequential information, and then generalize it to multiple layers.

Item Modeling with Heterogeneous Information   To alleviate the sparsity problem, we add item attribute information to the user–item bipartite graph and create a heterogeneous graph that includes different types of nodes. Inspired by HGAT [15], in this part, we present our proposed message passing layers which consider the heterogeneous information. Taking the item node as an example, it has different types of neighboring nodes such as users, categories, and cities. On one hand, different types of neighboring nodes may have different impacts on it. For example, the category of one item may be more informative than the user who interacted with it. On the other hand, different neighboring nodes of the same type could also have different importance. For example, different users may have different preference on one item. To capture both the different importance at both node level and type level, we design a dual-level attention mechanism when aggregating embeddings from neighboring nodes.

(a) Node-Level Attention. We design the node-level attention to learn the different importance of neighboring nodes with the same type and aggregate the representation of these neighbors to form a specific type embedding. Formally, given a specific item v and its neighboring node $v' \in \mathcal{N}_v^{\tau'}$ with type τ', the weight coefficient $\alpha_{vv'}$ of the specific node pair (v, v') can be formulated as follows:

$$\alpha_{vv'} = \frac{\exp(\sigma(\mathbf{a}_{\tau'}^{\top} \cdot [\mathbf{i}_v || \mathbf{h}_{v'}]))}{\sum_{k \in \mathcal{N}_v^{\tau'}} \exp(\sigma(\mathbf{a}_{\tau'}^{\top} \cdot [\mathbf{i}_v || \mathbf{h}_k]))}, \tag{5.36}$$

where $\sigma(\cdot)$ is the activation function, such like LeakReLU, $\mathbf{a}_{\tau'}$ is the attention vector for the type τ', $\mathbf{h}$ is the embedding of neighboring node, and $\|$ denotes the concatenate operation.

Then, for the item v, we can get the specific type embedding $\mathbf{g}_v^{\tau'}$ by aggregating neighboring nodes of the same type with corresponding coefficients as follows:

$$\mathbf{g}_v^{\tau'} = \sigma \left(\sum_{v' \in \mathcal{N}_v^{\tau'}} \alpha_{vv'} \cdot \mathbf{h}_{v'} \right). \tag{5.37}$$

(b) Type-Level Attention. For any type τ' belonging to the item v's neighboring node type set $\mathcal{T}$, we can get a type specific embedding $\mathbf{h}_v^{\tau'}$ following the Eq. (5.37). To capture different importance of different node types, we design a type-level attention defined as follows:

$$m_v^{\tau'} = V \cdot \tanh \left(\mathbf{w} \cdot \mathbf{g}_v^{\tau'} + b \right), \tag{5.38}$$

$$\beta_v^{\tau'} = \frac{\exp \left(m_v^{\tau'} \right)}{\sum_{\tau \in \mathcal{T}} \exp \left(m_v^{\tau} \right)}. \tag{5.39}$$

With the learned weights as coefficients, we can fuse these type embeddings $\mathbf{g}_v^{\tau'}$ to obtain the final item embedding $\tilde{\mathbf{i}}_v$ as follows:

$$\tilde{\mathbf{i}}_v = \sigma \left(\sum_{\tau' \in \mathcal{T}} \beta_v^{\tau'} \cdot \mathbf{g}_v^{\tau'} \right). \tag{5.40}$$

Notice that the above is an example of how to obtain item embedding with heterogeneous information; other typed nodes in heterogeneous graph such as attribute nodes can be modeled in the same way.

User Modeling with Fine-Grained Static and Dynamic Interest A great challenge for recommendation is how to accurately model user preferences. For traditional collaborative filtering or heterogeneous graph based recommendation methods, on one hand, they usually view an item as an entirety, which ignore the fact that users may have different preferences on different aspects of an item; on the other hand, they always neglect the sequential information of a user's interaction history, thus failing to capture the user's dynamic interests. Therefore, for user nodes, we present a carefully designed message passing layer to capture a user's fine-grained static interests and dynamic interests. More specifically, we propose an element-wise attention mechanism that assumes each dimension of the item embedding reflects a distinct aspect of the item. In addition, to capture a user's dynamic interests, we adopt a sequence-aware self-attention mechanism where each item

embedding is correlated with a position embedding, and self-attention is applied
to pay attention to important items.

(a) Element-Wise Attention. Here we present the details of element-wise attention
to capture user's fine-grained static preference. For a specific item i_j in user
u's interaction sequence $\mathcal{S}_u$, we can calculate a weight vector $\boldsymbol{\gamma}_j$ for different
aspects of item i_j as follows:

$$\boldsymbol{\gamma}_j = \tanh(\mathbf{W}_u \cdot \mathbf{i}_j + b), \tag{5.41}$$

where $\mathbf{W}_u \in \mathbb{R}^{d \times d}$, $\boldsymbol{\gamma}_j$ is the attention coefficients of different aspects, and a
large γ_j^k means that the kth aspect of item embedding $\mathbf{i}_j$ is strongly relevant to
the user's preference.

Then we aggregate with element-wise product between the weight coeffi-
cients $\boldsymbol{\gamma}_j$, and the user integrated item $\mathbf{i}_j$ to learn the embedding $\mathbf{u}_s$ reflects
user's fine-grained static interests:

$$\mathbf{u}_s = \sum_{j \in \mathcal{S}_u} \boldsymbol{\gamma}_j \odot \mathbf{i}_j. \tag{5.42}$$

(b) Sequence-Aware Self-attention. Motivated by the self-attention mechanism
widely used in NLP tasks such like machine translation[7, 41], we adopt a
sequence-aware self-attention mechanism. Specifically, each item $\hat{\mathbf{I}}_u$ is integrated with its position embedding and self-attention is used to pay attention to
critical items, which to learn the embedding $\mathbf{u}_d$ reflects user's dynamic interests
over time:

$$\text{ATTENTION}(\mathbf{Q}, \mathbf{K}, \mathbf{V}) = \text{softmax}\left(\frac{\mathbf{Q}\mathbf{K}^\top}{\sqrt{d}}\right) \cdot \mathbf{V}, \tag{5.43}$$

$$\mathbf{u}_d = \overset{H}{\underset{h=1}{\|}} \text{ATTENTION}\left(\hat{\mathbf{I}}_u \mathbf{W}^Q, \hat{\mathbf{I}}_u \mathbf{W}^K, \hat{\mathbf{I}}_u \mathbf{W}^V\right). \tag{5.44}$$

Equation (5.43) is the paradigm of self-attention, where ATTENTION() cal-
culates a weighted sum of all values $\mathbf{V}$ by queries $\mathbf{Q}$ and keys $\mathbf{K}$, and the
scale factor $\sqrt{d}$ is to avoid overly large values of the inner product result. In
Eq. (5.44), $\mathbf{W}^Q, \mathbf{W}^K, \mathbf{W}^V \in \mathbb{R}^{d \times d}$ is the projection matrices and we extend
the self-attention to multi-head attention by repeating it for H times and
concatenate the learned embeddings to get the final user dynamic interest
representation.

After getting the static interest embedding $\mathbf{u}_s$ and dynamic interest embedding $\mathbf{u}_d$, we combine them with a balance weight to get the final user embedding
$\tilde{\mathbf{u}}$:

$$\tilde{\mathbf{u}} = \lambda \mathbf{u}_d + (1 - \lambda)\mathbf{u}_s \tag{5.45}$$

(c) High-order Propagation. The above shows a single messaging passing layer with heterogeneous information and sequential information, which aggregates information from the first-order neighbors. To capture the high-order collaborative information, we can stack it to multiple layers in which each layer takes the last layer's output representation as its input.[1] After L-layer embedding propagation, we can get output embeddings of L different layers.

5.3.2.3 Optimization Objective

The embeddings of different layers may have different contributions in reflecting user preferences; following [44], we concatenate the representation of each layer to constitute the final embedding for both users and items:

$$\mathbf{u} = \tilde{\mathbf{u}}^1 \,||\, \tilde{\mathbf{u}}^2 \,||\cdots||\, \tilde{\mathbf{u}}^L, \quad \mathbf{i} = \tilde{\mathbf{i}}^1 \,||\, \tilde{\mathbf{i}}^2 \,||\cdots||\, \tilde{\mathbf{i}}^L. \tag{5.46}$$

Finally, we use the simple dot product to estimate the user's preference towards the target item:

$$\hat{y}(u, i) = \mathbf{u}^\top \mathbf{i}. \tag{5.47}$$

To optimize our model, we use the Bayesian Personalized Ranking (BPR) loss [32] as our loss function:

$$\mathcal{L} = \sum_{i \in S_u, j \notin S_u} -\ln \sigma(\hat{y}_{ui} - \hat{y}_{uj}) + \eta \,||\Theta||, \tag{5.48}$$

where $\sigma(\cdot)$ is the sigmoid function, Θ denotes all the trainable parameters and η is the regularization coefficient, S_u is the interaction sequence of the user u, and for each positive sample (u, i), we sample a negative sample j for training.

5.3.3 Experiments

5.3.3.1 Experimental Settings

Datasets To verify the effectiveness of our method, we conduct extensive experiments on three real-world datasets. MovieLens is a widely used benchmark dataset for recommendation task. In our experiments, we adopt a small version of 100K

---

[1] According to our experiments, fine-grained and dynamic user interest modeling layer is taken as the L-th layer for user modeling.

interactions ML100K and a larger version of 1M interactions ML1M. Yelp is a local business recommendation dataset which records the user ratings on local businesses.

Baselines We compare our method with three groups of recommendation baseline methods, namely collaborative filtering methods (BPR-MF [32], NeuMF [12], NGCF [43]), heterogeneous graph recommendation methods (NeuACF [11], HeRec [34]), and sequential recommendation methods (NARM [22], SR-GNN [13]).

Implementation Details For all the methods, we apply a grid search for hyper-parameters. For NGCF and SR-GNN, the layer of GNN is searched from 1 to 4. We implement our proposed model based on Tensorflow. The dimension of embeddings d is set as 64. For the self-attention network, the attention head number H is set as 8. The hyper-parameter λ to balance the weights of a user's dynamic interests and static interests is set as 0.5 and 0.2 for MovieLens and Yelp, respectively. In addition, the learning rate is 0.0005 for MovieLens and 0.00005 for Yelp. The coefficient of L2 normalization η for all the datasets is set to 10^{-5}. We set the depth of our proposed SHCF L as 4. We randomly initialize the model parameters with Xavier initializer, then use the Adam as the optimizer. To avoid over-fitting, we apply early stopping strategy and apply dropout (dropout rate is 0.1) in every layer of our proposed SHCF.

5.3.3.2 Performance Comparison

We first compare the recommendation performance of all the methods. For a fair comparison, the embedding dimension of all the methods is set as 64. Table 5.5 shows the experiment results of different methods. We have the following observations: (1) heterogeneous graph based recommendation methods generally perform better than traditional collaborative filtering methods. They especially have great improvements on the sparse dataset (i.e., Yelp), which illustrates that applying heterogeneous graph to incorporate side information for recommendation can alleviate the data sparsity problem and improve recommendation performance. (2) For the dense datasets ML100K and ML1M that users have adequate interaction behaviors (the average number of interactions per user is 103.9 and 165.3, respectively), sequential recommendation methods perform better than collaborative filtering methods and heterogeneous graph based recommendation methods. But for the sparse dataset Yelp where the average number of interactions per user decreases to 10.0, the performance of sequential recommendation methods significantly declines. It illustrates the limitation of sequential recommendation methods in modeling user embeddings based on only the user's own sequential item interactions without considering the collaborative information of similar users or items when the user does not have sufficient interaction records. (3) Our proposed model SHCF consistently outperforms all the baselines on all the datasets including the two dense datasets (i.e., ML100K and ML1M) and one sparse dataset (i.e., Yelp). These results verify the effectiveness of SHCF in modeling users and items in both sparse

Table 5.5 Recommendation performance of different models. The best result in each row is bold and the second best result is underlined. The improvements of our method over the second best models are shown in the last column

Dataset	Metrics	BPR-MF	NeuMF	NGCF	NeuACF	HeRec	NARM	SR-GNN	SHCF	Improve
ML100K	HR@5	0.4030	0.4057	0.4274	0.4337	0.4255	_0.5228_	0.5010	**0.5414**	3.56%
	NDCG@5	0.2747	0.2676	0.2889	0.2874	0.2798	_0.3659_	0.3510	**0.3859**	5.47%
	HR@10	0.5801	0.5689	0.5864	0.6034	0.6012	_0.6723_	0.6660	**0.7108**	5.73%
	NDCG@10	0.3312	0.3127	0.3402	0.3420	0.3325	0.4142	0.4048	**0.4401**	6.25%
	HR@15	0.6787	0.6706	0.6649	0.7084	0.6981	0.7529	_0.7598_	**0.7817**	2.88%
	NDCG@15	0.3573	0.3462	0.3611	0.3697	0.3521	_0.4354_	0.4298	**0.4592**	5.47%
	HR@20	0.7455	0.7595	0.7434	0.7720	0.7524	0.8038	_0.8048_	**0.8324**	3.43%
	NDCG@20	0.3731	0.3672	0.3796	0.3847	0.3721	_0.4475_	0.4396	**0.4693**	4.87%
ML1M	HR@5	0.4921	0.5092	0.5017	0.5050	0.4923	_0.6713_	0.6634	**0.6927**	3.18%
	NDCG@5	0.3376	0.3511	0.3437	0.3508	0.3455	0.5201	_0.5233_	**0.5299**	1.26%
	HR@10	0.6577	0.6803	0.6688	0.6684	0.6601	0.7603	_0.7699_	**0.7964**	3.19%
	NDCG@10	0.3910	0.4066	0.3977	0.4038	0.3982	0.5565	_0.5580_	**0.5639**	1.06%
	HR@15	0.7551	0.7761	0.7587	0.7593	0.7403	_0.8230_	0.8184	**0.8503**	3.32%
	NDCG@15	0.4168	0.4320	0.4216	0.4279	0.4194	0.5687	_0.5709_	**0.5785**	1.33%
	HR@20	0.8159	0.8369	0.8167	0.8232	0.8105	_0.8602_	0.8584	**0.8844**	2.81%
	NDCG@20	0.4311	0.4463	0.4353	0.4430	0.4328	0.5723	_0.5803_	**0.5968**	2.84%
Yelp	HR@5	0.3077	0.3571	_0.4097_	0.4094	0.3982	0.3490	0.3754	**0.4421**	7.91%
	NDCG@5	0.2086	0.2419	_0.2855_	0.2844	0.2765	0.2373	0.2565	**0.3100**	8.58%
	HR@10	0.4325	0.5018	_0.5584_	0.5553	0.5505	0.4900	0.5208	**0.5878**	5.27%
	NDCG@10	0.2488	0.2885	_0.3335_	0.3314	0.3311	0.2828	0.3035	**0.3572**	7.11%
	HR@15	0.5084	0.6006	0.6434	_0.6504_	0.6423	0.5851	0.6054	**0.6725**	3.40%
	NDCG@15	0.2689	0.3146	0.3544	_0.3565_	0.3499	0.3080	0.3259	**0.3796**	6.48%
	HR@20	0.5643	0.6717	0.7005	_0.7138_	0.6923	0.6496	0.6684	**0.7260**	1.71%
	NDCG@20	0.2821	0.3315	0.3686	_0.3715_	0.3603	0.3232	0.3408	**0.3922**	5.57%

and dense datasets by fully considering the high-order heterogeneous collaborative information and sequential information.

The more detailed methods and experiments description can be obtained in [24]. While the SHCF focuses on dealing with single-typed sequences, there are multiple interactions in real-world systems indeed, and the time span of recorded interactions becomes larger and larger, indicating not only short-term but also long-term heterogeneous preferences. The following section is to study the temporal heterogeneous interaction modeling.

5.4   Temporal Interaction

5.4.1   Overview

By modeling historical user–item interactions, recommender systems play a fundamental role in e-commerce [25, 39]. Existing methods [20, 49] are mainly capable of modelling short-term preferences from relatively recent interactions; capturing long-term preferences (e.g., preferred brands) from historical habits is also an important element of temporal dynamics [39]. However, these methods usually model short- and long-term preferences independently, ignoring the role of habits in driving the current, evolving demands. Taking Fig. 5.4a as an example, when browsing similar items (e.g., two schoolbags), users prefer to click those with attributes they habitually care (e.g., the brands). This presents the first research challenge: How to effectively model the complex temporal dynamics, coupling both historical habits and evolving demands? Another dimension overlooked by existing sequential models is the abundant heterogeneous structural information, as shown in Fig. 5.4b. This leads to the second research challenge: How to make full use of the temporal heterogeneous interactions to model the preferences of different types?

In this section, we propose THIGE to effectively learn user and item embeddings on temporal heterogeneous interaction graphs for next-item recommendation. THIGE first encodes heterogeneous interactions with temporal information. Building upon the temporal encoding, we take into account the influence of long-term habits on short-term demands, and design a habit-guided attention mechanism to couple short- and long-term preferences. To fully exploit the rich heterogeneous interactions to enhance multifaceted preferences, we further capture the latent relevance of varying types of interaction via heterogeneous self-attention mechanisms.

5.4.2   The THIGE Model

The overall framework is shown in Fig. 5.5. Specifically, we divide the historical interactions of a user into long- and short-term based on their timestamps. For

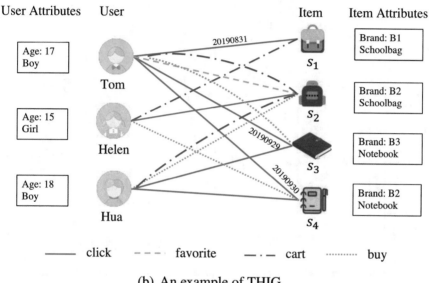

(a) The temporal evolution of interactions.

(b) An example of THIG.

Fig. 5.4 Toy example of next-item recommendation, from (**a**) a temporal sequence of interactions, to (**b**) a Temporal Heterogeneous Interaction Graph (THIG)

short-term preferences, we model users' sequences of recent interactions with gated recurrent units (GRU), to embed users' current demands $h_u^{(S)}$. For long-term preference, we model users' long-term interactions with a heterogeneous self-attention mechanism, to embed users' historical habits $h_u^{(L)}$. Different from the decoupled combination (e.g., simple concatenation) of long- and short-term embeddings in previous methods [20, 48], we propose to exploit the long-term historical habits to guide the learning of short-term demands using the habit-guided attention, which effectively captures the impact of habits on recent behaviors. Notice that Fig. 5.5 only shows the learning of user representations. For items, we do not distinguish their long- and short-term interactions, and only adopt a long-term model similar to that of users. The reason lies in the fact that there may be numerous users interacting with an item around a short period of time, and these users have no significant short-term sequential dependency.

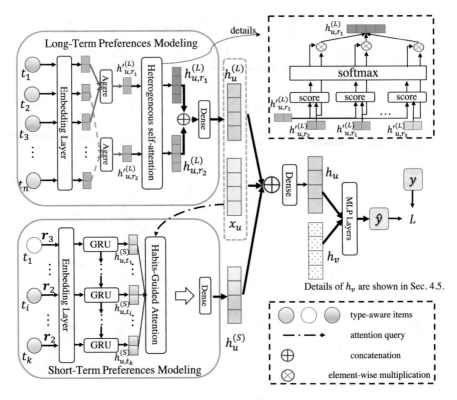

Fig. 5.5 Overall framework of user embedding in THIGE for next-item recommendation

5.4.2.1 Embedding Layer with Temporal Information

Each interacted item of a user is associated with not only attributes but also a timestamp. As shown in Fig. 5.5, the timestamps is in the form of $[t_1, t_2, \cdots, t_n]$.

Thus, the temporal embedding of an item v consists of both a static and a temporal component. The static component $\boldsymbol{x}_v = \boldsymbol{W}_{\phi(v)}\boldsymbol{a}_v$, where the input vector $\boldsymbol{a}_v \in \mathbb{R}^{d_\phi(v)}$ encodes the attributes of v, $\boldsymbol{W}_{\phi(v)} \in \mathbb{R}^{d \times d_\phi(v)}$ denotes the latent projection, $d_\phi(v)$ and d are the dimension of attributes and latent representation of v. Moreover, at time t, denoting Δt as the time span before the current time T and dividing the overall time span into B buckets, the temporal component of v is defined as $\boldsymbol{W}\xi(\Delta t)$, where $\xi(\Delta t) \in \mathbb{R}^B$ denotes the one-hot bucket representation of Δt, and $\boldsymbol{W} \in \mathbb{R}^{d_\mathcal{T} \times B}$ denotes the projection matrix, and $d_\mathcal{T}$ is the output dimension. Thus, the temporal embedding of an item v at time t is

$$\boldsymbol{x}_{v,t} = [\boldsymbol{W}\xi(\Delta t) \oplus \boldsymbol{x}_v], \tag{5.49}$$

where $\oplus$ denotes concatenation. Similarly, we generate the static representation of a user u as $\boldsymbol{x}_u \in \mathbb{R}^d$, and temporal representation of u at time t as $\boldsymbol{x}_{u,t}$.

To further consider the sequential evolution of heterogeneous interactions, we generate the ith interacted item embedding $\boldsymbol{x}_{v_i,t_i,r_i} = [\boldsymbol{x}_{v_i,t_i} \oplus \boldsymbol{r}_i]$ as the combination of the temporal embedding $\boldsymbol{x}_{v_i,t_i}$ and the corresponding type embedding $\boldsymbol{r}_i = \boldsymbol{W}_{\mathcal{R}} \boldsymbol{I}(r_i)$ where $\boldsymbol{I}(r_i)$ denotes the one-hot vector of r_i with dimension $|\mathcal{R}|$, $\boldsymbol{W}_{\mathcal{R}} \in \mathbb{R}^{d_{\mathcal{R}} \times |\mathcal{R}|}$ is the projection matrix and $d_{\mathcal{R}}$ is the latent dimension. For long-term preference modeling, we input the temporal embedding into type-aware aggregators to distinguish preferences of different types.

5.4.2.2 Short-Term Preference Modeling

Recent interactions of users usually indicate the evolving current demands. For instance, as shown in Fig. 5.4a, Tom's current demand has been evolved from bags to notebooks. In order to model the short-term and evolving preferences, we adopt gated recurrent units (GRU) [5], which can capture the dependency of recent interactions. Given a user u here, let his/her k recent interactions be $\{(v_i, t_i, r_i) \mid 1 \le i \le k\}$, where t_k is the most recent timestamp before the current time T. Subsequently, we encode the user preference at time t_i as $\boldsymbol{h}_{u,t_i}^{(S)}$, using a GRU based on the embedding of interaction (v_i, t_i, r_i) (i.e., $\boldsymbol{x}_{v_i,t_i,r_i}$) and his/her preference at t_{i-1}, as follows:

$$\boldsymbol{h}_{u,t_i}^{(S)} = \mathrm{GRU}\left(\boldsymbol{x}_{v_i,t_i,r_i}, \boldsymbol{h}_{u,t_{i-1}}^{(S)}\right), \quad \forall 1 < i \le k, \tag{5.50}$$

where $\boldsymbol{h}_{u,t_i}^{(S)} \in \mathbb{R}^d$. The time-dependent user embeddings $\{\boldsymbol{h}_{u,t_i}^{(S)} \mid 1 \le i \le k\}$ can be further aggregated to encode the current demand of user u.

However, the current and evolving demands of user are not only influenced by their recent transactions. Their long-term preferences, i.e., historical habits such as brands and lifestyle inclinations, often play a subtle but important role. Thus, we enhance the encoding of short-term preferences under the guidance of historical habits, in order to discover more fine-grained and personalized preferences. Specially, we propose a habit-guided attention mechanism to aggregate short-term user preferences, as follows:

$$\boldsymbol{h}_u^{(S)} = \sigma\left(W^{(S)} \cdot \sum_i a_{u,i} \boldsymbol{h}_{u,t_i}^{(S)} + b_s\right), \quad \forall 1 \le i \le k, \tag{5.51}$$

where $\boldsymbol{h}_u^{(S)} \in \mathbb{R}^d$ denotes the overall short-term preference of u, $W^{(S)} \in \mathbb{R}^{d \times d}$ denotes the projection matrix, σ is the activation function, and we adopt RELU here to ensure the non-linearity , b_s is the bias, and $a_{u,i}$ is the habit-guided weight:

$$a_{u,i} = \frac{\exp\left([\boldsymbol{h}_u^{(L)} \oplus \boldsymbol{x}_u]^T W_a \boldsymbol{h}_{u,t_i}^{(S)}\right)}{\sum_{j=1}^k \exp\left([\boldsymbol{h}_u^{(L)} \oplus \boldsymbol{x}_u]^T W_a \boldsymbol{h}_{u,t_j}^{(S)}\right)}, \tag{5.52}$$

where $h_u^{(L)} \in \mathbb{R}^d$ is the long-term preference of u which would have encoded the habits of u, and $W_a \in \mathbb{R}^{2d \times d}$ is a mapping to quantify the fine-grained relevance between the short-term preference of u and the long-term habits of u at different times. Therefore, how to encode the long-term habits $h_u^{(L)}$ in the context of heterogeneous interactions is the second key thesis of this work, as we will introduce next.

5.4.2.3   Long-Term Preference Modeling

Besides short-term preferences to encode current and evolving demands, users also exhibit long-term preferences to express personal and historical habits. In particular, there exist multiple types of heterogeneous interactions which have different relevance w.r.t. each other. For example, a "click" is more relevant to a "cart" or "buy" on the same item or similar items; "favorite" could be less relevant to "cart" or "buy", but is closely tied to the user's brand or lifestyle choices in the long run. Thus, different types of interactions entail both latent relevance and multifaceted preferences. Thereby, our goal is to fully encode the latent, fine-grained relevance of multifaceted long-term preferences.

Consider a user u, and his/her long-term interactions $\{(v_i, t_i, r_i) \mid 1 \leq i \leq n\}$ where $n \gg k$ (k is the count of recent interactions in short-term modeling). To differentiate the explicit interaction types, we first aggregate the embeddings of items which the user has interacted with under a specific type r:

$$h_{u,r}'^{(L)} = \sigma \left(W_r \cdot \text{aggre}(\{x_{v_i, t_i} \mid 1 \leq i \leq n, r_i = r\}) \right), \tag{5.53}$$

where $h_{u,r}'^{(L)} \in \mathbb{R}^d$ is the type-r long-term preferences of user u, $W_r \in \mathbb{R}^{d \times (d_T + d)}$ is the type-r learnable mapping, $\text{aggre}(\cdot)$ is an aggregator, and we utilize mean-pooling here.

While we can simply sum or concatenate the type-specific long-term preferences into an overall representation, there exists latent relevance among the types (e.g., "click" and "buy") and latent multifaceted preferences (e.g., brands and lifestyles). In this section, we design a heterogeneous self-attention mechanism to express the latent relevance of different-typed interactions and long-term multifaceted preferences. By concatenating all long-term preferences of different types as $H_u^{(L)} = \oplus_{r \in \mathcal{R}} h_{u,r}'^{(L)}$ with size d-by-$|\mathcal{R}|$, we first formulate the self-attention to capture the latent relevant of heterogeneous types in $\mathcal{R}$ w.r.t. each other:

$$h_{u,r}^{(L)} = \sum_{r' \in \mathcal{R}} \left(\frac{\exp\left(Q_{u,r}^T K_{u,r'}/\sqrt{d_a}\right)}{\sum_{r'' \in \mathcal{R}} \exp\left(Q_{u,r}^T K_{u,r''}/\sqrt{d_a}\right)} V_{u,r'} \right), \tag{5.54}$$

where $Q_u = W_Q H_u^{(L)}$, $K_u = W_K H_u^{(L)}$, $V_u = W_V H_u^{(L)}$, $W_Q, W_K \in \mathbb{R}^{d_a \times d}$, and $W_V \in \mathbb{R}^{d \times d}$ are the projection matrices, and d_a is the dimension of keys and queries.

Next, to express multifaceted preferences, we adopt a multi-head approach to model latent fine-grained facets. Specifically, the original embeddings of preferences are split into multi-heads and we adopt the self-attention for each head. The type-r long-term preference is concatenated from the h heads:

$$h_{u,r}^{(L)} = \oplus_{m=1}^{m=h} h_{u,r,m}^{(L)}, \tag{5.55}$$

where $h_{u,r,m}^{(L)}$ denotes the mth head based preference and there are h heads. The overall long-term preference can also be derived by fusing different types in $\mathcal{R}$:

$$h_u^{(L)} = \sigma \left(W^{(L)}(\oplus_{r \in \mathcal{R}} h_{u,r}^{(L)}) + b_l \right), \tag{5.56}$$

where $W^{(L)} \in \mathbb{R}^{d \times |\mathcal{R}|d}$ and b_l are the projection parameters. By now, both short- and long-term preferences have not been modeled. Taking the inherent attributes of users into consideration, the final representation of user u is calculated by

$$h_u = \sigma \left(W_u \left[x_u \oplus h_u^{(S)} \oplus h_u^{(L)} \right] + b_u \right), \tag{5.57}$$

where $h_u \in \mathbb{R}^d$ will be used for next-item prediction, and $W_u \in \mathbb{R}^{d \times 3d}$ and b_u are learnable parameters.

5.4.2.4 Preference Modeling of Items

The temporal interactions of an item are significantly different from those of a user. In practice, on a mass e-commerce platform, it is typical that many users interact with the same item around the same time constantly, without a meaningful sequential effect among different users. In other words, it is more reasonable to only model the general, long-term popularity of items. Thus, we model item representation $h_v^{(L)}$ similar to the long-term preference modeling of users in Eq. (5.56) with heterogeneous multi-head self-attention, and then encode the item representation as follows:

$$h_v = \sigma \left(W_v \left[x_v \oplus h_v^{(L)} \right] + b_v \right), \tag{5.58}$$

where $h_v \in \mathbb{R}^d$ is the final representation of item v for next-item prediction, and W_v and b_v are learnable parameters and x_v is the attribute vector of item v.

5.4.2.5 Optimization Objective

To deal with next-item recommendation, we predict $\hat{y}_{u,v}$ between user u and item v, indicating whether u will interact with v (under a given type) at the next time. Here we utilize a Multi-Layer Perception (MLP) [28]:

$$\hat{y}_{u,v} = \text{sigmoid}(\text{MLP}(\boldsymbol{h}_u \oplus \boldsymbol{h}_v)), \tag{5.59}$$

where $\boldsymbol{h}_u$ and $\boldsymbol{h}_v$ are the final representation of user u and item v, respectively. Model parameters can be optimized with the following cross-entropy loss:

$$L = - \sum_{\langle u,v \rangle} (1 - y_{u,v}) \log(1 - \hat{y}_{u,v}) + y_{u,v} \log(\hat{y}_{u,v}), \tag{5.60}$$

where $\langle u, v \rangle$ is a sample of user u and item v, and $y_{u,v} \in \{0, 1\}$ is the ground truth of the sample. We also optimize the L2 regularization of latent parameters to ensure the robustness.

5.4.3 Experiments

5.4.3.1 Experimental Settings

Datasets We evaluate the empirical performance of THIGE for next-item recommendation on three real-world datasets including Yelp, CloudTheme, and UserBehavior. For all datasets, we utilize the actual feedback of users as labels—among the candidates displayed to users.

Baselines We compare THIGE with six representative models, including sequential models (**DIEN** [49], **STAMP** [25]), long-term interest learning models (**SHAN** [48], **M3R** [39]), and heterogeneous GNNs (**MEIRec** [9], **GATNE** [4]).

Implementation Details For all baselines and our method, we set embedding size $d = 128$, $d_a = 128$, $d_{\mathcal{T}} = 16$, heads $h = 8$, the maximum iterations as 100, batch size as 128, learning rate as 0.001, and weight of regularization as 0.001 on all three datasets. The number of temporal buckets B is set as 60, 14, and 7 on the three datasets, respectively. For DIEN, MEIRec, and our THIGE, we set three-layers MLP with dimensions 64, 32, and 1. For our THIGE and all baselines learning long- or short-term preferences, we consider the last 10, 10, and 50 interactions as the short term, and sample up to 50, 50, and 200 historical interactions as the long term for Yelp, CloudTheme, and UserBehavior, respectively. We evaluate the performance of next-item recommendation with the metrics of F1, PR-AUC, and ROC-AUC.

5.4.3.2 Performance Comparison

We report the results of different methods for next-item recommendation in Table 5.6. In general, THIGE achieves the best performance on the three datasets,

Table 5.6 Performance of next-item recommendation (with standard deviation). The best result is in bold while the second best is underlined. PR. denotes PR-AUC and ROC. denotes ROC-AUC

Dataset	Metric	DIEN	STAMP	SHAN	M3R	MEIRec	GATNE	THIGE
Yelp	F1	39.52 (1.31)	40.37 (0.94)	40.17 (1.10)	33.49 (1.04)	42.86 (0.44)	42.21 (0.96)	**43.77** (0.66)
	PR.	30.04 (0.37)	31.36 (1.23)	32.35 (1.10)	26.40 (0.92)	32.69 (0.54)	33.39 (1.42)	**36.45** (1.66)
	ROC.	74.69 (0.57)	73.74 (1.15)	70.91 (1.14)	72.03 (1.33)	74.65 (0.23)	76.15 (0.64)	**79.23** (0.80)
CT.	F1	25.70 (1.25)	21.42 (0.91)	26.25 (1.09)	33.54 (1.67)	25.02 (0.98)	27.33 (0.50)	**37.17** (1.36)
	PR.	41.16 (0.22)	25.65 (0.44)	40.92 (1.09)	34.23 (0.95)	43.86 (0.42)	44.74 (0.20)	**51.94** (0.43)
	ROC.	68.41 (0.34)	52.97 (0.52)	67.48 (1.06)	62.92 (0.89)	69.98 (0.35)	71.22 (0.11)	**75.38** (0.33)
UB.	F1	67.32 (3.45)	63.06 (1.51)	58.84 (7.83)	61.37 (2.20)	66.48 (1.16)	**67.81** (1.14)	67.19 (0.98)
	PR.	63.38 (0.19)	59.09 (0.22)	63.86 (4.76)	57.68 (0.03)	64.94 (0.15)	65.42 (0.05)	**65.71** (0.09)
	ROC.	62.90 (0.23)	58.29 (0.40)	55.45 (3.98)	57.82 (0.09)	64.82 (0.16)	65.06 (0.08)	**65.39** (0.06)

outperform the second best method by 4.04% on Yelp, 5.84% on CloudTheme, and 0.51% on UserBehavior.

Compared with sequential models (DIEN, STAMP, SHAN, and M3R), the reason that THIGE is superior is twofold. First, THIGE designs a more effective way to integrate long- and short-term preferences, such that the current demands are explicitly guided by historical habits. Second, it also considers different types of interactions between users and items, leading to better performance.

Compared with GNN-based models (MEIRec and GATNE), the main improvement of THIGE comes from jointly modeling historical habits and evolving demands. Moreover, MEIRec models heterogeneous interactions in an entirely decoupled manner, whereas GATNE and THIGE achieve better performance by modeling their latent relevance. It is also not surprising that heterogeneous GNN-based methods typically outperform sequential models, as the former accounts for multi-typed interactions whereas the latter only models single-typed interactions.

The whole details and experimental analysis of THIGE are introduced in [17].

5.5   Conclusion

Focusing on the preserving dynamics on real-world heterogeneous graphs, this chapter introduces three representative models including DyHNE, SHCF, and THIGE to model the incremental heterogeneous structures, the type-aware sequential evolution of entities as well as the multiple long-term habits and short-term demands. Specifically, this chapter designs a matrix perturbation based DyHNE to model the changes between different semantic-level snapshots and learns the dynamic embedding of heterogeneous nodes in Sect. 5.2. And then, as the dynamical interactions are in the form of type-aware sequences, we present the heterogeneous graph neural collaborative filtering model to make full use of high-order heterogeneous collaborative singles and sequential information for sequential recommendation in Sect. 5.3. Furthermore, with the consideration that heterogeneous dynamics express the multiple historical habits and current demands of entities, this chapter proposes a unified model which integrates both heterogeneous long-/short-term preferences to handle the problem of next-item recommendation in Sect. 5.4.

References

1. Aggarwal, C., Subbian, K.: Evolutionary network analysis: a survey. ACM Comput. Surv. **47**(1), 10 (2014)
2. Belkin, M., Niyogi, P.: Laplacian eigenmaps and spectral techniques for embedding and clustering. In: NeurIPS, pp. 585–591 (2002)
3. Cai, H., Zheng, V.W., Chang, K.C.C.: A comprehensive survey of graph embedding: problems, techniques, and applications. IEEE Trans. Knowl. Data Eng. **30**(9), 1616–1637 (2018)

4. Cen, Y., Zou, X., Zhang, J., Yang, H., Zhou, J., Tang, J.: Representation learning for attributed multiplex heterogeneous network. Preprint. arXiv:1905.01669 (2019)
5. Cho, K., Van Merriënboer, B., Gulcehre, C., Bahdanau, D., Bougares, F., Schwenk, H., Bengio, Y.: Learning phrase representations using RNN encoder-decoder for statistical machine translation. Preprint. arXiv:1406.1078 (2014)
6. Cui, P., Wang, X., Pei, J., Zhu, W.: A survey on network embedding. IEEE Trans. Knowl. Data Eng. **31**(5), 833–852 (2018)
7. Devlin, J., Chang, M.W., Lee, K., Toutanova, K.: Bert: Pre-training of deep bidirectional transformers for language understanding. Preprint. arXiv:1810.04805 (2018)
8. Dong, Y., Chawla, N.V., Swami, A.: metapath2vec: Scalable representation learning for heterogeneous networks. In: Proceedings of the 23rd ACM SIGKDD International Conference on Knowledge Discovery and Data Mining, pp. 135–144 (2017)
9. Fan, S., Zhu, J., Han, X., Shi, C., Hu, L., Ma, B., Li, Y.: Metapath-guided heterogeneous graph neural network for intent recommendation. In: KDD '19: Proceedings of the 25th ACM SIGKDD International Conference on Knowledge Discovery & Data Mining, pp. 2478–2486 (2019)
10. Golub, G.H., Van Loan, C.F.: Matrix Computations. JHU Press, Baltimore (2012)
11. Han, X., Shi, C., Wang, S., Philip, S.Y., Song, L.: Aspect-level deep collaborative filtering via heterogeneous information networks. In: IJCAI'18: Proceedings of the 27th International Joint Conference on Artificial Intelligence, pp. 3393–3399 (2018)
12. He, X., Liao, L., Zhang, H., Nie, L., Hu, X., Chua, T.S.: Neural collaborative filtering. In: Proceedings of the 26th International Conference on World Wide Web, pp. 173–182 (2017)
13. Hidasi, B., Karatzoglou, A., Baltrunas, L., Tikk, D.: Session-based recommendations with recurrent neural networks. Preprint. arXiv:1511.06939 (2015)
14. Hidasi, B., Quadrana, M., Karatzoglou, A., Tikk, D.: Parallel recurrent neural network architectures for feature-rich session-based recommendations. In: RecSys, pp. 241–248 (2016)
15. Hu, L., Yang, T., Shi, C., Ji, H., Li, X.: Heterogeneous graph attention networks for semi-supervised short text classification. In: Proceedings of the 2019 Conference on Empirical Methods in Natural Language Processing and the 9th International Joint Conference on Natural Language Processing (EMNLP-IJCNLP), pp. 4823–4832 (2019)
16. Jacob, Y., Denoyer, L., Gallinari, P.: Learning latent representations of nodes for classifying in heterogeneous social networks. In: WSDM '14: Proceedings of the 7th ACM International Conference on Web Search and Data Mining, pp. 373–382 (2014)
17. Ji, Y., Yin, M., Fang, Y., Yang, H., Wang, X., Jia, T., Shi, C.: Temporal heterogeneous interaction graph embedding for next-item recommendation. In: Proceedings of The European Conference on Machine Learning and Principles and Practice of Knowledge Discovery in Databases, pp. 314–329 (2020)
18. Kang, W.C., McAuley, J.: Self-attentive sequential recommendation. In: 2018 IEEE International Conference on Data Mining (ICDM), pp. 197–206 (2018)
19. Kato, T.: Perturbation Theory for Linear Operators, vol. 132. Springer Science & Business Media, Berlin (2013)
20. Le, D.T., Lauw, H.W., Fang, Y.: Modeling contemporaneous basket sequences with twin networks for next-item recommendation. In: Proceedings of the Twenty-Seventh International Joint Conference on Artificial Intelligence (2018)
21. Li, J., Dani, H., Hu, X., Tang, J., Chang, Y., Liu, H.: Attributed network embedding for learning in a dynamic environment. In: Proceedings of the 2017 ACM on Conference on Information and Knowledge Management, pp. 387–396 (2017)
22. Li, J., Ren, P., Chen, Z., Ren, Z., Lian, T., Ma, J.: Neural attentive session-based recommendation. In: Proceedings of the 2017 ACM on Conference on Information and Knowledge Management, pp. 1419–1428 (2017)
23. Li, X., Wu, Y., Ester, M., Kao, B., Wang, X., Zheng, Y.: Semi-supervised clustering in attributed heterogeneous information networks. In: Proceedings of the 26th International Conference on World Wide Web, pp. 1621–1629 (2017)

24. Li, C., Hu, L., Shi, C., Song, G., Yuanfu, L.: Sequence-aware heterogeneous graph neural collaborative filtering. In: Proceedings of the 2021 SIAM International Conference on Data Mining (SDM) (2021)
25. Liu, Q., Zeng, Y., Mokhosi, R., Zhang, H.: Stamp: short-term attention/memory priority model for session-based recommendation. In: Proceedings of the 24th ACM SIGKDD International Conference on Knowledge Discovery & Data Mining, pp. 1831–1839 (2018)
26. Lu, Y., Wang, X., Shi, C., Yu, P.S., Ye, Y.: Temporal network embedding with micro-and macro-dynamics. In: Proceedings of the 28th ACM International Conference on Information and Knowledge Management, pp. 469–478 (2019)
27. Ng, A.Y., Jordan, M.I., Weiss, Y.: On spectral clustering: analysis and an algorithm. In: Advances in Neural Information Processing Systems, pp. 849–856 (2002)
28. Pal, S.K., Mitra, S.: Multilayer perceptron, fuzzy sets, and classification. IEEE Trans. Neural Netw. 3(5), 683–697 (1992)
29. Parlett, B.N.: The symmetric eigenvalue problem, vol. 20. SIAM (1998)
30. Perozzi, B., Al-Rfou, R., Skiena, S.: DeepWalk: online learning of social representations. In: Proceedings of the 20th ACM SIGKDD International Conference on Knowledge Discovery and Data Mining, pp. 701–710 (2014)
31. Quadrana, M., Karatzoglou, A., Hidasi, B., Cremonesi, P.: Personalizing session-based recommendations with hierarchical recurrent neural networks. In: Proceedings of the Eleventh ACM Conference on Recommender Systems, pp. 130–137 (2017)
32. Rendle, S., Freudenthaler, C., Gantner, Z., Schmidt-Thieme, L.: BPR: Bayesian personalized ranking from implicit feedback. Preprint. arXiv:1205.2618 (2012)
33. Shang, J., Qu, M., Liu, J., Kaplan, L.M., Han, J., Peng, J.: Meta-path guided embedding for similarity search in large-scale heterogeneous information networks. Preprint. arXiv:1610.09769 (2016)
34. Shi, C., Hu, B., Zhao, W.X., Yu, P.S.: Heterogeneous information network embedding for recommendation. IEEE Trans. Knowl. Data Eng. 31(2), 357–370 (2019)
35. Shi, C., Li, Y., Zhang, J., Sun, Y., Philip, S.Y.: A survey of heterogeneous information network analysis. IEEE Trans. Knowl. Data Eng. 29(1), 17–37 (2016)
36. Sun, Y., Barber, R., Gupta, M., Aggarwal, C.C., Han, J.: Co-author relationship prediction in heterogeneous bibliographic networks. In: 2011 International Conference on Advances in Social Networks Analysis and Mining, pp. 121–128 (2011)
37. Sun, F., Liu, J., Wu, J., Pei, C., Lin, X., Ou, W., Jiang, P.: Bert4rec: sequential recommendation with bidirectional encoder representations from transformer. In: Proceedings of the 28th ACM International Conference on Information and Knowledge Management, pp. 1441–1450 (2019)
38. Tang, J., Qu, M., Wang, M., Zhang, M., Yan, J., Mei, Q.: Line: large-scale information network embedding. In: Proceedings of the 24th International Conference on World Wide Web, pp. 1067–1077 (2015)
39. Tang, J., Belletti, F., Jain, S., Chen, M., Beutel, A., Xu, C., H Chi, E.: Towards neural mixture recommender for long range dependent user sequences. In: The World Wide Web Conference, pp. 1782–1793 (2019)
40. Trefethen, L.N., Bau III, D.: Numerical Linear Algebra, vol. 50. SIAM, Philadelphia (1997)
41. Vaswani, A., Shazeer, N., Parmar, N., Uszkoreit, J., Jones, L., Gomez, A.N., Kaiser, Ł., Polosukhin, I.: Attention is all you need. In: Neural Information Processing Systems (NIPS 2017), pp. 5998–6008 (2017)
42. Wang, H., Zhang, F., Hou, M., Xie, X., Guo, M., Liu, Q.: Shine: signed heterogeneous information network embedding for sentiment link prediction. In: Proceedings of the Eleventh ACM International Conference on Web Search and Data Mining, pp. 592–600 (2018)
43. Wang, X., He, X., Wang, M., Feng, F., Chua, T.S.: Neural graph collaborative filtering. In: Proceedings of the 42nd International ACM SIGIR Conference on Research and Development in Information Retrieval, pp. 165–174 (2019)
44. Wang, X., Ji, H., Shi, C., Wang, B., Ye, Y., Cui, P., Yu, P.S.: Heterogeneous graph attention network. In: The World Wide Web Conference, pp. 2022–2032 (2019)

45. Wang, X., Lu, Y., Shi, C., Wang, R., Cui, P., Mou, S.: Dynamic heterogeneous information network embedding with meta-path based proximity. IEEE Trans. Knowl. Data Eng. 1–1 (2020)
46. Xu, C., Zhao, P., Liu, Y., Sheng, V.S., Xu, J., Zhuang, F., Fang, J., Zhou, X.: Graph contextualized self-attention network for session-based recommendation. In: Proceedings of the Twenty-Eighth International Joint Conference on Artificial Intelligence, vol. 19, pp. 3940–3946 (2019)
47. Yin, Y., Ji, L.X., Zhang, J.P., Pei, Y.L.: DHNE: network representation learning method for dynamic heterogeneous networks. IEEE Access 7, 134,782–134,792 (2019)
48. Ying, H., Zhuang, F., Zhang, F., Liu, Y., Xu, G., Xie, X., Xiong, H., Wu, J.: Sequential recommender system based on hierarchical attention networks. In: IJCAI International Joint Conference on Artificial Intelligence (2018)
49. Zhou, G., Mou, N., Fan, Y., Pi, Q., Bian, W., Zhou, C., Zhu, X., Gai, K.: Deep interest evolution network for click-through rate prediction. In: Proceedings of the AAAI Conference on Artificial Intelligence, vol. 33, pp. 5941–5948 (2019)
50. Zhu, D., Cui, P., Zhang, Z., Pei, J., Zhu, W.: High-order proximity preserved embedding for dynamic networks. IEEE Trans. Knowl. Data Eng. 30(11), 2134–2144 (2018)

Chapter 6
Emerging Topics of Heterogeneous Graph Representation

Abstract Heterogeneous graph (HG) embedding, aiming to project HG into a low-dimensional space, has attracted considerable research attention. We have introduced some kinds of HG embedding methods, and there are also some other essential topics on HG embeddings. In this chapter, we will introduce three novel HG embedding methods. Specifically, to learn semantic-preserving and robust node representations, we study the problem of adversarial learning on HG. Also, we work with large-scale heterogeneous interaction graphs and focus on the problem of importance sampling on HG embedding. Moreover, we explore the intrinsic spaces of HG and propose a hyperbolic space based HG embedding method.

6.1 Introduction

Recently, some efforts begin to combine graph embedding methods with varied machine learning technologies to obtain better graph representations. These machine learning technologies include generative adversarial networks (GANs) [12, 25], importance sampling [5, 18], and hyperbolic representation [20, 22]. Specifically, GANs hinge on the idea of adversarial learning, in which discriminators and generators compete with each other to learn better underlying data distributions. Besides, importance sampling aims to reduce the computational and memory costs in machine learning. Moreover, hyperbolic spaces are non-Euclidean spaces, which are more suitable to model data with a hierarchical structure.

Some homogeneous graph embedding methods combined with these technologies have shown their effectiveness in modeling graphs. For example, GAN-based homogeneous graph embedding methods [6, 28] leverage generative and discriminative components to learn graph representation. The generative component aims to learn the underlying connectivity distribution in the graph, while the discriminative component aims to predict the probability of edge existence between nodes. Besides, to make large-scale graph representation learning possible, a natural idea is to sample a small set of important nodes from the whole graph. The importance sampling based homogeneous graph embedding methods have shown their powerful ability in large-scale graph representation. Moreover, some researchers begin to embed graphs into low-dimensional hyperbolic spaces. They find that embedding a graph in hyperbolic spaces would have a low distortion when the graph has a hierarchical structure [11, 22], and the hierarchical relation between nodes can be reflected by analyzing hyperbolic embedding [22].

In this chapter, we introduce three emerging HG embedding methods. First, **HG e**mbedding with **GAN**-based adversarial learning (named **HeGAN** [16]) is designed to learn semantic-preserving and robust node representations, which leverages the relation-aware discriminator and generator to fit the heterogeneous setting. Second, to reduce the cost of large-scale HG embedding, **Heter**ogeneous importance **Samp**ling (named **HeteSamp** [19]) is proposed, which introduces both type-dependent and type-fusion samplers with self-normalized and adaptive estimators to ensure the model efficiency. Third, assuming the underlying spaces of HG are hyperbolic spaces, **H**yperbolic **H**eterogeneous **N**etwork **E**mbedding (named **HHNE**) [31] is proposed to preserve the structure and semantic information of HG in hyperbolic spaces.

6.2   Adversarial Learning

6.2.1   Overview

GANs [12, 25] have been developed for learning robust latent representations in various applications [29, 34]. In recent researches, GANs only study homogeneous graphs [6, 23, 28, 35], such as the bibliographic data in Fig. 6.1a, so they do not consider the heterogeneity of nodes and edges, resulting in unsatisfactory performance on HG. In existing methods, positive (real) and negative (fake) nodes can only be differentiated by the network structure, which is shown in Fig. 6.1b. Therefore, it is

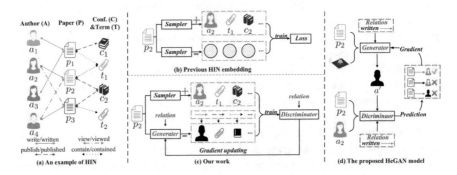

Fig. 6.1 Overview of HeGAN. (**a**) A toy example of HG for bibliographic data. (**b**) An example of previous HG embedding. (**c**) Relation-aware discriminator and generator in our work. (**d**) The framework of HeGAN for adversarial learning on HGs

urgent to design new formal discriminator and generator to distinguish and simulate the real and fake semantic-rich nodes involving various relationships. Generators are essentially picking an existing node from the original network according to the learned distribution, without the ability to generalize to "unseen" nodes. Not surprisingly, they do not generate the most representative fake nodes, as such nodes may not even appear in the network. Thus, it is important to design a generator that can efficiently produce latent fake samples.

We exploit the heterogeneity of HG in adversarial setting in order to learn semantic-preserving and robust node representations. Different from previous efforts [2, 6, 23, 28, 35], HeGAN raises two major novelties to address the challenges of adversarial learning on HG. It is not only relation-aware to capture rich semantics, but also equipped with a generalized generator that is effective and efficient. It proposes a new form of discriminator and generator, as illustrated in Fig. 6.1c. To a given relation, the discriminator can tell whether a node pair is real or fake, whereas the generator can produce fake node pairs that mimic real pairs. To further improve the effectiveness and efficiency of sample generation, we propose a generalized generator, which is able to directly sample latent nodes from a continuous distribution.

6.2.2 The HeGAN Model

6.2.2.1 Model Framework

HeGAN is based on GAN so that it consists of a discriminator and a generator. It is a novel framework for HG embedding with GAN-based adversarial learning. The discriminator and generator are relation-aware in HeGAN. The discriminator is able to tell apart the specified relationship between nodes, and the generator will try to generate nodes which has certain relation with other nodes. The node pair is considered true in the following cases: (i) It is a positive pair based on network topology; (ii) The pair is formed under the correct relation. HeGAN provides a generalized generator, which is able to directly sample latent nodes from a continuous distribution, such that (i) no softmax computation is necessary; and (ii) fake samples are not restricted to the existing nodes.

Existing studies typically model the distribution of nodes using some forms of softmax over all nodes in the original network. In terms of effectiveness, their fake samples are limited to the nodes in the network, whereas the most representative fake samples may fall "in between" the existing nodes in the embedding space. For the bibliographic data in Fig. 6.1a, given a paper p_2, it can only choose fake samples from $\mathcal{V}$, such as a_1 and a_3. However, both may not be adequately similar to real samples such as a_2. Towards a better sample generation, HeGAN introduces a generalized generator that can produce latent nodes not appearing in the embedding space. The generator can sample fake nodes directly without using a softmax. The framework of HeGAN is in Fig. 6.1d.

6.2.2.2 Relation-Aware Discriminator

In an HG, the relation-aware discriminator $D(e_v|u, r; \theta^D)$ evaluates the relation r between the pair of nodes u and v. Specifically, $u \in \mathcal{V}$ is a given node and $r \in \mathcal{R}$ is a given relation from a HG $\mathcal{G}$, e_v is the embedding of a sample node v (which can be real or fake), and θ^D denotes the model parameters of D. In essence, D outputs a probability that the sample v is connected to u under the relation r. We quantify this probability as:

$$D(\mathbf{e}_v|u, r; \theta^D) = \frac{1}{1 + \exp\left(-\mathbf{e}_u^{D^\top} \mathbf{M}_r^D \mathbf{e}_v\right)}, \tag{6.1}$$

where $\mathbf{e}_v \in \mathbb{R}^{d \times 1}$ is the embedding of the sample v, $\mathbf{e}_u^D \in \mathbb{R}^{d \times 1}$ is the learnable embedding of node u, and $\mathbf{M}_r^D \in \mathbb{R}^{d \times d}$ is a learnable relation matrix for relation r. $\theta^D = \{\mathbf{e}_u^D : u \in \mathcal{V}, \mathbf{M}_r^D : r \in \mathcal{R}\}$ forms the model parameters of D, i.e., the union of all node embeddings and relation matrices learnt by D. Naturally, the probability should be high when v is a positive sample related to u through r, or low when it is a negative sample. In general, a sample v forms a triple $\langle u, v, r \rangle$ together with the given u and r, and each triple belongs to one of the three cases below with regard to its polarity. Each case also contributes to one part of the discriminator loss, inspired by the idea of conditional GAN [25].

Case 1:   Connected under a given relation. That is, nodes u and v are indeed connected through the right relation r on the HG $\mathcal{G}$, such as $\langle a_2, p_2, \text{write} \rangle$ shown in Fig. 6.1a. Such a triple is considered positive and can be modeled by the below loss.

$$\mathcal{L}_1^D = \mathbb{E}_{\langle u,v,r \rangle \sim P_\mathcal{G}} - \log D\left(\mathbf{e}_v^D | u, r\right). \tag{6.2}$$

Here we draw the positive triple from the observed $\mathcal{G}$, denoted as $\langle u, v, r \rangle \sim P_\mathcal{G}$.

Case 2:   Connected under an incorrect relation. That is, u and v are connected on the HG under a wrong relation $r' \neq r$, such as $\langle a_2, p_2, \text{view} \rangle$. The discriminator is expected to mark them as negative, as their connectivity does not match the desired semantics carried by the given relation r. We define this part of loss as follows:

$$\mathcal{L}_2^D = \mathbb{E}_{\langle u,v \rangle \sim P_\mathcal{G}, r' \sim P_{\mathcal{R}'}} - \log\left(1 - D\left(\mathbf{e}_v^D | u, r'\right)\right). \tag{6.3}$$

Here, we still draw the pair of nodes $\langle u, v \rangle$ from $\mathcal{G}$, but the negative relation r' is drawn from $\mathcal{R}' = \mathcal{R} \setminus \{r\}$ uniformly.

Case 3:   Fake node from a relation-aware generator. That is, given a node $u \in \mathcal{V}$, it can form a fake pair with the node v supplied by the generator $G(u, r; \theta^G)$, such as $\langle a', p_2, \text{write} \rangle$ in Fig. 6.1d. The generator is also relation-aware: It attempts to generate a fake node's embedding that mimics the real nodes connected to u

under the correct relation r. Again, the discriminator aims to identify this triple as negative, which can be formulated as follows:

$$\mathcal{L}_3^D = \mathbb{E}_{\langle u,r\rangle \sim P_{\mathcal{G}}, \mathbf{e}_v' \sim G(u,r;\theta^G)} - \log\left(1 - D(\mathbf{e}_v'|u, r)\right). \tag{6.4}$$

Note that the fake sample v's embedding $\mathbf{e}_v'$ is drawn from the generator G's learnt distribution. On the other hand, the discriminator D simply treats $\mathbf{e}_v'$ as non-learnable input, and only optimizes its own parameters θ^D.

The loss function of the discriminator is comprehensive by the above three parts:

$$\mathcal{L}^D = \mathcal{L}_1^D + \mathcal{L}_2^D + \mathcal{L}_3^D + \lambda^D \|\theta^D\|_2^2, \tag{6.5}$$

where $\lambda_D > 0$ controls the regularization term to avoid over-fitting. The parameters θ^D of the discriminator can be optimized by minimizing $\mathcal{L}^D$.

6.2.2.3 Relation-Aware Generalized Generator

Given a node $u \in \mathcal{V}$ and a relation $r \in \mathcal{R}$, the generator $G(u, r; \theta^G)$ aims to generate a fake node v likely to connect to u in the context of relation r. In other words, v should be as close as possible to a real node on $\mathcal{G}$, in which has $\langle u, v, r\rangle \sim P_{\mathcal{G}}$. On the other hand, the generator is generalized, which means the fake node v can be latent and not found in $\mathcal{V}$.

To meet the two requirements, the generator should also employ relation-specific matrices (for relation-awareness), and generate samples from an underlying continuous distribution (for generalization). One of the suitable distributions is Gaussian distribution:

$$\mathcal{N}\left(\mathbf{e}_u^{G\top}\mathbf{M}_r^G, \sigma^2\mathbf{I}\right), \tag{6.6}$$

where $\mathbf{e}_u^G \in \mathbb{R}^{d\times 1}$ and $\mathbf{M}_r^G \in \mathbb{R}^{d\times d}$ denote the node embedding of $u \in \mathcal{V}$ and the relation matrix of $r \in \mathcal{R}$ for the generator. In other words, it is a Gaussian distribution with mean $\mathbf{e}_u^{G\top}\mathbf{M}_r^G$ and covariance $\sigma^2\mathbf{I} \in \mathbb{R}^{d\times d}$ for some choices of σ. Intuitively, the mean represents a fake node is likely to be connected to u by relation r, and the covariance represents potential deviations. As neural networks have shown strong ability in modeling complex structure [15], we integrate the multi-layer perceptron (MLP) into the generator for enhancing the expression of the fake samples. Hence, our generator is formulated as follows:

$$G(u, r; \theta^G) = f(\mathbf{W}_L \cdots f(\mathbf{W}_1\mathbf{e} + \mathbf{b}_1) + \mathbf{b}_L), \tag{6.7}$$

where we draw $\mathbf{e}$ from the distribution $\mathcal{N}(\mathbf{e}_u^{G\top}\mathbf{M}_r^G, \sigma^2\mathbf{I})$. Here $\mathbf{W}_*$ and $\mathbf{b}_*$ respectively denote the weight matrix and the bias vector for each layer, and f is an activation function. The parameter set of the generator is thus $\theta^G = \{\mathbf{e}_u^G : u \in \mathcal{V}, \mathbf{M}_r^G : r \in \mathcal{R}, \mathbf{W}_*, \mathbf{b}_*\}$, i.e., the union of all node embeddings and relation matrices, as well as the parameters of MLP.

As motivated earlier, the generator aims to fool the discriminator by generating close-to-real fake samples, such that the discriminator gives high score to them.

$$\mathcal{L}^G = \mathbb{E}_{\langle u,r \rangle \sim P_{\mathcal{G}}, \mathbf{e}_v' \sim G(u,r;\theta^G)} - \log D(\mathbf{e}_v'|u, r) + \lambda^G \|\theta^G\|_2^2, \qquad (6.8)$$

where $\lambda^G > 0$ controls the regularization term. The parameters θ^G of the generator can be optimized by minimizing $\mathcal{L}^G$.

Algorithm 6.1 Model training for HeGAN

Input: HG $\mathcal{G}$, number of generator and discriminator trainings per epoch n_G, n_D, number of samples n_s

1: Initialize θ_G and θ_D for G and D, respectively
2: **while** not converge **do**
3:     **for** $n = 0; n < n_D$ **do**   ▷ **Discriminator training**
4:         Sample a batch of triples, i.e., $\langle u, v, r \rangle \sim P_{\mathcal{G}}$
5:         Generate n_s fake nodes $\mathbf{e}_v' \sim G(u, r; \theta^G)$ for each $\langle u, r \rangle$
6:         Sample n_s relations $r' \sim P_{\mathcal{R}'}$ for each $\langle u, v \rangle$
7:         Update θ^D according to Eq. (6.5)
8:     **end for**
9:     **for** $n = 0; n < n_G$ **do**   ▷ **Generator training**
10:        Sample a batch of triples, i.e., $\langle u, v, r \rangle \sim P_{\mathcal{G}}$
11:       Generate n_s fake nodes $\mathbf{e}_v' \sim G(u, r; \theta^G)$ for each $\langle u, r \rangle$
12:       Update θ^G according to Eq. (6.8)
13:     **end for**
14: **end while**
15: **return** θ^G and θ^D

6.2.2.4 Model Training

We adopt the iterative optimization strategy to train HeGAN. The iterative optimization methods follow the steps below: At each iteration, θ^G is fixed first, and then the generator generates fake samples to optimize θ^D and thus improve the discriminator. Next, θ^D is fixed, followed by optimizing θ^G in order to produce increasingly better fake samples as evaluated by the discriminator. The above steps repeat until the model converges. The model training for HeGAN is outlined in Algorithm 6.1.

6.2.3 Experiments

6.2.3.1 Experimental Settings

We conduct extensive experiments on three benchmark datasets [10, 15], namely DBLP (with four types of nodes: *Author, paper, conference*, and *term*), Yelp (*user, business, service, star*, and *reservation*), AMiner (*author, paper, conference*, and *reference*). We organize them into HGs, as summarized in Table 6.1. We perform node classification and link prediction on three datasets.

For the baselines, we consider three categories of network embedding methods: Traditional (Deepwalk [24], LINE [27]), GAN-based (GraphGAN [28], ANE [6]), and HG (HERec-HNE [26], HIN2vec [10], Metapath2vec [8]) embedding algorithms.

6.2.3.2 Node Classification

We use 80% of the labeled nodes to train a logistic regression classifier, and test the classifier on the remaining 20% of the nodes. We report *Accuracy* on the test set in Table 6.2.

In node classification, HeGAN consistently outperforms the best baseline with statistical significance. It is also worth noting that our performance margins over the best baseline become smaller compared to the node clustering task, since in classification all methods are helped by the supervision, narrowing their gaps.

6.2.3.3 Link Prediction

In this task, we predict *user-business* links on Yelp, and *author-paper* links on DBLP and AMiner. We randomly hide 20% of such links from the original network as the ground truth positives, and randomly sample disconnected node pairs of the given form as negative instances. The ground truth serves as our test set. We adopt two ways to perform link prediction, namely, *logistic regression* and *inner product*. The evaluation metrics consist of *Accuracy, AUC*, and *F1*, as shown in Table 6.3.

We observe that, the performance of HeGAN over the best baseline is much larger with inner product than with logistic regression. It is hypothesized that HeGAN innately learns a much better structure and semantics preserving embed-

Table 6.1 Description of datasets

Datasets	#Nodes	#Edges	#Node types	#Labels
DBLP	37, 791	170, 794	4	4
Yelp	3913	38, 680	5	3
Aminer	312, 776	599, 951	4	6

Table 6.2 Performance comparison on node classification (bold: best; underline: runner-up)

Methods	DBLP			Yelp			AMiner		
	Micro-F1	Macro-F1	Accuracy	Micro-F1	Macro-F1	Accuracy	Micro-F1	Macro-F1	Accuracy
Deepwalk	0.9201	0.9242	0.9298	0.8262	0.7551	0.8145	0.9519	0.9460	0.9529
LINE-1st	0.9239	0.9213	0.9285	0.8229	0.7440	0.8126	0.9776	0.9713	0.9788
LINE-2nd	0.9144	0.9172	0.9236	0.7591	0.5518	0.7571	0.9469	0.9341	0.9471
GraphGAN	0.9198	0.9210	0.9286	0.8098	0.7268	0.7820	-	-	-
ANE	0.9143	0.9153	0.9189	0.8232	0.7623	0.7932	0.9256	0.9203	0.9221
HERec-HNE	0.9214	0.9228	0.9299	0.7962	0.7713	0.7912	0.9801	0.9726	0.9784
HIN2vec	0.9141	0.9115	0.9224	0.8352	0.7610	0.8200	0.9799	0.9775	0.9801
Metapath2vec	0.9288	0.9296	0.9360	0.7953	0.7884	0.7839	0.9853	0.9860	0.9857
HeGAN	**0.9381****	**0.9375****	**0.9421****	**0.8524****	**0.8031****	**0.8432****	**0.9864***	**0.9873***	**0.9883***

Table 6.3 Performance comparison on link prediction. (bold: best; underline: runner-up)

	Methods	DBLP			Yelp			AMiner		
		Acc	AUC	F1	Acc	AUC	F1	Acc	AUC	F1
Logistic Regression	Deepwalk	0.5441	0.5630	0.5208	0.7161	0.7825	0.7182	0.4856	0.5182	0.4618
	LINE-1st	0.6546	0.7121	0.6685	0.7226	0.7971	0.7099	0.5983	0.6413	0.6080
	LINE-2nd	0.6711	0.6500	0.6208	0.6335	0.6745	0.6499	0.5604	0.5114	0.4925
	GraphGAN	0.5241	0.5330	0.5108	0.7123	0.7625	0.7132	–	–	–
	ANE	0.5123	0.5430	0.5280	0.6983	0.7325	0.6838	0.5023	0.5280	0.4938
	HERec-HNE	0.7123	0.7823	0.6934	0.7087	0.7623	0.6923	0.7089	0.7776	0.7156
	HIN2vec	0.7180	0.7948	0.7006	0.7219	0.7959	0.7240	0.7142	0.7874	0.7264
	Metapath2vec	0.5969	0.5920	0.5698	0.7124	0.7798	0.7106	0.7069	0.7623	0.7156
	HeGAN	**0.7290****	**0.8034****	**0.7119****	**0.7240****	**0.8075****	**0.7325****	**0.7198****	**0.7957****	**0.7389****
Inner Product	Deepwalk	0.5474	0.7231	0.6874	0.5654	0.8164	0.6953	0.5309	0.6064	0.6799
	LINE-1st	0.6647	0.7753	0.7363	0.6769	0.7832	0.7199	0.6113	0.6899	0.7123
	LINE-2nd	0.4728	0.4797	0.6325	0.4193	0.7347	0.5909	0.5000	0.4785	0.6666
	GraphGAN	0.5532	0.6825	0.6214	0.5702	0.7725	0.6894	–	–	–
	ANE	0.5218	0.6543	0.6023	0.5432	0.7425	0.6324	0.5421	0.6123	0.6623
	HERec-HNE	0.5123	0.7473	0.6878	0.5323	0.6756	0.7066	0.6063	0.6912	0.6798
	HIN2vec	0.5775	0.8295	0.6714	0.6273	**0.8340**	0.4194	0.5348	0.6934	0.6824
	Metapath2vec	0.4775	0.6926	0.6287	0.5124	0.6324	0.6702	0.6243	0.7123	0.6953
	HeGAN	**0.7649****	**0.8712****	**0.7837****	**0.7391****	0.8298	**0.7705****	**0.6505****	**0.7431****	**0.7752****

ding space than the baseline methods, since the inner product only relies on the learnt representations without resorting to any external supervision.

The complete method and more experiments can be found in [16].

6.3 Importance Sampling

6.3.1 Overview

To make large-scale graph representation learning possible, researchers have proposed several sampling strategies on Graph Neural Networks (GNNs), including node-wise neighborhood sampling [14, 33] and layer-wise neighborhood sampling [5, 18, 37] to accelerate the training process of GNNs by reducing the number of edges in computation. The former is to sample neighbors for each node while the latter is to sample neighbors from the whole graph. Unfortunately, both the node- and layer-wise sampling only deal with homogeneous graphs, which are inadequate in many real-world scenarios such as E-commerce graphs, as they (1) deal with homogeneous graphs only; (2) take all the nodes of a graph as initial candidates; (3) still incur rapidly increasing cost with more layers. There are two challenges we have to face with as follows.

First, how to design an effective sampler that works with the heterogeneous neighborhoods? While a sampler can be easily defined on a per-node, per-type basis, it is not clear how we can sample neighborhoods in a batch, given that the candidates consist of different types of nodes. Two alternatives could work, namely, **type-dependent** and **type-fusion** sampling. In the former, a sampler is deployed for each neighbor type, and these type-based samplers sample from their respective sub-neighborhood on their own. In the latter, a single sampler is deployed, which treats the entire common neighborhood of the batch as its candidates. Intuitively, the type-fusion sampler would work better, as it takes the influences of total types of interactions into account and models the sampling distribution over all types jointly.

Second, how to design the corresponding effective estimators with the sampled heterogeneous neighborhoods? With the neighbors being sampled based on the global importance in a batch, traditional importance sampling could introduce unwanted variance because of the imbalance between the local importance to the given target node and the global importance to the given batch. To address this challenge, we propose **self-normalized estimators** and **adaptive estimators**, respectively. The former is to adjust the estimators by self-normalizing importance of sampled neighborhoods while the latter is to automatically learn the global importance (i.e., the importance distribution of candidate neighborhoods) by modeling both structural and attributed information.

Finally, we test the proposed sampling strategies on two public datasets, and the experimental results demonstrate that the proposed framework can achieve statistically significant improvements. Compared with the full model without any sampling, we achieve a memory cost reduction by up to 92.48% and a time cost reduction by up to 85.95%, while maintaining the same level of accuracy on two real-world datasets.

6.3.2 The HeteSamp Model

This section introduces the architecture of the proposed HeteSamp model, which consists of the general **H**eterogeneous **I**nteraction **G**raph **E**mbedding (named **HIGE**) and the heterogeneous sampling framework. The heterogeneous sampling framework adopts type-dependent and type-fusion strategies to reduce the expensive time cost and high computational complexity. Notice that, we name the large-scale heterogeneous graph as heterogeneous interaction graph to emphasize the rapidly increasing scale. Moreover, as HeteSamp consists of multiple samplers and estimators, we name the specific strategies according to their adopted sampler and estimator.

6.3.2.1 The General HIGE

To tackle with HIGs, a general idea is to reconstruct node embeddings by propagating information from its heterogeneous neighbors, and backwardly propagate gradients [4, 30, 36], as shown in Fig. 6.2a. Specifically, given a node v_i with type $\phi(v_i)$ and its edges $\{e_{v_i,v_j,r} | v_j \in \mathcal{N}_{v_i,r}, r \in \mathcal{R}\}$, the aggregated information of node v_i from type-r neighborhoods is

$$g'_{v_i,r} = \sum_{v_j \in \mathcal{N}_{v_i,r}} \frac{1}{|\mathcal{N}_{v_i,r}|} w(v_i, v_j, r) h_{v_j}, \tag{6.9}$$

where $g'_{v_i,r}$ is the aggregated information, $\mathcal{N}_{v_i,r}$ is the type-r neighborhoods of node v_i, $v_j \in \mathcal{N}_{v_i,r}$ is a specific type-r neighbor of node v_i, $w(v_i, v_j, r)$ denotes the weight of edge $e_{v_i,v_j,r}$, $h_{v_j} \in \mathbb{R}^d$ denotes the embedding of node v_j, d is the

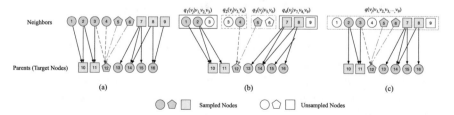

Fig. 6.2 The sampling processes of type-dependent and type-fusion sampling. (**a**) Batch-wise information propagation. (**b**) Type-dependent sampling. (**c**) Type-fusion sampling

dimension of $\boldsymbol{h}_{v_j}$. $w(v_i, v_j, r)$ between node v_i and v_j is calculated by

$$w(v_i, v_j, r) = \sigma(\boldsymbol{x}_{v_i,v_j,r}\boldsymbol{\lambda}_r + b_r), \tag{6.10}$$

where σ is an activation function, $\boldsymbol{x}_{v_i,v_j,r} \in \mathbb{R}^{d_r}$ is the edge features between node v_i and v_j, d_r is the length of type-r edge features, and $\boldsymbol{\lambda}_r \in \mathbb{R}^{d_r \times 1}$ and $b_r \in \mathbb{R}$ are the weight and bias parameters shared by type-r edges.

We then take all semantics into account by aggregating information from different-typed neighborhoods and constructing the global embedding $\boldsymbol{h}'_{v_i} \in \mathbb{R}^d$ of node v_i as

$$\boldsymbol{h}'_{v_i} = \sigma\left(concat\left(\boldsymbol{g}'_{v_i,r_0}, \boldsymbol{g}'_{v_i,r_1}, \cdots, \boldsymbol{g}'_{v_i,r_{|\mathcal{R}|}}\right) \boldsymbol{W}_{\phi(v_i)} + b_{\phi(v_i)}\right), \tag{6.11}$$

where σ is an activation function, $concat(\cdot)$ is the concatenation option, $\boldsymbol{W}_{\phi(v_i)} \in \mathbb{R}^{|\mathcal{R}|d \times d}$ denotes the projection matrix of type-$\phi(v_i)$ nodes, $b_{\phi(v_i)}$ denotes the bias, and $|\mathcal{R}|$ denotes the total number of edge types $\mathcal{R}$ in the heterogeneous graph. The computational complexity of the general model is linear with the scale of edges and nodes on heterogeneous graphs.

6.3.2.2   Batch-Wise Heterogeneous Sampling

To reduce computational overhead and memory cost of the training process, a naïve idea is to sample several neighbors rather than aggregating information from all neighborhoods. Previous works [14, 33] on homogeneous works usually adopt node-wise sampling to sample several neighbors per node for learning. Based on the global and local structural information, current works [5, 18, 37] propose layer-wise sampling strategies which consider the whole neighborhood as the candidates. However, these strategies are to deal with homogeneous graphs but ignore the abundant semantics of heterogeneous nodes and edges. Moreover, these strategies can only deal with small graphs because the overhead for node-wise and layer-wise sampling during optimization could be quite expensive, or even unaffordable.

In this section, Eq. (6.9) can be re-written as

$$\boldsymbol{g}'_{v_i,r} = \mathbb{E}_q\left[\frac{p(v_j|v_i, r)}{q(v_j|\boldsymbol{B}_k)}w(v_i, v_j, r)\boldsymbol{h}_{v_j}\right], \tag{6.12}$$

where $\boldsymbol{B}_k$ denotes the target nodes in the kth mini-batch, $q(v_j|\boldsymbol{B}_k)$ denotes the corresponding sampling probability in the kth mini-batch, or in other words, the importance in this batch, and $p(v_j|v_i, r)$ equals to $\frac{1}{|\mathcal{N}_{v_i,r}|}$. Thereby, we attempt to make heterogeneous sampling for each batch. The batch-wise heterogeneous sampling probability is defined as

$$\hat{v}_j \sim q(\hat{v}_j|v_1, v_2, \cdots, v_{|\boldsymbol{B}_k|}) \quad v_j \in \{\mathcal{N}_{v_i,r}|r \in \mathcal{R}, v_i \in \boldsymbol{B}_k\}, \tag{6.13}$$

where $\hat{v}_j$ is the sampled neighbor. As shown in Fig. 6.2a, we sample instances from the union neighborhood of all the target nodes in a batch, where the union neighborhood denotes the union of the individual neighbors of each target node. Since sampling is now done for each batch instead of each target node, batch-wise sampling is often a good choice to reduce computational overhead and memory cost of the training process. Compared with layer-wise sampling methods [5, 18, 37] which require the total nodes as candidates and have to load the whole graph structure before training, our batch-wise heterogeneous sampling focuses on the neighborhood union in the current batch. Compared with node-wise sampling which sample neighborhoods for each target node, our batch-wise heterogeneous sampling contains the advantages of reducing the overhead of sampling.

6.3.2.3 Type-Dependent Sampling Strategy

By now, we have defined the general batch-wise heterogeneous sampling. Different from traditional importance sampling, the neighborhoods of each batch connect with target nodes based on different-typed interactions. A straightforward idea is to design a sampler for each type, respectively. For example, in Fig. 6.2b, there are four types of candidates, and we sample neighbors from each type of candidates respectively by using one sampler. By adopting type-dependent sampling, we consider $q(v_j|v_1, v_2, \cdots, v_{|B_k|})$ as a set of $\{q_r(v_j|\cdot)|r \in \mathbb{R}\}$, and sample type-$r$ neighborhoods with the corresponding sampler $q_r(v_j|\cdot)$. The remaining problem for type-dependent sampling is how to design the exact form of this sampler so as to keep low variance for efficient training. Here we define the average information μ_{q_r} of $g'_{i,r}$ as

$$\mu_{q_r} = \frac{1}{|B_k|} \sum_{v_i \in B_k} g'_{v_i,r} = \frac{1}{|B_k|} \sum_{v_j} q_r(v_j|\cdot) \sum_{v_i \in B_k} \frac{p(v_j|v_i, r)w(v_i, v_j, r)h_{v_j}}{q_r(v_j|\cdot)},$$

(6.14)

where $|B_k|$ is the number of type-r samples. Then, the variance Var_{q_r} of μ_{q_r} is calculated by

$$Var_{q_r}(\mu_{q_r}) = \frac{1}{|B_k|} \mathbb{E}_{q_r} \frac{\left[\mu_{q_r} q_r(v_j|\cdot) - \sum_{v_i \in B_k} p(v_j|v_i, r)w(v_i, v_j, r)h_{v_i}\right]^2}{q_r(v_j|\cdot)^2}.$$

(6.15)

To ensure minimizing the variance, a better sampler is shown as

$$q_r(v_j) = \frac{\sum_{v_i \in B_k} p(v_j|v_i, r)^2}{\sum_{v_j} \sum_{v_i \in B_k} p(v_j|v_i, r)^2}.$$

(6.16)

6.3.2.4   Type-Fusion Sampling Strategy

Essentially, the type-dependent strategy pays attention to the influence of same-type neighborhoods, without jointly considering the effect of heterogeneous types. Thus, they only reduce the individual variance of each type, lacking a global picture on the overall variance of the batch. To address this weakness, we propose a type-fusion sampling strategy which considers the entire neighborhoods as the candidates.

The aggregation in Eq. (6.11) is to gather information from different-typed neighborhoods, and can be rewritten as

$$g'_{v_i} = \sum_{r \in \mathcal{R}} \sum_{v_j \in \mathcal{N}_{v_i,r}} p(v_j|v_i, r) w(v_i, v_j, r) h_{v_j} W_r, \qquad (6.17)$$

where $W_r \in \mathbb{R}^{d \times d}$ denotes the relation-wise projection matrix. Under the type-fusion strategy, the average information in k-th batch over all types is given by

$$\mu_q = \frac{1}{|B_k|} \sum_{v_i \in B_k} g'_{v_i} = \frac{1}{|B_k|} \sum_{v_j} q(v_j|\cdot) \sum_{r \in \mathcal{R}} \sum_{v_i \in B_k} \frac{p(v_j|v_i, r) w(v_i, v_j, r) h_{v_j} W_r}{q(v_j|\cdot)}. \qquad (6.18)$$

As shown in Fig. 6.2c, the different-typed neighborhoods are sampled according to the type-fusion sampling distribution. Similar to that in Sect. 6.3.2.3, the corresponding sampler is defined as follows:

$$q_s(v_j) = \frac{\sum_{r \in \mathcal{R}} \sum_{v_i} p(v_j|v_i, r)^2}{\sum_{r \in \mathcal{R}} \sum_{v_j} \sum_{v_i \in B_k} p(v_j|v_i, r)^2}, \qquad (6.19)$$

where $q_s(v_j)$ is the structure-based type-fusion sampler.

While the above sampling strategies only consider link-based weight as the importance of nodes but ignoring the attributed-based importance on edges, we further pay attention to the edge features within HIGs and propose the corresponding sampler as follows:

$$q_a(v_j) = \frac{\sum_{r \in \mathcal{R}} \sum_{v_i} p(v_j|v_i, r) f(x(v_i, v_j, r))}{\sum_{r \in \mathcal{R}} \sum_{v_j} \sum_{v_i \in B_k} p(v_j|v_i, r) f(x(v_i, v_j, r))}, \qquad (6.20)$$

where $q_a(v_j)$ denotes the adaptive sampler, $f(x(v_i, v_j, r))$ denotes the importance of edge features which is calculated by $\sigma(x(v_i, v_j, r) W_{x,r})$, and $W_{x,r} \in \mathbb{R}^{d_r \times 1}$ is the type-r learnable parameter.

6.3.2.5 Heterogeneous Self-Normalized and Adaptive Estimators

In type-dependent or type-fusion strategies, the batch-wise estimator is

$$\hat{g}_{v_i} = \frac{1}{n} \sum_{r \in \mathcal{R}} \sum_{v_j \in \hat{\mathcal{N}}_{B_k,r}} \frac{p(\hat{v}_j|\cdot)}{q(\hat{v}_j|\cdot)} w(v_i, \hat{v}_j, r) \mathbf{h}_{\hat{v}_j} \mathbf{W}_r, \tag{6.21}$$

where $\hat{g}_{v_i}$ is the approximated information, n is the number of samples, $\hat{\mathcal{N}}_{B_k,r}$ is the sampled type-r neighborhoods in this batch, $p(\hat{v}_j|\cdot)$ is equal to $\frac{1}{|\mathcal{N}_{v_i,r}|}$, and $q(\hat{v}_j|\cdot)$ denotes the heterogeneous samplers, such as q_r, q_s, and q_a. However, this could increase the variance resulted from the imbalance of $p(\hat{v}_j|\cdot)$ and $q(\hat{v}_j|\cdot)$.

To address such problem, for structure-based $q_s(v_j)$, a promising way is to balance the weights based on self-normalized importance, similar to that in [18]. The corresponding estimator can be computed as

$$\hat{g}_{sn,v_i} = \sum_{r \in \mathcal{R}} \sum_{\hat{v}_j \in \hat{\mathcal{N}}_{v_i,r}} \frac{\pi(\hat{v}_j)}{\sum_{\hat{v}'_j \in \hat{\mathcal{N}}_{v_i,r}} \pi(\hat{v}'_j)} w_{v_i,v_j,r} \mathbf{h}_j \mathbf{W}_r, \tag{6.22}$$

where $\hat{g}_{sn,v_i}$ denotes the self-normalized information, $\pi(v_j) = \frac{p(v_j|\cdot)}{q(v_j|\cdot)}$, $\hat{\mathcal{N}}_r$ is the sampled neighborhoods of type r. Besides, for the adaptive sampling, we add the variance to the loss function and explicitly minimize the variance during training to achieve a better $q_a(v_j)$.

6.3.2.6 Optimization Framework

The overall loss function consists of four parts

$$L_k = L_{task,k} + \alpha L_{ep,k} + \beta \Omega(\mathbf{\Theta}) + \xi Var_{q_a,k}(\hat{\mu}_q), \tag{6.23}$$

where L_k is the loss value in k-th batch, $L_{task,k}$ is the loss from supervised learning in the same batch, $L_{ep,k}$ is the embedding propagation loss with sampling in the k-th batch, $Var_{q_a,k}(\hat{\mu}_q)$ is the variance of sampled information, $\Omega(\mathbf{\Theta})$ is the regularization of all latent parameters, and α, β, and ξ are three hyper-parameters. Notice that, for type-dependent sampling and structure-based type-fusion sampling, ξ is 0.

6.3.3 Experiments

6.3.3.1 Experimental Settings

Datasets We evaluate the empirical performance of our method on two real-world heterogeneous graphs, including the small Aminer graph consisting of 41,523 nodes and 199,429 edges, and the large Alibaba graph consisting of 4,527,222 nodes and 49,785,900 edges.

Baseline We first compare our various importance sampling strategies with the general HIGE model and HAN [30] without any sampling. We also compare our model with state-of-the-art sampling algorithms on GCNs including Fast-GCN [18] and AS-GCN [5] to showcase our advantage of effectiveness and scalability, whereas the sampling-based models not only achieve superior accuracy, but also scale to large graphs. Finally, to study the performance of our proposed variance reduction, we also substitute our variance reduction sampler with a uniform sampler.

Parameter Settings For all baselines and our methods, we set $\beta = 0.1$, $\gamma = 0.1$, $\psi = 0.1$, $\alpha = 0.4$. The maximum iteration is 100 for Aminer and 5 for Alibaba. We adopt Micro-F1 and Macro-F1 as the evaluation metrics for node classification on Aminer while adopting F1 and AUC as the evaluation metrics for link prediction on Alibaba. All the metrics are positively related to the performance of methods.

6.3.3.2 Empirical Validation

We perform empirical validation on the small Aminer graph for node classification and the much larger Alibaba interaction graph for link prediction. Notice that the non-HIGE-based baselines, GraphsAGE, Fast-GCN and AS-GCN are to deal with node classification, and these baselines can only work on Aminer; they cannot perform on Alibaba graph when the task is link prediction and this graph is too large. Furthermore, they are likely to suffer from insufficient memory on the large-scale Alibaba graph even if we implement the modified models for the task of link prediction.

6.3.3.3 Effectiveness

We report the results in Tables 6.4 and 6.5, respectively. We progressively sample more neighbors per batch, and the sampling rate ranges from about 3%–24% on smaller Aminer and 1.25%–10% on the larger Alibaba graph. We make the following observations.

First, VarR-TF-AS generally achieves the best performance, whereas VarR-TF-SN comes as a close competitor. In particular, VarR-TF-AS performs as well as or even better than the original HIGE. This phenomenon is reasonable. On the one hand, HIGE aggregates information from all neighbors which may introduce

Table 6.4 Micro/Macro-F1 scores for node classification on Aminer. Excluding HIGE and HAN, the best method is bolded and the second best is underlined

	Micro-F1				Macro-F1			
Sampling size	128	256	512	1024	128	256	512	1024
Per batch	~ 3%	~ 6%	~ 12%	~ 24%	~ 3%	~ 6%	~ 12%	~ 24%
HIGE-Nil	0.1990				0.1961			
HIGE	0.9646				0.9593			
HAN	0.9512				0.9508			
GraphSAGE	0.2022	0.2039	0.2125	0.2119	0.1989	0.2023	0.2036	0.2073
Fast-GCN	0.2117	0.2244	0.2318	0.2361	0.1850	0.1898	0.2116	0.2119
AS-GCN	0.2361	0.2307	0.2390	0.2390	0.2005	0.2028	0.2004	0.2068
Unif-TD	0.3043	0.4123	0.4256	0.4221	0.2638	0.3907	0.3553	0.3962
Unif-TF	0.1780	0.2229	0.3785	0.6296	0.1266	0.1952	0.3081	0.5727
VarR-TD	0.8086	0.7990	0.8971	0.9123	0.7913	0.7894	0.8945	0.9133
VarR-TF	0.9461	_0.9671_	**0.9712**	0.9675	0.9424	_0.9651_	**0.9696**	_0.9664_
VarR-TF-SN	**0.9659**	0.9643	0.9612	_0.9684_	_0.9637_	0.9629	_0.9631_	0.9603
VarR-TF-AS	_0.9650_	**0.9676**	_0.9705_	**0.9687**	**0.9649**	**0.9667**	0.9602	**0.9671**

Table 6.5 F1 and AUC scores for purchase prediction on the Alibaba graph. Excluding HIGE and HAN, the best method is bolded and the second best is underlined

	F1				ROC-AUC			
Sampling size	512	1024	2048	4096	512	1024	2048	4096
Per batch	~ 1.25%	~ 2.5%	~ 5%	~ 10%	~ 1.25%	~ 2.5%	~ 5%	~ 10%
HIGE-Nil	0.3994				0.5134			
HIGE	0.5663				0.7715			
HAN	0.5618				0.7704			
Unif-TD	0.4017	0.4226	0.4352	0.4371	0.5768	0.5826	0.5908	0.5924
Unif-TF	0.4008	0.4122	0.4125	0.4451	0.5731	0.5790	0.5862	0.5977
VarR-TD	0.4841	0.5003	0.5274	0.5682	0.6207	0.6504	0.6925	0.7475
VarR-TF	_0.5769_	_0.5908_	0.5709	_0.5833_	0.7648	**0.7671**	0.7653	_0.7796_
VarR-TF-SN	**0.5780**	0.5883	_0.5802_	0.5798	_0.7669_	0.7625	_0.7660_	0.7742
VarR-TF-AS	0.5729	**0.5913**	**0.5806**	**0.5844**	**0.7674**	_0.7641_	**0.7676**	**0.7799**

noisy information while our sampling strategies are to sample valuable neighbors for training. On the other hand, we optimize the sampler by reducing the variance loss which enhances the similarity limitation between the sampled neighbors and its target nodes.

Second, compared to the corresponding uniform samplers, the variance reduction samplers guarantee more stable estimators and thus produce better results. In the VarR-* methods, type-fusion strategies outperform type-dependent methods, as the former consider all types jointly rather than independently and reduce the variance of the whole batch rather than a single type.

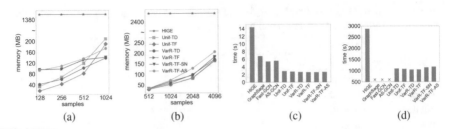

Fig. 6.3 Average memory cost and running time per iteration of training. (**a**) Aminer Memory. (**b**) Alibaba Memory. (**c**) Aminer Time. (**d**) Alibaba Time

Third, our proposed general HIGE can perform competitively with or even better than HAN. However, HAN has higher time complexity. For Alibaba dataset, the time cost of HAN (about 1 day) is quite larger than HIGE (about 5 h).

Fourth, all sampling strategies for HIGE perform significantly better than GCN-based models and GraphsAGE. On the one hand, sampling for HIGE takes the graph heterogeneity into consideration, whereas the two GCN-based models do not make use of such information. On the other hand, the sampling size or number of layers of Fast-GCN and AS-GCN may be not enough. However, even under current settings, Fast-GCN and AS-GCN are already several times slower than our method, as we shall see in the efficiency study.

6.3.3.4 Efficiency

We first investigate the efficiency of our sampling strategies in memory cost. As shown in Fig. 6.3a, b, our models incur by up to 92.48% less memory. Further note that the differences in both the number of edges and memory cost are more prominent on the larger Alibaba graph, indicating excellent scalability of our sampling strategies. Second, in terms of the running time, as shown in Fig. 6.3c, d, VarR-TF-AS and VarR-TF-SN require less time than HIGE to attain close performance on the two datasets. Compared to HIGE, our models incur by up to 84.39% less time cost. Notice that when dealing with large-scale Alibaba dataset, the time cost is quite larger (about 1 day) than HIGE (about 5 h).

The specific designs of HeteSamp can be found in [19].

6.4 Hyperbolic Representation

6.4.1 Overview

HG embedding has attracted considerable research attention recently. Most HG embedding methods choose Euclidean spaces to represent HG graphs, which is because Euclidean spaces are the natural generalization of our intuition-friendly, and visual three-dimensional space. However, a fundamental problem is that what are the appropriate or intrinsic underlying spaces of HGs? Therefore, we wonder whether Euclidean spaces are the intrinsic spaces of HGs?

Recently, hyperbolic spaces have gained momentum in the context of network science [20]. Hyperbolic spaces are spaces of constant negative curvature [3]. A superiority of hyperbolic spaces is that they expand faster than Euclidean spaces [22]. Therefore, it is easy to model complex data with low-dimensional embedding in hyperbolic spaces. Due to the characteristic of hyperbolic spaces, [20] assumes hyperbolic spaces underlie complex network and finds that data with power-law structure is suitable to be modeled in hyperbolic spaces. Also, some researchers begin to embed different data in hyperbolic spaces. For instance, [7] embeds text in hyperbolic spaces. Nickel and Kiela [22] and Ganea et al. [11] learn the hyperbolic embeddings of homogeneous networks. However, it is unknown whether HGs are suitable to be embedded in hyperbolic spaces.

In this section, we analyze the relation distribution in HGs and propose HHNE, which is able to preserve the structure and semantic information in hyperbolic spaces. HHNE leverages the meta-path guided random walk to generate heterogeneous neighborhoods to capture the structure and semantic relations in HGs. Then the proximity between nodes is measured by the distance in hyperbolic spaces. Also, HHNE is able to maximize the proximity between the neighborhood nodes while minimize the proximity between the negative sampled nodes. The optimization strategy of HHNE is derived to optimize hyperbolic embeddings iteratively.

6.4.2 The HHNE Model

6.4.2.1 Model Framework

HHNE leverages the meta-path guided random walk to obtain neighbors for each node to capture the structure and semantic relations in HGs. Also, HHNE learns the embeddings by maximizing the proximity between the neighborhood nodes and minimizing the proximity between the negative sampled nodes. Moreover, we derive the optimization strategy of HHNE to upgrade the hyperbolic embeddings.

6.4.2.2 Hyperbolic HG Embedding

To design HG embedding methods in hyperbolic spaces, we make use of the Poincaré ball model to describe hyperbolic spaces. Let $\mathbb{D}^d = \{x \in \mathbb{R}^d : \|x\| < 1\}$ be the *open* d-dimensional unit ball. The Poincaré ball model is defined by the manifold $\mathbb{D}^d$ equipped with the following Riemannian metric tensor $g_x^{\mathbb{D}}$:

$$g_x^{\mathbb{D}} = \lambda_x^2 g^{\mathbb{E}} \quad \text{where } \lambda_x := \frac{2}{1 - \|x\|^2}, \tag{6.24}$$

where $x \in \mathbb{D}^d$, $g^{\mathbb{E}} = \mathbf{I}$ denotes the Euclidean metric tensor.

HHNE aims to learn the representation of nodes to preserve the structure and semantic correlations in hyperbolic spaces. Given an HG $G = (V, E, T, \phi, \psi)$ with $|T_V| > 1$, HHNE is interested in learning the embeddings $\Theta = \{\theta_i\}_{i=1}^{|V|}$, $\theta_i \in \mathbb{D}^d$. HHNE preserves the structure by facilitating the proximity between a node and its neighborhoods. HHNE uses meta-path guided random walks [8] to obtain heterogeneous neighborhoods of a node. In meta-path guided random walks, the node sequences are restrained by the node types which are defined by meta-paths. Specifically, let t_{v_i} and t_{e_i} as the types of node v_i and edge e_i, respectively, given a meta-path $\mathcal{P} = t_{v_1} \xrightarrow{t_{e_1}} \ldots t_{v_i} \xrightarrow{t_{e_i}} \ldots \xrightarrow{t_{e_{n-1}}} t_{v_n}$, the transition probability at step i is defined as follows:

$$p(v^{i+1}|v_{t_{v_i}}^i, \mathcal{P}) = \begin{cases} \frac{1}{|N_{t_{v_{i+1}}}(v_{t_{v_i}}^i)|} & (v^{i+1}, v_{t_{v_i}}^i) \in E, \phi(v^{i+1}) = t_{v_{i+1}} \\ 0 & \text{otherwise}, \end{cases} \tag{6.25}$$

where $v_{t_{v_i}}^i$ is node $v \in V$ with type t_{v_i}, and $N_{t_{v_{i+1}}}(v_{t_{v_i}}^i)$ denotes the $t_{v_{i+1}}$ type of neighborhood of node $v_{t_{v_i}}^i$. The meta-path guided random walk strategy ensures that the semantic relationships between different types of nodes can be properly incorporated into HHNE.

In order to preserve the proximity between nodes and its neighborhoods in hyperbolic spaces, HHNE uses distances in Poincaré ball model to measure their proximity. Given nodes embeddings $\theta_i, \theta_j \in \mathbb{D}^d$, the distance in Poincaré ball is given by:

$$d_{\mathbb{D}}(\theta_i, \theta_j) = \cosh^{-1}\left(1 + 2\frac{\|\theta_i - \theta_j\|^2}{(1 - \|\theta_i\|^2)(1 - \|\theta_j\|^2)}\right). \tag{6.26}$$

It is worth noting that as the Poincaré ball model is defined in metric spaces, the distance in Poincaré ball meets the triangle inequality and can well preserve the transitivity in HG. Then, HHNE calculates the probability of having the heterogeneous neighbor c_t, given a node v

$$p(v|c_t; \Theta) = \sigma[-d_{\mathbb{D}}(\theta_v, \theta_{c_t})],$$

where $\sigma(x) = \frac{1}{1+\exp(-x)}$. Then the object of HHNE is to maximize the probability as follows:

$$\arg\max_{\Theta} \sum_{v \in V} \sum_{c_t \in C_t(v)} \log p(v|c_t; \Theta). \tag{6.27}$$

To achieve efficient optimization, HHNE leverages the negative sampling proposed in [21], which basically samples a small number of negative objects to enhance the influence of positive objects. For a given node v, HHNE aims to maximize the proximity between v and its neighborhood c_t while minimizes the proximity between v and its negative sampled node n. Therefore, the objective function Eq. (6.27) can be rewritten as follows:

$$\mathcal{L}(\Theta) = \log \sigma[-d_{\mathbb{D}}(\theta_{c_t}, \theta_v)] + \sum_{m=1}^{M} \mathbb{E}_{n^m \sim P(n)}\{\log \sigma[d_{\mathbb{D}}(\theta_{n^m}, \theta_v)]\}, \qquad (6.28)$$

where $P(n)$ is a pre-defined distribution from which a negative node n^m is drawn for M times. HHNE builds the node frequency distribution by drawing nodes regardless of their types.

6.4.2.3 Optimization

As the parameters of the model live in a Poincaré ball which has a Riemannian manifold structure, the back-propagated gradient is a Riemannian gradient. It means that the Euclidean gradient-based optimization, such as $\theta_i \leftarrow \theta_i + \eta \nabla_{\theta_i}^E \mathcal{L}(\Theta)$, makes no sense as an operation in the Poincaré ball, because the addition operation is not defined in this manifold. Instead, HHNE can optimize Eq. (6.28) via a Riemannian stochastic gradient descent (RSGD) optimization method [1]. In particular, let $\mathcal{T}_{\theta_i} \mathbb{D}^d$ denote the tangent space of a node embedding $\theta_i \in \mathbb{D}^d$, and HHNE can compute the Riemannian gradient $\nabla_{\theta_i}^R \mathcal{L}(\Theta) \in \mathcal{T}_{\theta_i} \mathbb{D}^d$ of $\mathcal{L}(\Theta)$. Using RSGD, HHNE can be optimized by maximizing Eq. (6.28), and a node embedding can be updated in the form of:

$$\theta_i \leftarrow \exp_{\theta_i}(\eta \nabla_{\theta_i}^R \mathcal{L}(\Theta)), \qquad (6.29)$$

where $\exp_{\theta_i}(\cdot)$ is exponential map in the Poincaré ball. The exponential map is given by Ganea et al. [11]:

$$\exp_{\theta_i}(s)$$
$$= \frac{\lambda_{\theta_i}\left(\cosh(\lambda_{\theta_i}\|s\|) + \langle \theta_i, \frac{s}{\|s\|}\rangle \sinh(\lambda_{\theta_i}\|s\|)\right)}{1 + (\lambda_{\theta_i} - 1)\cosh(\lambda_{\theta_i}\|s\|) + \lambda_{\theta_i}\langle \theta_i, \frac{s}{\|s\|}\rangle \sinh(\lambda_{\theta_i}\|s\|)}\theta_i \qquad (6.30)$$
$$+ \frac{\frac{1}{\|s\|}\sinh(\lambda_{\theta_i}\|s\|)}{1 + (\lambda_{\theta_i} - 1)\cosh(\lambda_{\theta_i}\|s\|) + \lambda_{\theta_i}\langle \theta_i, \frac{s}{\|s\|}\rangle \sinh(\lambda_{\theta_i}\|s\|)}s.$$

As the Poincaré ball model is a conformal model of hyperbolic spaces, i.e., $g_x^{\mathbb{D}} = \lambda_x^2 g^{\mathbb{E}}$, the Riemannian gradient ∇^R is obtained by rescaling the Euclidean gradient

∇^E by the inverse of the metric tensor, i.e., $\frac{1}{g_x^{\mathbb{D}}}$:

$$\nabla_{\theta_i}^R \mathcal{L} = \left(\frac{1}{\lambda_{\theta_i}}\right)^2 \nabla_{\theta_i}^E \mathcal{L}. \tag{6.31}$$

Furthermore, the gradients of Eq. (6.28) can be derived as follows:

$$\frac{\partial \mathcal{L}}{\partial \theta_{u^m}} = \frac{4}{\alpha \sqrt{\gamma^2 - 1}} \left[\mathbb{I}_v[u^m] - \sigma(-d_{\mathbb{D}}(\theta_{c_t}, \theta_{u^m}))\right]$$
$$\cdot \left[\frac{\theta_{c_t}}{\beta_m} - \frac{\|\theta_{c_t}\|^2 - 2\langle\theta_{c_t}, \theta_{u^m}\rangle + 1}{\beta_m^2}\theta_{u^m}\right], \tag{6.32}$$

$$\frac{\partial \mathcal{L}}{\partial \theta_{c_t}} = \sum_{m=0}^M \frac{4}{\beta_m \sqrt{\gamma^2 - 1}} \left[\mathbb{I}_v[u^m] - \sigma(-d_{\mathbb{D}}(\theta_{c_t}, \theta_{u^m}))\right]$$
$$\cdot \left[\frac{\theta_{u^m}}{\alpha} - \frac{\|\theta_{u^m}\|^2 - 2\langle\theta_{c_t}, \theta_{u^m}\rangle + 1}{\alpha^2}\theta_{c_t}\right], \tag{6.33}$$

where $\alpha = 1 - \|\theta_{c_t}\|^2$, $\beta_m = 1 - \|\theta_{u^m}\|^2$, $\gamma = 1 + \frac{2}{\alpha\beta}\|\theta_{c_t} - \theta_{u^m}\|^2$ and when $m = 0$, $u^0 = v$. $\mathbb{I}_v[u]$ is an indicator function to indicate whether u is v. Then, HHNE can be updated by using Eqs. (6.32)–(6.33) iteratively.

6.4.3 Experiments

6.4.3.1 Experimental Setup

Datasets The basic statistics of the two HGs used in our experiments are shown in Table 6.6.

Baselines HHNE is compared with the following state-of-the-art methods: (1) the homogeneous graph embedding methods, i.e., DeepWalk [24], LINE [27], and node2vec [13]; (2) the heterogeneous graph embedding methods, i.e., metapath2vec [8]; (3) the hyperbolic homogeneous graph embedding methods, i.e., PoincaréEmb [22].

Table 6.6 Statistics of datasets

DBLP	# A	# P	# V	# P-A	# P-V
	14,475	14,376	20	41,794	14,376
MovieLens	# A	# M	# D	# M-A	# M-D
	11,718	9160	3510	64,051	9160

Parameter Settings For random walk based methods DeepWalk, node2vec, meta-path2vec, and HHNE, we set neighborhood size as 5, walk length as 80, walks per node as 40. For LINE, metapath2vec, PoincaréEmb, and HHNE, we set the number of negative samples as 10. For methods based on meta-path guided random walks, we use "APA" for relation "P-A" in network reconstruction and link prediction experiments in DBLP; "APVPA" for relation "P-V" in above experiments in DBLP; "AMDMA" for all relation in above experiments in MovieLens. In visualization experiment, in order to focus on analyzing the relation of "A" and "P", we use "APA".

6.4.3.2 Network Reconstruction

A good HG embedding method should ensure that the learned embeddings can preserve the original HG structure. The reconstruction error in relation to the embedding dimension is then a measure for the capacity of the model. More specifically, we use network embedding methods to learn feature representations. Then for each type of links in the HG, we enumerate all pairs of objects that can be connected by such a link and calculate their proximity [17], i.e., the distance in Poincaré ball model for HHNE and PoincaréEmb. Finally, we use the AUC [9] to evaluate the performance of each embedding method. For example, for link type "write," we calculate all pairs of authors and papers in DBLP and compute the proximity for each pair. Then using the links between authors and papers in real DBLP network as ground truth, we compute the AUC value for each embedding method.

The results are shown in Table 6.7. As we can see, HHNE consistently performs the best in all the tested HGs. The results demonstrate that HHNE can effectively preserve the original network structure and reconstruct the network, especially on the reconstruction of P-V and M-D edges. Also, please note that HHNE achieves very promising results when the embedding dimension is very small. This suggests that regarding hyperbolic spaces underlying HG is reasonable and hyperbolic spaces have strong ability of modeling network when the dimension of spaces is small.

6.4.3.3 Link Prediction

Link prediction aims to infer the unknown links in an HG given the observed HG structure, which can be used to test the generalization performance of a network embedding method. The experimental setting is similar to [32]. For each type of edge, 20% of edges are removed randomly from the network while ensuring that the rest network structure is still connected. The proximity of all pair of nodes is calculated in the test. AUC is used as the evaluation metric.

From the results in Table 6.8, HHNE outperforms the baselines upon all the dimensionality, especially in the low dimensionality. The results can demonstrate the generalization ability of HHNE. In DBLP dataset, the results of HHNE in

Table 6.7 AUC scores for network reconstruction. The best result in each row is bold

Dataset	Edge	Dimension	Deepwalk	LINE(1st)	LINE(2nd)	node2vec	metapath2vec	PoincaréEmb	HHNE
DBLP	P-A	2	0.6933	0.5286	0.6740	0.7107	0.6686	0.8251	**0.9835**
		5	0.8034	0.5397	0.7379	0.8162	0.8261	0.8769	**0.9838**
		10	0.9324	0.6740	0.7541	0.9418	0.9202	0.8921	**0.9887**
		15	0.9666	0.7220	0.7868	0.9719	0.9500	0.8989	**0.9898**
		20	0.9722	0.7457	0.7600	0.9809	0.9623	0.9024	**0.9913**
		25	0.9794	0.7668	0.7621	0.9881	0.9690	0.9034	**0.9930**
	P-V	2	0.7324	0.5182	0.6242	0.7595	0.7286	0.5718	**0.8449**
		5	0.7906	0.5500	0.6349	0.8019	0.9072	0.5529	**0.9984**
		10	0.8813	0.7070	0.6333	0.8922	0.9691	0.6271	**0.9985**
		15	0.9353	0.7295	0.6343	0.9382	0.9840	0.6446	**0.9985**
		20	0.9505	0.7369	0.6444	0.9524	0.9879	0.6600	**0.9985**
		25	0.9558	0.7436	0.6440	0.9596	0.9899	0.6760	**0.9985**
MoiveLens	M-A	2	0.6320	0.5424	0.6378	0.6402	0.6404	0.5231	**0.8832**
		5	0.6763	0.5675	0.7047	0.6774	0.6578	0.5317	**0.9168**
		10	0.7610	0.6202	0.7739	0.7653	0.7231	0.5404	**0.9211**
		15	0.8244	0.6593	0.7955	0.8304	0.7793	0.5479	**0.9221**
		20	0.8666	0.6925	0.8065	0.8742	0.8189	0.5522	**0.9239**
		25	0.8963	0.7251	0.8123	0.9035	0.8483	0.5545	**0.9233**
	M-D	2	0.6626	0.5386	0.6016	0.6707	0.6589	0.6213	**0.9952**
		5	0.7263	0.5839	0.6521	0.7283	0.7230	0.7266	**0.9968**
		10	0.8246	0.6114	0.6969	0.8308	0.8063	0.7397	**0.9975**
		15	0.8784	0.6421	0.7112	0.8867	0.8455	0.7378	**0.9972**
		20	0.9117	0.6748	0.7503	0.9186	0.8656	0.7423	**0.9982**
		25	0.9345	0.7012	0.7642	0.9402	0.8800	0.7437	**0.9992**

Table 6.8 AUC scores for link prediction. The best result in each row is bold

Dataset	Edge	Dimension	Deepwalk	LINE(1st)	LINE(2nd)	node2vec	metapath2vec	PoincaréEmb	HHNE
DBLP	P-A	2	0.5813	0.5090	0.5909	0.6709	0.6536	0.6742	**0.8777**
		5	0.7370	0.5168	0.6351	0.7527	0.7294	0.7381	**0.9041**
		10	0.8250	0.5427	0.6510	0.8469	0.8279	0.7699	**0.9111**
		15	0.8664	0.5631	0.6582	0.8881	0.8606	0.7743	**0.9111**
		20	0.8807	0.5742	0.6644	0.9037	0.8740	0.7806	**0.9106**
		25	0.8878	0.5857	0.6782	0.9102	0.8803	0.7830	**0.9117**
	P-V	2	0.7075	0.5160	0.5121	0.7369	0.7059	0.8257	**0.9331**
		5	0.7197	0.5663	0.5216	0.7286	0.8516	0.8878	**0.9409**
		10	0.7292	0.5873	0.5332	0.7481	0.9248	0.9113	**0.9619**
		15	0.7325	0.5896	0.5425	0.7583	0.9414	0.9142	**0.9625**
		20	0.7522	0.5891	0.5492	0.7674	0.9504	0.9185	**0.9620**
		25	0.7640	0.5846	0.5512	0.7758	0.9536	0.9192	**0.9612**
MoiveLens	M-A	2	0.6278	0.5053	0.5712	0.6349	0.6168	0.5535	**0.7715**
		5	0.6353	0.5636	0.5874	0.6402	0.6212	0.5779	**0.8255**
		10	0.6680	0.5914	0.6361	0.6700	0.6332	0.5984	**0.8312**
		15	0.6791	0.6184	0.6442	0.6814	0.6382	0.5916	**0.8319**
		20	0.6868	0.6202	0.6596	0.6910	0.6453	0.5988	**0.8318**
		25	0.6890	0.6256	0.6700	0.6977	0.6508	0.5995	**0.8309**
	M-D	2	0.6258	0.5139	0.6501	0.6299	0.6191	0.5856	**0.8520**
		5	0.6482	0.5496	0.6607	0.6589	0.6332	0.6290	**0.8967**
		10	0.6976	0.5885	0.7499	0.7034	0.6687	0.6518	**0.8984**
		15	0.7163	0.6647	0.7756	0.7241	0.6702	0.6715	**0.9007**
		20	0.7324	0.6742	0.7982	0.7412	0.6746	0.6821	**0.9000**
		25	0.7446	0.6957	0.8051	0.7523	0.6712	0.6864	**0.9018**

10 dimensionality exceed all the baselines in higher dimensionality results. In MovieLens dataset, HHNE with only 2 dimensionality surpasses baselines in all dimensionality. Besides, both LINE(1st) and PoincaréEmb preserve proximities of node pairs linked by an edge, while LINE(1st) embed network into Euclidean spaces and Poincaré embed network into hyperbolic spaces. PoincaréEmb performs better than LINE(1st) in most cases, especially in dimensionality lower than 10, suggesting the superiority of embedding network into hyperbolic spaces. Because HHNE can preserve high-order network structure and handle different types of nodes in HG, HHNE is more effective than PoincaréEmb.

More detailed introduction of HHNE can be found in [31].

6.5   Conclusion

In this chapter, we have introduced three emerging HG embedding topics, as well as the related HG embedding methods. For adversarial learning, HeGAN is designed for learning semantic-preserving and robust HG embedding based on the adversarial principle. Besides, for importance sampling, HeteSamp studies the problem of accelerating large-scale HG embedding with importance sampling. Moreover, for hyperbolic representation, HHNE makes the efforts to embed HGs in hyperbolic spaces, and the optimization strategies are derived to optimize the hyperbolic embedding. We hope more HG embedding methods with deeper insights can be proposed to discover the abundant semantic of HG in the future.

References

1. Bonnabel, S., et al.: Stochastic gradient descent on Riemannian manifolds. IEEE Trans. Automat. Contr. **58**(9), 2217–2229 (2013)
2. Cai, X., Han, J., Yang, L.: Generative adversarial network based heterogeneous bibliographic network representation for personalized citation recommendation. In: Thirty-Second AAAI Conference on Artificial Intelligence, pp. 5747–5754 (2018)
3. Cannon, J.W., Floyd, W.J., Kenyon, R., Parry, W.R., et al.: Hyperbolic geometry. Flavors Geom. **31**, 59–115 (1997)
4. Cen, Y., Zou, X., Zhang, J., Yang, H., Zhou, J., Tang, J.: Representation learning for attributed multiplex heterogeneous network. In: Proceedings of the 25th ACM SIGKDD International Conference on Knowledge Discovery and Data Mining (KDD), pp. 1358–1368 (2019)
5. Chen, J., Ma, T., Xiao, C.: FastGCN: Fast learning with graph convolutional networks via importance sampling. In: Proceedings of the Conference ICLR (2018). arXiv preprint arXiv:1801.10247
6. Dai, Q., Li, Q., Tang, J., Wang, D.: Adversarial network embedding. In: Proceedings of the AAAI Conference on Artificial Intelligence (AAAI), pp. 2167–2174 (2018)
7. Dhingra, B., Shallue, C., Norouzi, M., Dai, A., Dahl, G.: Embedding text in hyperbolic spaces. In: Proceedings of the Twelfth Workshop on Graph-Based Methods for Natural Language Processing (TextGraphs), pp. 59–69 (2018)

8. Dong, Y., Chawla, N.V., Swami, A.: metapath2vec: Scalable representation learning for heterogeneous networks. In: Proceedings of the 23rd ACM SIGKDD International Conference on Knowledge Discovery and Data Mining (KDD), pp. 135–144 (2017)

9. Fawcett, T.: An introduction to roc analysis. Pattern Recogn. Lett. **27**(8), 861–874 (2006)

10. Fu, T.y., Lee, W.C., Lei, Z.: Hin2vec: Explore meta-paths in heterogeneous information networks for representation learning. In: Proceedings of the 2017 ACM on Conference on Information and Knowledge Management (CIKM), pp. 1797–1806 (2017)

11. Ganea, O., Becigneul, G., Hofmann, T.: Hyperbolic entailment cones for learning hierarchical embeddings. In: International Conference on Machine Learning (ICML), pp. 1646–1655 (2018)

12. Goodfellow, I., Pouget-Abadie, J., Mirza, M., Xu, B., Warde-Farley, D., Ozair, S., Courville, A., Bengio, Y.: Generative adversarial nets. In: Advances in Neural Information Processing Systems (NeurIPS), pp. 2672–2680 (2014)

13. Grover, A., Leskovec, J.: node2vec: Scalable feature learning for networks. In: Proceedings of the 22nd ACM SIGKDD International Conference on Knowledge Discovery and Data Mining (KDD), pp. 855–864 (2016)

14. Hamilton, W.L., Ying, Z., Leskovec, J.: Inductive representation learning on large graphs. In: Proceedings of the 31st International Conference on Neural Information Processing Systems, pp. 1025–1035 (2017)

15. He, X., Liao, L., Zhang, H., Nie, L., Hu, X., Chua, T.S.: Neural collaborative filtering. In: Proceedings of the 26th International Conference on World Wide Web (WWW), pp. 173–182 (2017)

16. Hu, B., Fang, Y., Shi, C.: Adversarial learning on heterogeneous information networks. In: Teredesai, A., Kumar, V., Li, Y., Rosales, R., Terzi, E., Karypis, G. (eds.) Proceedings of the 25th ACM SIGKDD International Conference on Knowledge Discovery and Data Mining (KDD), pp. 120–129 (2019)

17. Huang, Z., Mamoulis, N.: Heterogeneous information network embedding for meta path based proximity. arXiv preprint arXiv:1701.05291 (2017)

18. Huang, W., Zhang, T., Rong, Y., Huang, J.: Adaptive sampling towards fast graph representation learning. In: Advances in Neural Information Processing Systems (NeurIPS), pp. 4563–4572 (2018)

19. Ji, Y., Yin, M., Yang, H., Zhou, J., Zheng, V.W., Shi, C., Fang, Y.: Accelerating large-scale heterogeneous interaction graph embedding learning via importance sampling. ACM Trans. Knowl. Discov. Data **15**(1), 1–23 (2020)

20. Krioukov, D., Papadopoulos, F., Kitsak, M., Vahdat, A., Boguná, M.: Hyperbolic geometry of complex networks. Phys. Rev. E **82**(3), 036106 (2010)

21. Mikolov, T., Sutskever, I., Chen, K., Corrado, G.S., Dean, J.: Distributed representations of words and phrases and their compositionality. In: Advances in Neural Information Processing Systems (NIPS), pp. 3111–3119 (2013)

22. Nickel, M., Kiela, D.: Poincaré embeddings for learning hierarchical representations. Adv. Neural Inf. Proces. Syst. **30**, 6338–6347 (2017)

23. Pan, S., Hu, R., Long, G., Jiang, J., Yao, L., Zhang, C.: Adversarially regularized graph autoencoder for graph embedding. In: Proceedings of the Twenty-Seventh International Joint Conference on Artificial Intelligence (IJCAI-18), pp. 2609–2615 (2018)

24. Perozzi, B., Al-Rfou, R., Skiena, S.: Deepwalk: Online learning of social representations. In: Proceedings of the 20th ACM SIGKDD International Conference on Knowledge Discovery and Data Mining (KDD), pp. 701–710 (2014)

25. Reed, S., Akata, Z., Yan, X., Logeswaran, L., Schiele, B., Lee, H.: Generative adversarial text to image synthesis. In: International Conference on Machine Learning (ICML), pp. 1060–1069 (2016)

26. Shi, C., Hu, B., Zhao, X., Yu, P.: Heterogeneous information network embedding for recommendation. IEEE Trans. Knowl. Data Eng. **31**(2), 357–370 (2018)

27. Tang, J., Qu, M., Wang, M., Zhang, M., Yan, J., Mei, Q.: Line: Large-scale information network embedding. In: Proceedings of the 24th International Conference on World Wide Web (WWW), pp. 1067–1077 (2015)
28. Wang, H., Wang, J., Wang, J., Zhao, M., Zhang, W., Zhang, F., Xie, X., Guo, M.: GraphGAN: Graph representation learning with generative adversarial nets. In: Proceedings of the AAAI Conference on Artificial Intelligence (AAAI), pp. 2508–2515 (2018)
29. Wang, J., Yu, L., Zhang, W., Gong, Y., Xu, Y., Wang, B., Zhang, P., Zhang, D.: IRGAN: A minimax game for unifying generative and discriminative information retrieval models. In: Proceedings of the 40th International ACM SIGIR Conference on Research and Development in Information Retrieval (SIGIR), pp. 515–524 (2017)
30. Wang, X., Ji, H., Shi, C., Wang, B., Ye, Y., Cui, P., Yu, P.S.: Heterogeneous graph attention network. In: The World Wide Web Conference (WWW), pp. 2022–2032 (2019)
31. Wang, X., Zhang, Y., Shi, C.: Hyperbolic heterogeneous information network embedding. In: Proceedings of the AAAI Conference on Artificial Intelligence (AAAI), pp. 5337–5344 (2019)
32. Xu, L., Wei, X., Cao, J., Yu, P.S.: Embedding of embedding (EOE): joint embedding for coupled heterogeneous networks. In: Proceedings of the Tenth ACM International Conference on Web Search and Data Mining (WSDM), pp. 741–749 (2017)
33. Ying, R., He, R., Chen, K., Eksombatchai, P., Hamilton, W.L., Leskovec, J.: Graph convolutional neural networks for web-scale recommender systems. In: Proceedings of the 24th ACM SIGKDD International Conference on Knowledge Discovery and Data Mining (KDD), pp. 974–983 (2018)
34. Yu, L., Zhang, W., Wang, J., Yu, Y.: SeqGAN: Sequence generative adversarial nets with policy gradient. In: Proceedings of the AAAI Conference on Artificial Intelligence (AAAI), pp. 2852–2858 (2017)
35. Yu, W., Zheng, C., Cheng, W., Aggarwal, C.C., Song, D., Zong, B., Chen, H., Wang, W.: Learning deep network representations with adversarially regularized autoencoders. In: Proceedings of the 24th ACM SIGKDD International Conference on Knowledge Discovery and Data Mining (KDD), pp. 2663–2671 (2018)
36. Zheng, V.W., Sha, M., Li, Y., Yang, H., Fang, Y., Zhang, Z., Tan, K., Chang, K.C.: Heterogeneous embedding propagation for large-scale e-commerce user alignment. In: Proceedings of the 2018 IEEE International Conference on Data Mining (ICDM), pp. 1434–1439 (2018)
37. Zou, D., Hu, Z., Wang, Y., Jiang, S., Sun, Y., Gu, Q.: Layer-dependent importance sampling for training deep and large graph convolutional networks. In: Advances in Neural Information Processing Systems (NeurIPS), pp. 11247–11256 (2019)

Chapter 7
Heterogeneous Graph Representation for Recommendation

Abstract With the rapid development of web services, various kinds of useful auxiliary data (a.k.a., side information) become available in recommender systems. To characterize these complex and heterogeneous auxiliary data, heterogeneous graph (HG) representation methods have been widely adopted due to the flexibility in modeling data heterogeneity. In this chapter, we introduce three HG representation based recommendation systems solving the unique challenges existing in diverse real-world scenarios, including Top-N recommendation (MCRec), cold-start recommendation (MetaHIN), and bibliographic recommendation (ASI). In the field of HG representation for recommendation, methods mainly contain three key components: HG constructions, HG representation learning and recommendation based on the HG representation.

7.1 Introduction

In recent years, recommender systems, which help users discover items of interest from a large resource collection, have been playing an increasingly important role in various online services [15], such as item recommendation and collaborator recommendation. Traditional recommendation methods (e.g., matrix factorization) mainly aim to learn an effective prediction function for recovering and completing interaction matrix. With the rapid development of web services, various kinds of auxiliary data (e.g., side information) become available in recommender systems. Although auxiliary data is likely to contain useful information for recommendation, it is difficult to model and utilize these heterogeneous and complex information in recommender systems.

As a promising direction, heterogeneous graph has been proposed as a powerful information modeling method [26, 27, 30]. Due to its flexibility in modeling data heterogeneity, heterogeneous graph (HG) has been adopted in recommender systems to characterize rich auxiliary data. Under the HG based representation, the recommendation problem can be considered as a similarity search task over the HG [30]. Such a recommendation setting is called as HG based recommendation. HG

© The Author(s), under exclusive license to Springer Nature Singapore Pte Ltd. 2022   175
C. Shi et al., *Heterogeneous Graph Representation Learning and Applications*,
Artificial Intelligence: Foundations, Theory, and Algorithms,
https://doi.org/10.1007/978-981-16-6166-2_7

based recommendation has been widely adopted in recommender systems due to its
excellence in modeling complex context information [8, 25, 39]. Although the existing HG based recommendation methods have achieved performance improvement
to some extent, they still meet unique challenges existing in diverse applications.

In this chapter, we introduce three HG representation based recommendation
systems solving the challenges in diverse real-world scenarios, including Top-N
recommendation, cold-start recommendation and bibliographic recommendation.
First, to leverage rich meta-path based context for top-N recommendation, Meta-
path based Context for Recommendation (named MCRec) is designed as a novel
deep neural network with the co-attention mechanism, explicitly learning the
representation of meta-path and their complex relationships. Second, to better
address the cold-start problem in recommendation, a Meta-learning approach
to cold-start recommendation on Heterogeneous Information Networks (named
MetaHIN) is proposed to capture richer semantics, by exploiting the power of
meta-learning at the model level and HINs at the data level simultaneously. Third,
to capture relationships between authors in bibliographic recommendation, Author
Set Identification model (named ASI) first studies the problem of author set
identification, which is to identify an author set related to an anonymous paper.

7.2 Top-N Recommendation

7.2.1 Overview

The existing heterogeneous graph based recommendation methods can be cate-
gorized into two types. The first type leverages path based semantic relatedness
as direct features for recommendation relevance [8, 25, 39], and the second type
performs some transformation on path based similarities for learning effective
transformed features [39, 42]. These two types of methods both extract meta-path
based features for improving the characterization of two-way user–item interactions,
as illustrated in Fig. 7.1, while these existing methods have two major shortcomings.
First, these models seldom learn an explicit representation for path or meta-path in
the recommendation task. Second, they do not consider the mutual effect between
the meta-path and the involved user–item pair in an interaction.

A basic idea for the problems is to leverage rich meta-path information from
heterogeneous graph for top-N recommendation in a more principled way. Our main
idea is to: (1) Learn explicit representations for meta-path based context tailored
for the recommendation task. (2) Characterize a three-way interaction of the form:
⟨ user, meta-path, item⟩. However, the solution is challenging. We have to consider
three key problems: (1) How to design the base architecture that is suitable for the
complicated heterogeneous graph based interaction scenarios. (2) How to generate
meaningful path instances for constructing high-quality meta-path based context.

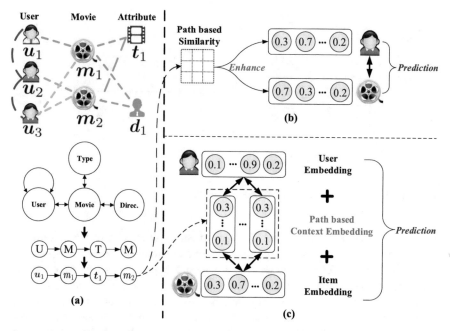

Fig. 7.1 The illustration for heterogeneous graph based recommendation setting (network schema, meta-path, path instance) and the comparison between our model and previous methods (two-way interaction vs. three-way meta-path based interaction). (**a**) Heterogeneous Information Network and Network Schema. (**b**) Previous Models. (**c**) Our Model

(3) How to capture the mutual effect between the involved user–item pair and meta-path based context in an interaction.

In this section, we introduce a novel deep neural network with the co-attention mechanism by leveraging rich meta-path based context, which is able to learn interaction-specific representations for users, items, and meta-path context. We present the proposed model that leverages **M**eta-path based **C**ontext for **Rec**ommendation, called **MCRec**. To our knowledge, it is the first time that meta-path based context has been explicitly modeled in a three-way neural interaction model for top-*N* recommendation in heterogeneous graph. More details about MCRec are given in the next section.

7.2.2   The MCRec Model

7.2.2.1   Model Framework

Differing existing heterogeneous graph based recommendation models, which only learn the representations for users and items, we explicitly incorporate meta-paths as the context in an interaction between a user and an item. Instead of modeling the

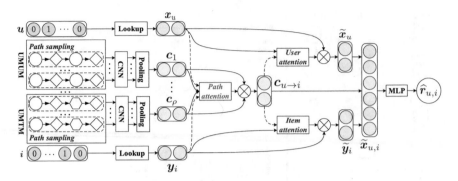

Fig. 7.2 The overall architecture of the proposed model

two-way interaction $\langle user, item \rangle$, we aim to characterize a three-way interaction $\langle user,$ meta-paths, $item \rangle$. We present the overall architecture for the proposed model in Fig. 7.2. As we can see, for learning a better interaction function that generates the recommendations, we learn the representations (i.e., embedding) for users, items, and their interaction contexts. Besides the components for learning user and item embeddings, the most important part lies in the embedding of meta-path based context. We first use a priority based sampling technique to select high-quality path instances. Hence, the meta-path based context is first modeled into a low-dimensional embedding using a hierarchical neural network. With the initially learned embeddings for users, items, and meta-path based context, the co-attention mechanism further improves the three representations through alternative enhancement. Due to the incorporation of meta-path based context, our model is expected to yield a better performance and also improve the interpretability for recommendation results.

7.2.2.2 Characterizing Meta-path Based Context for Interaction

Sampling Path Instances via Priority Based Random Walk Existing heterogeneous graph embedding models mainly adopt a meta-path guided random walk strategy to generate path instances [7], relying on a uniform sampling over the outgoing nodes. Intuitively, at each step, the walker should wander to a neighbor of a higher "priority" score with a larger probability, since such an outgoing node can reflect more reliable semantics by forming a closer link. Hence, we propose to use a similar pretrain technique to measure the priority degree of each candidate outgoing node. First, we train the feature-based matrix factorization framework SVDFeature [5] on all the available historical interaction records to learn a potential vector for each node that has a history of user–item interactions. We can incorporate the entities from heterogeneous graph related to an interaction as the context of a training instance. With the learned latent factors, we can compute the pairwise similarities between two consecutive nodes along a path instance and then average these similarities for ranking the candidate path instances. Finally, given a meta-path, we only keep top K path instances with the highest average similarities.

Meta-path Based Context Embedding After obtaining path instances from multiple meta-paths, we focus on how to model these meta-path based contexts as an informative embedding. Our method naturally follows a hierarchical structure: embedding a single path instance $\rightarrow$ embedding a single meta-path $\rightarrow$ embedding the aggregated meta-paths.

For path instance embedding, formally, given a path p from some meta-path ρ, let $\mathbf{X}^p \in \mathbb{R}^{L \times d}$ denote the embedding matrix formed by concatenating node embeddings, where L is the length of the path instance and d is the embedding dimension for entities. We adopt the commonly used convolution neural network (CNN) to deal with sequences of variable lengths. The structure of CNN consists of a convolution layer and a max pooling layer. We learn the embedding of a path instance p using CNN as follows:

$$\mathbf{h}_p = CNN(\mathbf{X}^p; \boldsymbol{\theta}), \tag{7.1}$$

where $\mathbf{X}^p$ denotes the matrix of the path instance p and $\boldsymbol{\theta}$ denotes all the related parameters in CNNs.

For meta-path embedding, since a meta-path can produce multiple path instances, we further apply the max pooling operation to derive the embedding for a meta-path. Let $\{\mathbf{h}_p\}_{p=1}^K$ denote the embeddings for the K selected path instances from meta-path ρ. The embedding $\mathbf{c}_\rho$ for meta-path ρ can be given by

$$\mathbf{c}_\rho = \text{max-pooling}\left(\{\mathbf{h}_p\}_{p=1}^K\right). \tag{7.2}$$

Our max pooling operation is carried out over K path instance embeddings, which aims to capture the important dimension features from multiple path instances.

For simple average embedding for meta-path based context, we apply the average pooling operation to derive the embedding for modeling the aggregate meta-path based context

$$\mathbf{c}_{u \rightarrow i} = \frac{1}{|\mathcal{M}_{u \rightarrow i}|} \sum_{\rho \in \mathcal{M}_{u \rightarrow i}} \mathbf{c}_\rho, \tag{7.3}$$

where $\mathbf{c}_{u \rightarrow i}$ is the embedding for meta-path based context and $\mathcal{M}_{u \rightarrow i}$ is the set of the considered meta-paths for the current interaction. In this naive embedding method, each meta-path indeed receives equal attention, and the representation of meta-path based context fully depends on the generated path instances. It fails to take the involved user and item into consideration, which lacks the ability of capturing varying semantics from meta-paths in different interaction scenarios.

7.2.2.3 Improving Embeddings for Interaction Via Co-Attention Mechanism

Inspired by the recent progress of attention mechanism made in computer vision and natural language processing [21, 38], we propose a novel co-attention mechanism to improve the embeddings of users, items, and meta-paths.

Attention for Meta-path Based Context Since distinct meta-paths may have different semantics in an interaction, we learn the interaction-specific attention weights over meta-paths conditioned on the involved user and item. Given the user embedding $\mathbf{x}_u$, item embedding $\mathbf{y}_i$, the context embedding $\mathbf{c}_\rho$ for a meta-path ρ, we adopt a two-layer architecture to implement the attention

$$\boldsymbol{\alpha}_{u,i,\rho}^{(1)} = f\left(\mathbf{W}_u^{(1)}\mathbf{x}_u + \mathbf{W}_i^{(1)}\mathbf{y}_i + \mathbf{W}_\rho^{(1)}\mathbf{c}_\rho + \mathbf{b}^{(1)}\right), \tag{7.4}$$

$$\boldsymbol{\alpha}_{u,i,\rho}^{(2)} = f\left(\mathbf{w}^{(2)\top}\boldsymbol{\alpha}_{u,i,\rho}^{(1)} + b^{(2)}\right), \tag{7.5}$$

where $\mathbf{W}_*^{(1)}$ and $\mathbf{b}^{(1)}$ denote the weight matrix and the bias vector for the first layer, and the $\mathbf{w}^{(2)}$ and $b^{(2)}$ denote the weight vector and the bias for the second layer. $f(\cdot)$ is set to the ReLU function.

The final meta-path weights are obtained by normalizing the above attentive scores over all the meta-paths using the softmax function,

$$\alpha_{u,i,\rho} = \frac{\exp\left(\alpha_{u,i,\rho}^{(2)}\right)}{\sum_{\rho'\in\mathcal{M}_{u\to i}}\exp\left(\alpha_{u,i,\rho'}^{(2)}\right)}, \tag{7.6}$$

which can be interpreted as the contribution of the meta-path ρ to the interaction between u and i. After we obtain the meta-path attention scores $\alpha_{u,i,\rho}$, the new embedding for aggregate meta-path context can be given as the following weighted sum:

$$\mathbf{c}_{u\to i} = \sum_{\rho\in\mathcal{M}_{u\to i}} \alpha_{u,i,\rho}\cdot\mathbf{c}_\rho, \tag{7.7}$$

where $\mathbf{c}_\rho$ is the learned embedding for the meta-path ρ in Eq. (7.2). Since the attention weights $\{\alpha_{u,i,\rho}\}$ are generated for each interaction, they are interaction-specific and able to capture varying interaction context.

Attention for Users and Items Given a user and an item, the meta-path connecting them provides important interaction context, which is likely to affect the original representations of users and items. Giving original user and item latent embeddings $\mathbf{x}_u$ and $\mathbf{y}_i$, and the meta-path based context embedding $\mathbf{c}_{u\to i}$ for the interaction between u and i, we use a single-layer network to compute the attention vectors $\boldsymbol{\beta}_u$

and $\boldsymbol{\beta}_i$ for user u and item i as

$$\boldsymbol{\beta}_u = f(\mathbf{W}_u \mathbf{x}_u + \mathbf{W}_{u \to i} \mathbf{c}_{u \to i} + \mathbf{b}_u), \tag{7.8}$$

$$\boldsymbol{\beta}_i = f(\mathbf{W}_i' \mathbf{y}_i + \mathbf{W}_{u \to i}' \mathbf{c}_{u \to i} + \mathbf{b}_i'), \tag{7.9}$$

where $\mathbf{W}_*$ and $\mathbf{b}_u$ denote the weight matrix and bias vector for user attention layer, and $\mathbf{W}_*'$ and $\mathbf{b}_i'$ denote the weight matrix and bias vector for item attention layer. Similarly, $f(\cdot)$ is set to the ReLU function. Then, the final representations of user and item are computed by using an element-wise product "$\odot$" with the attention vectors:

$$\tilde{\mathbf{x}}_u = \boldsymbol{\beta}_u \odot \mathbf{x}_u, \tag{7.10}$$

$$\tilde{\mathbf{y}}_i = \boldsymbol{\beta}_i \odot \mathbf{y}_i. \tag{7.11}$$

The attention vectors $\boldsymbol{\beta}_u$ and $\boldsymbol{\beta}_i$ are used for improving the original user and item embeddings conditioned on the calibrated meta-path based context $\mathbf{c}_{u \to i}$ (Eq. (7.7)).

By combining the two parts of attention components, our model improves the original representations for users, items, and meta-path based context in a mutual enhancement way. We call such an attention mechanism co-attention. To our knowledge, few heterogeneous graph based recommendation methods are able to learn explicit representations for meta-paths, especially in an interaction-specific way.

7.2.2.4 Overall Architecture

Until now, given an interaction between user u and item i, we have the embeddings for user u, item i, and the meta-path connecting them. We combine the three embedding vectors into a unified representation of the current interaction as below:

$$\widetilde{\mathbf{x}}_{u,i} = \tilde{\mathbf{x}}_u \oplus \mathbf{c}_{u \to i} \oplus \tilde{\mathbf{y}}_i, \tag{7.12}$$

where "$\oplus$" denotes the vector concatenation operation, $\mathbf{c}_{u \to i}$ (Eq. (7.7)) denotes the embedding of the meta-path based context for $\langle u, i \rangle$, $\tilde{\mathbf{x}}_u$ (Eq. (7.10)), and $\tilde{\mathbf{y}}_i$ (Eq. (7.11)) denotes the improved embeddings of user u and item i, respectively. $\widetilde{\mathbf{x}}_{u,i}$ encodes the information of an interaction from three aspects: the involved user, the involved item, and the corresponding meta-path based context. Following [11], we feed $\widetilde{\mathbf{x}}_{u,i}$ into a MLP component in order to implement a non-linear function for modeling complicated interactions:

$$\hat{r}_{u,i} = \text{MLP}(\widetilde{\mathbf{x}}_{u,i}). \tag{7.13}$$

MLP component involves two hidden layers with ReLU as the activation function and an output layer with the sigmoid function. With the premise that neural network models can learn more abstractive features of data via using a small number of hidden units for higher layers [10], we empirically implement a tower structure for the MLP component, halving the layer size for each successive higher layer.

Defining a proper objective function for model optimization is a key step for learning a good recommendation model. Traditional point-wise recommendation models for the rating prediction task usually adopt the squared error loss [16]. However, in our task, we only have implicit feedback available. Following [11, 32], we learn the parameters of our model with negative sampling and the objective for an interaction $\langle u, i \rangle$ can be formulated as follows:

$$\ell_{u,i} = -\log \hat{r}_{u,i} - E_{j \sim P_{neg}}[\log(1 - \hat{r}_{u,j})], \tag{7.14}$$

where the first term models the observed interaction, and the second term models the negative feedback drawn from the noise distribution P_{neg}. In MERec, we set the distribution P_{neg} as uniform distribution, which is flexible to extend to other biased distributions, i.e., popularity based distribution.

7.2.3 Experiments

7.2.3.1 Experimental Settings

Datasets In experiments, we employ three real datasets from different domains, namely MovieLens[1] movie dataset, LastFM[2] music dataset, and Yelp[3] business dataset. The detailed descriptions of the three datasets are shown in Table 7.1. The selected meta-paths for each dataset are reported in Table 7.2.

Baselines In this section, we consider two kinds of representative recommendation methods: CF-based methods (ItemKNN [24], BPR [22], MF [16], and NeuMF [11]) only utilizing implicit feedback, and HIN-based methods utilizing rich heterogeneous information (SVDFeature$_{hete}$ [5], SVDFeature$_{mp}$, HeteRS [20], and FMG$_{rank}$ [42]). To examine the effectiveness of our priority based sampling strategy and co-attention mechanism, we prepare three variants of MCRec (MCRec$_{rand}$, MCRec$_{avg}$, and MCRec$_{mp}$). MCRec$_{rand}$ employs the random meta-path guided sampling strategy for path generation. MCRec$_{avg}$ employs the naive context embedding strategy for meta-paths. MCRec$_{mp}$ reserves the attention components for meta-paths and removes the attention component for users and items.

---

[1] https://grouplens.org/datasets/movielens/.

[2] https://www.last.fm.

[3] http://www.yelp.com/dataset-challenge.

Table 7.1 Statistics of the three datasets. The first row of each dataset corresponds to the number of users, items, and interactions

Datasets	Relations (A–B)	#A	#B	#A–B
MovieLens	User–Movie	943	1682	100, 000
	User–User	943	943	47, 150
	Movie–Movie	1682	1682	82, 798
	Movie–Genre	1682	18	2861
LastFM	User–Artist	1892	17, 632	92, 834
	User–User	1892	1892	18, 802
	Artist–Artist	17, 632	17, 632	153, 399
	Artist–Tag	17, 632	11, 945	184, 941
Yelp	User–Business	16, 239	14, 284	198, 397
	User–User	16, 239	16, 239	158, 590
	Business–City (Ci)	14, 267	47	14, 267
	Business–Category (Ca)	14, 180	511	40, 009

Table 7.2 The selected meta-paths used in each dataset

Dataset	Meta-paths
MovieLens	UMUM, UMGM, UUUM, UMMM
LastFM	UATA, UAUA, UUUA, UUA
Yelp	UBUB, UBCaB, UUB, UBCiB

Parameter Settings For our method MCRec, we set the batch size to 256, the learning rate to 0.001, the regularization parameter to 0.0001, the CNN filter size to 3, the dimension of user and item embeddings to 128, the dimension of predictive factors to 32, and the number of sampled path instances to 5. For MF and NeuMF, we follow the optimal configuration in [11]. Moreover, we use 10% training data as the validation set to optimize the parameters for the other methods.

Evaluation Metrics The top-*N* recommendation task usually adopts similar evaluation metrics. Following [11, 39], we use Precision at rank K (Prec@K), Recall at rank K (Recall@K), and Normalized Discounted Cumulative Gain at rank K (NDCG@K) as the evaluation metrics. The final results are first averaged over all the test items of a user and then averaged over all the users. For stability, we perform ten runs using different random-splitting training/test sets and report the average results.

7.2.3.2 Comparisons and Analysis

To evaluate the performance, we randomly split the entire user implicit feedback records of each dataset into training and test sets, i.e., we use 80% feedback records

Table 7.3 Results of effectiveness experiments on three datasets. We use "*" to mark the best performance from the baselines for each comparison. We use "#" to indicate the improvement of MCRec over the best performance from the baselines is significant based on paired t-test at the significance level of 0.01. Here we simplify Prec@10 (%) to P@10, Recall@10 (%) to R@10, and NDCGG@10 (%) to N@10. The best results of all methods are indicated in bold

Model	MovieLens			LastFM			Yelp		
	P@10	R@10	N@10	P@10	R@10	N@10	P@10	R@10	N@10
ItemKNN	25.8	15.4	56.9	41.6	45.1	79.8	13.9	54.2	53.8
BRP	30.1	19.5	64.6	41.3	44.9	81.0	14.7	55.0	55.5
MF	32.5	20.5	65.1	43.6	46.3	79.2	15.0	53.5	53.2
NeuMF	32.9*	20.9	65.9	45.4	46.8	81.0	15.0	58.6	57.1
SVDFeature$_{hete}$	31.7	20.2	64.5	45.8	48.4	82.9*	14.0	56.1	52.9
SVDFeature$_{mp}$	31.1	19.3	65.4	43.9	46.5	81.2	15.2	59.3	59.7*
HeteRS	24.9	16.7	59.7	42.8	44.9	80.3	14.2	56.1	56.0
FMG$_{rank}$	32.6	21.7*	66.8*	46.3*	49.2*	82.6	15.4*	59.5*	58.6
MCRec$_{rand}$	32.2	21.0	66.5	45.4	48.0	80.0	15.1	58.4	57.2
MCRec$_{avg}$	32.7	21.1	66.3	46.5	49.1	83.1	16.0	59.3	60.2
MCRec$_{mp}$	34.0	22.0	68.3	46.6	49.2	84.3	16.6	63.0	62.3
MCRec	**34.5**#	**22.6**#	**69.0**#	**48.1**#	**50.7**#	**85.3**#	**16.9**#	**63.3**#	**63.0**#

to predict the remaining 20% feedback records.[4] We randomly sample 50 negative samples that have no interaction records with the target user. Then, we rank the list consisting of the positive item and 50 negative items.

The comparison results of our proposed model and baselines on three datasets are reported in Table 7.3. There are some observations and analysis. (1) Our complete model MCRec is consistently better than all the baselines on the three datasets. The results indicate the effectiveness of MCRec on the task of top-N recommendation, which has adopted a more principled way to leverage heterogeneous context information for improving recommendation performance. (2) Considering the three variants of MCRec, we can find that the overall performance order is as follows: MCRec > MCRec$_{mp}$ > MCRec$_{avg}$ > MCRec$_{rand}$. The results show that the co-attention mechanism is able to better utilize the meta-path based context for recommendation. First, the importance of each meta-path should depend on a specific interaction instead of being treated equal (i.e., MCRec$_{avg}$). Second, meta-paths provide important context for the interaction between users and items, which has a potential influence on the learned representations of users and items. Ignoring such influence may not be able to achieve the optimal performance for utilizing meta-path based context information (i.e., MCRec$_{mp}$). In addition, although MCRec$_{rand}$ achieves competitive performance compared to baselines, it is worse than the complete MCRec. Our complete model adopts the priority based

---

[4] We hold out 10% training data as the validation set for parameter tuning.

sampling strategy to generate path instances, while MCRec$_{rand}$ adopts a random sampling strategy.

The more detailed method description and experiment validation can be seen in [13].

7.3   Cold-Start Recommendation

7.3.1   Overview

In recommender systems, the interaction data of new users or new items are often of high sparsity, leading to the so-called cold-start issue [44] in which it becomes challenging to learn effective user or item representations. To alleviate this problem, at the data level, heterogeneous information network (HIN) [26] has been leveraged to enrich user–item interactions with complementary heterogeneous information. As shown in Fig. 7.3a, a toy HIN can be constructed for movie recommendation, which captures how the movies are related with each other via actors and directors, in addition to the existing user–movie interactions. On the HIN, higher-order graph structures like meta-paths [30], a relation sequence connecting two objects, can effectively capture semantic contexts. For instance, the meta-path User–Movie–Actor–Movie or UMAM encodes the semantic context of "movies starring the same actor as a movie rated by the user". Together with the content-based methods, HIN-based methods [13, 40] also assume a data-level strategy to alleviate the cold-start problem, as illustrated in Fig. 7.3b.

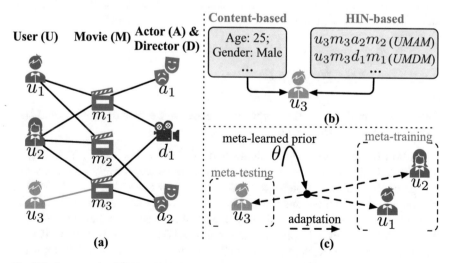

Fig. 7.3   An example of HIN and existing data or model-level alleviation for cold-start recommendation. (**a**) An example of HIN. (**b**) Data-level alleviation. (**c**) Model-level alleviation

On another line, at the model level, the recent episodic meta-learning paradigm [9] has offered insights into modeling new users or items with scarce interaction data [34]. Meta-learning focuses on deriving general knowledge (i.e., a prior) across different learning tasks, so as to rapidly adapt to a new learning task with the prior and a small amount of training data. To some extent, cold-start recommendation can be formulated as a meta-learning problem, where each task is to learn the preferences of one user. From the tasks of existing users, the meta-learner learns a prior with strong generalization capacity during meta-training, such that it can be easily and quickly adapted to the new tasks of cold-start users with scarce interaction data during meta-testing. As illustrated in Fig. 7.3c, the cold-start user u_3 (with only one movie rating) can be adapted from the prior θ in meta-testing, where the prior is derived by learning how to adapt to existing users u_1 and u_2 in meta-training.

In this section, we propose to address the cold-start recommendation at both data and model levels, in which learning the preference of each user is regarded as a task in meta-learning, and a HIN is exploited to augment data. One is to augment the task for each user with multifaceted semantic contexts. That is, in a task of a specific user, besides considering the items directly interacted with the user, we also introduce items that are semantically related to the user via higher-order graph structures, i.e., meta-paths. These related items form the semantic contexts of each task, which can be further differentiated into multiple facets as implied by different meta-paths. The other is to propose a co-adaptation meta-learner, which is equipped with both semantic-wise adaptation and task-wise adaptation. Specifically, the semantic-wise adaptation learns a unique semantic prior for each facet. While the semantic priors are derived from different semantic spaces, they are regulated by a global prior to capture the general knowledge of encoding contexts on a HIN. Furthermore, the task-wise adaptation is designed for each task (i.e., user), which updates the preference of each user from the various semantic priors, such that tasks sharing the same facet of semantic contexts can hinge on a common semantic prior.

7.3.2   The MetaHIN Model

Before we introduce the framework of MetaHIN, we first define the cold-start problem on HINs as follows [26].

Definition 1 (Cold-Start Recommendation) Given a HIN $G = \{V, E, O, R\}$, let $V_U, V_I \subset V$ denote the set of user and item objects, respectively. Given a set of ratings between users and items, i.e., $\mathcal{R} = \{r_{u,i} \geq 0 : u \in V_U, i \in V_I, \langle u, i \rangle \in E\}$, we aim to predict the unknown rating $r_{u,i} \notin \mathcal{R}$ between user u and item i. In particular, if u is a new user with only a handful of existing ratings, i.e., $|\{r_{u',i} \in \mathcal{R} : u' = u\}|$ is small, it is known as user cold-start (UC); correspondingly, if i is a new item, it is known as item cold-start (IC); if both u and i are new, it is known as user–item cold-start (UIC).

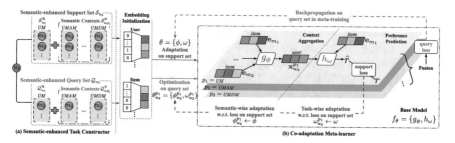

Fig. 7.4 Illustration of the meta-training procedure of a task in MetaHIN. (**a**) Semantic-enhanced task constructor, where the support and query sets are augmented with meta-path based heterogeneous semantic contexts. (**b**) Co-adaptation meta-learner, with semantic- and task-wise adaptations on the support set, while the global prior θ is optimized on the query set. During meta-testing, each task follows the same procedure except updating the global prior

7.3.2.1   Model Framework

As illustrated in Fig. 7.4, the proposed MetaHIN consists of two components: semantic-enhanced task constructor in Fig. 7.4a and co-adaptation meta-learner in Fig. 7.4b. First, we design a semantic-enhanced task constructor to augment the support and query sets of user tasks with heterogeneous semantic contexts, which comprise of items related to the user through meta-paths on a HIN. The semantic contexts are multifaceted in nature, such that each meta-path represents a different facet of heterogeneous semantics. Second, compared to task-wise adaptation, we perform semantic-wise adaptation, in order to adapt the global prior θ to finer-grained semantic priors for different facets (i.e., meta-paths) in a task. The global prior θ captures the general knowledge of encoding contexts for recommendation and can be materialized in the form of a base model f_θ. Thus, our co-adaptation meta-learner performs both semantic- and task-wise adaptions on the support set and further optimizes the global prior on the query set.

7.3.2.2   Semantic-Enhanced Task Constructor

Given a user u with task $\mathcal{T}_u = (\mathcal{S}_u, \mathcal{Q}_u)$, the semantic-enhanced support set is defined as

$$\mathcal{S}_u = \left(\mathcal{S}_u^{\mathcal{R}}, \mathcal{S}_u^{\mathcal{P}} \right), \tag{7.15}$$

where $\mathcal{S}_u^{\mathcal{R}}$ is a set of items that has been rated by user u, and $\mathcal{S}_u^{\mathcal{P}}$ represents the semantic contexts based on a set of meta-paths $\mathcal{P}$.

For new users in cold-start scenarios, the set of rated items $\mathcal{S}_u^{\mathcal{R}}$ is usually small, i.e., a new user only has a few ratings. For meta-training tasks, we follow previous work [17] to construct $\mathcal{S}_u^{\mathcal{R}}$ by sampling a small subset of items rated by u, i.e., $\{i \in V_I : r_{u,i} \in \mathcal{R}\}$, in order to simulate new users.

On the other hand, the semantic contexts $\mathcal{S}_u^{\mathcal{P}}$ are employed to encode multi-faceted semantics into the task. Specifically, assume a set of meta-paths $\mathcal{P}$, such that each path $p \in \mathcal{P}$ starts with User–Item and ends with Item with a length up to l. For example, in Fig. 7.3a, $\mathcal{P} = \{UM, UMAM, UMDM, UMUM\}$ if we set $l = 3$. For each user–item interaction $\langle u, i \rangle$, we define the semantic context of $\langle u, i \rangle$ induced by meta-path p as follows:

$$C_{u,i}^p = \{j : j \in \text{items reachable along } p \text{ starting from } u - -i\}. \tag{7.16}$$

For instance, the semantic context of $\langle u_2, m_2 \rangle$ induced by UMAM is $\{m_2, m_3, \ldots\}$. Since in each task u may interact with multiple items, we build the p-induced semantic context for the task $\mathcal{T}_u$ as

$$\mathcal{S}_u^p = \bigcup_{i \in \mathcal{S}_u^{\mathcal{R}}} C_{u,i}^p. \tag{7.17}$$

Finally, accounting for all meta-paths in $\mathcal{P} = \{p_1, p_2, \ldots, p_n\}$, the semantic contexts $\mathcal{S}_u^{\mathcal{P}}$ of task $\mathcal{T}_u$ are formulated as

$$\mathcal{S}_u^{\mathcal{P}} = \left(\mathcal{S}_u^{p_1}, \mathcal{S}_u^{p_2}, \ldots, \mathcal{S}_u^{p_n}\right). \tag{7.18}$$

In essence, $\mathcal{S}_u^{\mathcal{P}}$ is the set of items that are reachable from user u via all items he/she has rated along the meta-paths, which incorporates multifaceted semantic contexts such that each meta-path represents one facet. As shown in Fig. 7.4a, following the meta-path UMAM, the reachable items of user u_2 are $\{m_2, m_3, \ldots\}$, which are the movies starring the same actor of movies that u_2 has rated in the past. That is, the semantic context induced by UMAM incorporates movies starring the same actor as a facet of user preferences, which makes sense since the user might be a fan of an actor and prefers most movies played by the actor.

Likewise, we can construct the semantic-enhanced query set $\mathcal{Q}_u = (\mathcal{Q}_u^{\mathcal{R}}, \mathcal{Q}_u^{\mathcal{P}})$. In particular, $\mathcal{Q}_u^{\mathcal{R}}$ contains items rated by u for calculating the task loss in meta-training, or items with hidden rating for making predictions in meta-testing; $\mathcal{Q}_u^{\mathcal{P}}$ captures the semantic contexts induced by meta-paths $\mathcal{P}$. Note that in a task $\mathcal{T}_u$, the items with ratings in the support and query sets are mutually exclusive, i.e., $\mathcal{S}_u^{\mathcal{R}} \cap \mathcal{Q}_u^{\mathcal{R}} = \emptyset$.

7.3.2.3   Co-Adaptation Meta-Learner

The co-adaptation meta-learner can learn fine-grained prior knowledge through semantic- and task-wise adaptations with semantic-enhanced tasks. The global prior can be abstracted as a base model to encode the general knowledge of how to learn with contexts on HINs, which can be further adapted to different semantic facets within a task.

Base Model As shown in Fig. 7.4b, the base model f_θ involves context aggregation g_ϕ to derive user embeddings, and preference prediction h_ω to estimate the rating score, i.e., $f_\theta = (h_\omega, g_\phi)$.

In context aggregation, the user embeddings are aggregated from his/her contexts, which are his/her related items via direct interactions or meta-paths (i.e., semantic contexts), since user preferences are reflected in items. Following [17], we initialize the user and item embeddings based on their features (or an embedding look up if there are no features), say $\mathbf{e}_u \in \mathbb{R}^{d_U}$ for user u and $\mathbf{e}_i \in \mathbb{R}^{d_I}$ for item i where d_U, d_I are the embedding dimensions. Subsequently, we obtain user u's embedding $\mathbf{x}_u$ as follows:

$$\mathbf{x}_u = g_\phi(u, C_u) = \sigma\left(\text{MEAN}(\{\mathbf{W}\mathbf{e}_j + \mathbf{b} : j \in C_u\})\right), \qquad (7.19)$$

where C_u denotes the set of items related to user u via direct interactions (i.e., the rated items) or meta-paths (i.e., their induced semantic contexts), $\text{MEAN}(\cdot)$ is mean-pooling, and σ is the activation function (we use LeaklyReLU). Here g_ϕ is the context aggregation function parameterized by $\phi = \{\mathbf{W} \in \mathbb{R}^{d \times d_I}, \mathbf{b} \in \mathbb{R}^d\}$, which are trainable to distill semantic information for user preferences. $\mathbf{x}_u$ can be further concatenated with u's initial embedding $\mathbf{e}_u$, when user features are available.

In preference prediction, given user u's embedding $\mathbf{x}_u$ and item i's embedding $\mathbf{e}_i$, we estimate the rating of user u on the item i as

$$\hat{r}_{ui} = h_\omega(\mathbf{x}_u, \mathbf{e}_i) = \text{MLP}(\mathbf{x}_u \oplus \mathbf{e}_i), \qquad (7.20)$$

where MLP is a two-layer multi-layer perceptron, and $\oplus$ denotes concatenation. Here h_ω is the rating prediction function parameterized by ω, which contains the weights and biases in MLP. Finally, we minimize the following loss for user u to learn his/her preferences:

$$\mathcal{L}_u = \frac{1}{|\mathcal{R}_u|} \sum_{i \in \mathcal{R}_u} (r_{ui} - \hat{r}_{ui})^2, \qquad (7.21)$$

where $\mathcal{R}_u = \{i : r_{ui} \in \mathcal{R}\}$ denotes the set of items rated by u, and r_{ui} is the actual rating of u on item i.

Note that the base model $f_\theta = (g_\phi, h_\omega)$ is a supervised model for recommendation, which typically requires a large number of example ratings to achieve reasonable performance, which is not upheld in the cold-start scenario. As motivated, we recast the cold-start recommendation as a meta-learning problem. Specifically, we abstract the base model $f_\theta = \{g_\phi, h_\omega\}$ as encoding the prior knowledge $\theta = \{\phi, \omega\}$ of how to learn user preferences from contexts on HINs. Next, we detail the proposed co-adaptation meta-learner to learn the prior knowledge.

Co-adaptation The goal of the co-adaptation meta-learner is to learn the prior knowledge $\theta = (\phi, \omega)$, which can quickly adapt to a new user task with just a few example ratings. As discussed in Fig. 7.4a, each task is augmented with multifaceted semantic contexts. Thus, the prior should not only encode the global knowledge shared across tasks, but also become capable of generalizing to different semantic facets within each task. To this end, we enhance the meta-learner with semantic- and task-wise adaptations.

For semantic-wise adaptation, the semantic-enhanced support set $\mathcal{S}_u$ of the task $\mathcal{T}_u$ is associated with semantic contexts induced by different meta-paths (e.g., UMAM and UMDM in Fig. 7.4), where each meta-path represents one semantic facet. The semantic-wise adaptation evaluates the loss based on the semantic context induced by a meta-path p (i.e., $\mathcal{S}_u^p$). With one (or a few) gradient descent step w.r.t. the p-specific loss, the global context prior ϕ, which encodes how to learn with contexts on a HIN, is adapted to the semantic space induced by the meta-path p.

Formally, given a task $\mathcal{T}_u$ of user u, the support set $\mathcal{S}_u = (\mathcal{S}_u^{\mathcal{R}}, \mathcal{S}_u^{\mathcal{P}})$ is augmented with semantic contexts $\mathcal{S}_u^{\mathcal{P}}$, comprising various facets $\mathcal{S}_u^{p_i}$ induced by different meta-paths p_i as in Eq. (7.18). Given a meta-path $p \in \mathcal{P}$, user u's embedding in the semantic space of p is

$$\mathbf{x}_u^p = g_\phi\left(u, \mathcal{S}_u^p\right). \tag{7.22}$$

In this semantic space of p, we can further calculate the loss on the support set of rated items $\mathcal{S}_u^{\mathcal{R}}$ in task $\mathcal{T}_u$ as

$$\mathcal{L}_{\mathcal{T}_u}\left(\omega, \mathbf{x}_u^p, \mathcal{S}_u^{\mathcal{R}}\right) = \frac{1}{|\mathcal{S}_u^{\mathcal{R}}|} \sum_{i \in \mathcal{S}_u^{\mathcal{R}}} \left(r_{ui} - h_\omega\left(\mathbf{x}_u^p, \mathbf{e}_i\right)\right)^2, \tag{7.23}$$

where $h_\omega(\mathbf{x}_u^p, \mathbf{e}_i)$ represents the predicted rating of user u on item i in the meta-path p-induced semantic space.

Next, we adapt the global context prior ϕ w.r.t. the loss in each semantic space of p in task $\mathcal{T}_u$ with one gradient descent step, to obtain the semantic prior ϕ_u^p. Thus, the meta-learner learns more fine-grained prior knowledge for various semantic facets:

$$\phi_u^p = \phi - \alpha \frac{\partial \mathcal{L}_{\mathcal{T}_u}\left(\omega, \mathbf{x}_u^p, \mathcal{S}_u^{\mathcal{R}}\right)}{\partial \phi} = \phi - \alpha \frac{\partial \mathcal{L}_{\mathcal{T}_u}\left(\omega, \mathbf{x}_u^p, \mathcal{S}_u^{\mathcal{R}}\right)}{\partial \mathbf{x}_u^p} \frac{\partial \mathbf{x}_u^p}{\partial \phi}, \tag{7.24}$$

where α is the semantic-wise learning rate, and $\mathbf{x}_u^p = g_\phi(u, \mathcal{S}_u^p)$ is a function of ϕ.

For task-wise adaptation, in the semantic space of meta-path p with adapted semantic prior ϕ_u^p, the task-wise adaptation further adapts the global prior ω, which encodes how to learn rating predictions of u, to the task $\mathcal{T}_u$ with one (or a few) gradient descent step.

The semantic prior ϕ_u^p subsequently updates user u' embeddings in the semantic space of p on the support set to $\mathbf{x}_u^{p\langle S \rangle} = g_{\phi_u^p}(u, \mathcal{S}_u^p)$, which further transforms the global prior ω to the same space:

$$\omega^p = \omega \odot \kappa \left(\mathbf{x}_u^{p\langle S \rangle} \right), \qquad (7.25)$$

where $\odot$ is the element-wise product and $\kappa(\cdot)$ serves as a transformation function realized with a fully connected layer. Intuitively, ω is gated into the current p-induced semantic space. We then adapt ω^p to the task $\mathcal{T}_u$ with one gradient descent step:

$$\omega_u^p = \omega^p - \beta \frac{\partial \mathcal{L}_{\mathcal{T}_u} \left(\omega^p, \mathbf{x}_u^{p\langle S \rangle}, \mathcal{S}_u^{\mathcal{R}} \right)}{\partial \omega^p}, \qquad (7.26)$$

where β is the task-wise learning rate.

With the semantic- and task-wise adaptations, we have adapted the global prior θ to the semantic- and task-specific parameters $\theta_u^p = \{\phi_u^p, \omega_u^p\}$ in the p-induced semantic space of task $\mathcal{T}_u$. Given a set of meta-paths $\mathcal{P}$, the meta-learner is trained by optimizing the performance of the adapted parameters θ_u^p on the query set $\mathcal{Q}_u$ in all semantic spaces of $\mathcal{P}$ across all meta-training tasks. That is, as shown in Fig. 7.4b, the global prior $\theta = (\phi, \omega)$ will be optimized through backpropagation of the query loss:

$$\min_{\theta} \sum_{\mathcal{T}_u \in \mathcal{T}^{\text{tr}}} \mathcal{L}_{\mathcal{T}_u} \left(\omega_u, \mathbf{x}_u, \mathcal{Q}_u^{\mathcal{R}} \right), \qquad (7.27)$$

where ω_u and $\mathbf{x}_u$ are fused from multiple semantic spaces (i.e., meta-paths in $\mathcal{P}$). Specifically,

$$\omega_u = \sum_{p \in \mathcal{P}} a_p \omega_u^p, \quad \mathbf{x}_u = \sum_{p \in \mathcal{P}} a_p \mathbf{x}_u^{p\langle Q \rangle}, \qquad (7.28)$$

where $a_p = \text{softmax}(-\mathcal{L}_{\mathcal{T}_u}(\omega_u^p, \mathbf{x}_u^{p\langle Q \rangle}, \mathcal{Q}_u^{\mathcal{R}}))$ is the weight of the p-induced semantic space, and $\mathbf{x}_u^{p\langle Q \rangle} = g_{\phi_u^p}(u, \mathcal{Q}_u^p)$ is u's embedding aggregated on the query set. Since the loss value reflects the model performance [3], it is intuitive that the larger the loss value in a semantic space, the smaller the corresponding weight should be.

In summary, the co-adaption meta-learner aims to optimize the global prior θ across several tasks, in such a way that the query loss of each meta-training task $\mathcal{T}_u$ using the adapted parameters $\{\theta_u^p : p \in \mathcal{P}\}$ can be minimized (i.e., "learning to learn"); it does not directly update the global prior using task data. In particular, with the co-adaption mechanism, we adapt the parameters not only to each task, but also to each semantic facet within a task.

7.3.3   Experiments

7.3.3.1   Experimental Settings

Datasets   We conduct experiments on three benchmark datasets, namely, DBook[5], MovieLens[6], and Yelp[7], from publicly accessible repositories.

Baselines   We compare our proposed MetaHIN with three categories of methods. (1) Traditional methods, including FM [23], NeuMF [11], and GC-MC [1]. As they cannot handle HINs, we take the heterogeneous information (e.g., actor) as the features of users or items. (2) HIN-based methods, including mp2vec [7] and HERec [28]. Both methods are based on meta-paths, and we utilize the same set of meta-paths as in our method. (3) Cold-start methods, including content-based DropoutNet [35], as well as meta-learning based MeteEmb [19] and MeLU [17]. Since they do not handle HINs either, we input the heterogeneous information as user or item features following the original papers. We follow [17] to train the non-meta-learning baselines with the union of rated items in all support and query sets from meta-training tasks. To handle new users or items, we fine-tune the trained models with support sets and evaluate on query sets in meta-testing tasks.

Evaluation Metrics   We adopt three widely used evaluation protocols [17, 28, 36], namely, mean absolute error (MAE), root mean square error (RMSE), and normalized discounted cumulative gain at rank K (nDCG@K). Here we use $K = 5$.

7.3.3.2   Comparisons and Analysis

In this experiment, we empirically compare MetaHIN to several state-of-the-art baselines, in three cold-start scenarios and the traditional non-cold-start scenario. Table 7.4 demonstrates the performance comparison between all methods w.r.t. four recommendation scenarios.

Cold-Start Scenarios   The first three parts of Table 7.4 present three cold-start scenarios (UC, IC, and UIC). Overall, our MetaHIN consistently yields the best performance among all methods on three datasets. For instance, MetaHIN improves over the best baseline w.r.t. MAE by 3.05–5.26%, 2.89–5.55%, and 2.22–5.19% on three datasets, respectively. Among different baselines, traditional methods (e.g., MF, NeuMF, and GC-MC) are least competitive despite incorporating heterogeneous information as content features. Such treatment of heterogeneous information is not ideal as higher-order graph structures are lost. HIN-based methods perform better due to the incorporation of such structures (i.e., meta-paths). Nevertheless,

---

[5] https://book.douban.com.

[6] https://grouplens.org/datasets/movielens/.

[7] https://www.yelp.com/dataset/challenge.

Table 7.4 Experimental results in four recommendation scenarios and on three datasets. A smaller MAE or RMSE value and a larger nDCG@5 value indicate a better performance. The best method is bolded, and second best is underlined

Scenario	Model	DBook			MovieLens			Yelp		
		MAE ↓	RMSE ↓	nDCG@5 ↑	MAE ↓	RMSE ↓	nDCG@5 ↑	MAE ↓	RMSE ↓	nDCG@5 ↑
Existing items for new users (User Cold-start or UC)	FM	0.7027	0.9158	0.8032	1.0421	1.3236	0.7303	0.9581	1.2177	0.8075
	NeuMF	0.6541	0.8058	0.8225	0.8569	1.0508	0.7708	0.9413	1.1546	0.7689
	GC-MC	0.9061	0.9767	0.7821	1.1513	1.3742	0.7213	0.9321	1.1104	0.8034
	mp2vec	0.6669	0.8391	0.8144	0.8793	1.0968	0.8233	0.8972	1.1613	0.8235
	HERec	0.6518	0.8192	0.8233	0.8691	0.9916	0.8389	0.8894	1.0998	0.8265
	DropoutNet	0.8311	0.9016	0.8114	0.9291	1.1721	0.7705	0.8557	1.0369	0.7959
	MeteEmb	0.6782	0.8553	0.8527	0.8261	1.0308	0.7795	0.8988	1.0496	0.7875
	MeLU	0.6353	0.7733	0.8793	0.8104	0.9756	0.8415	0.8341	1.0017	0.8275
	MetaHIN	**0.6019**	**0.7261**	**0.8893**	**0.7869**	**0.9593**	**0.8492**	**0.7915**	**0.9445**	**0.8385**
New items for existing users (Item Cold-start or IC)	FM	0.7186	0.9211	0.8342	1.3488	1.8503	0.7218	0.8293	1.1032	0.8122
	NeuMF	0.7063	0.8188	0.7396	0.9822	1.2042	0.6063	0.9273	1.1009	0.7722
	GC-MC	0.9081	0.9702	0.7634	1.0433	1.2753	0.7062	0.8998	1.1043	0.8023
	mp2vec	0.7371	0.9294	0.8231	1.0615	1.3004	0.6367	0.7979	1.0304	0.8337
	HERec	0.7481	0.9412	0.7827	0.9959	1.1782	0.7312	0.8107	1.0476	0.8291
	DropoutNet	0.7122	0.8021	0.8229	0.9604	1.1755	0.7547	0.8116	1.0301	0.7943
	MeteEmb	0.6741	0.7993	0.8537	0.9084	1.0874	0.8133	0.8055	0.9407	0.8092
	MeLU	0.6518	0.7738	0.8882	0.9196	1.0941	0.8041	0.7567	0.9169	0.8451
	MetaHIN	**0.6252**	**0.7469**	**0.8902**	**0.8675**	1.0462	**0.8341**	**0.7174**	**0.8696**	**0.8551**

(continued)

Table 7.4 (continued)

Scenario	Model	DBook			MovieLens			Yelp		
		MAE ↓	RMSE ↓	nDCG@5 ↑	MAE ↓	RMSE ↓	nDCG@5 ↑	MAE ↓	RMSE↓	nDCG@5 ↑
New items for new users (User–Item Cold-start or UIC)	FM	0.8326	0.9587	0.8201	1.3001	1.7351	0.7015	0.8363	1.1176	0.8278
	NeuMF	0.6949	0.8217	0.8566	0.9686	1.2832	0.8063	0.9860	1.1402	0.7836
	GC-MC	0.7813	0.8908	0.8003	1.0295	1.2635	0.7302	0.8894	1.1109	0.7923
	mp2vec	0.7987	1.0135	0.8527	1.0548	1.2895	0.6687	0.8381	1.0993	0.8137
	HERec	0.7859	0.9813	0.8545	0.9974	1.1012	0.7389	0.8274	0.9887	0.8034
	DropoutNet	0.8316	0.8489	0.8012	0.9635	1.1791	0.7617	0.8225	0.9736	0.8059
	MeteEmb	0.7733	0.9901	0.8541	0.9122	1.1088	0.8087	0.8285	0.9476	0.8188
	MeLU	0.6517	0.7752	0.8891	0.9091	1.0792	0.8106	0.7358	0.8921	0.8452
	MetaHIN	**0.6318**	**0.7589**	**0.8934**	**0.8586**	**1.0286**	**0.8374**	**0.7195**	**0.8695**	**0.8521**
Existing items for existing users (Non-cold-start)	FM	0.7358	0.9763	0.8086	1.0043	1.1628	0.6493	0.8642	1.0655	0.7986
	NeuMF	0.6904	0.8373	0.7924	0.9249	1.1388	0.7335	0.7611	0.9731	0.8069
	GC-MC	0.8056	0.9249	0.8032	0.9863	1.2238	0.7147	0.8518	1.0327	0.8023
	mp2vec	0.6897	0.8471	0.8342	0.8788	1.1006	0.7091	0.7924	1.0191	0.8005
	HERec	0.6794	0.8409	0.8411	0.8652	1.0007	0.7182	0.7911	0.9897	0.8101
	DropoutNet	0.7108	0.7991	0.8268	0.9595	1.1731	0.7231	0.8219	1.0333	0.7394
	MeteEmb	0.7095	0.8218	0.7967	0.8086	1.0149	0.8077	0.7677	0.9789	0.7740
	MeLU	0.6519	0.7834	0.8697	0.8084	0.9978	0.8433	0.7382	0.9028	0.8356
	MetaHIN	**0.6393**	**0.7704**	**0.8859**	**0.7997**	**0.9491**	**0.8499**	**0.6952**	**0.8445**	**0.8477**

supervised learning methods generally cannot perform effectively given limited training data for new users and items.

On the other hand, meta-learning methods typically cope better in such cases. In particular, the best baseline is consistently MeLU or MeteEmb. However, they still underperform our MetaHIN in all scenarios. The reason might be that both of them only integrate heterogeneous information as content features, without capturing multifaceted semantics derived from higher-order structures like meta-paths. In contrast, in MetaHIN, we perform semantic- and task-wise co-adaptions, to effectively adapt to not only tasks, but also different semantic facets within a task.

Non-cold-start Scenario In the last part of Table 7.4, we investigate the traditional recommendation scenario. Our MetaHIN is still robust, outperforming all the baselines. While this is a traditional scenario, the datasets are still very sparse in general. Thus, incorporating the semantic-rich HINs can often alleviate the sparsity challenge at the data level. MetaHIN further addresses the problem at the model level with the co-adaptation meta-learner and thus can better deal with sparse data. Of course, compared to cold-start scenarios, MetaHIN's performance lift over the baselines tends to be smaller as the sparsity issue is not as severe.

The more detailed method description and experiment validation can be seen in [18].

7.4 Author Set Recommendation

7.4.1 Overview

Heterogeneous bibliographic network [29] has also received more and more attention in recent years. As an important related task, the problem of author identification has been extensively studied, which aims to rank potential authors for an anonymous paper based on public information. The existing studies mainly employ the network structure or semantic content of the paper to predict the correlation between the paper and the author, while they usually ignore relationships among authors. Generally, in many scenarios such as finding a potential author group for a given paper, the relationships among authors are very significant. Therefore, in this section, we propose to study a new problem called author set identification. We illustrate the problem setting in Fig. 7.5, in which the heterogeneous bibliographic network and network schema are given as the input. The goal is to learn a model that can identify the optimal author set for a new anonymous paper. The problem of author set identification is to acquire an author set with a strong relationship, while the traditional problem only gets an author ranking for the target node of anonymous paper.

A basic idea for the problem is to find a set of closely connected authors that are related to an anonymous paper. Therefore, we need to characterize the

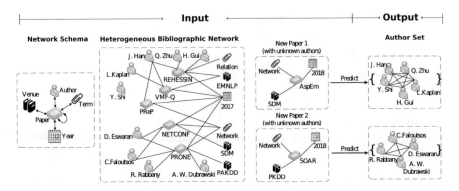

Fig. 7.5 The problem of author set identification in heterogeneous bibliographic networks

relationship between anonymous paper and authors, as well as that between authors. However, it is non-trivial to take both relationships into account simultaneously. Moreover, the number of subsets of authors is enormous especially when the number of authors is large. Hence, it is also very difficult to select the optimal one from all subsets. There are two challenges in this problem. (1) How can we model interactions between anonymous paper and authors, meanwhile preserving rich inherent structural information among authors in heterogeneous bibliographic network. (2) How can we find an optimal set of closely connected authors that are related to the anonymous paper.

In this section, we propose a novel **A**uthor **S**et **I**dentification approach called **ASI**. In order to tackle the first challenge, we propose to only emphasize on two types of nodes including anonymous paper and candidate authors. Therefore, ASI first constructs a paper–author interactive network denoted by weighted paper-ego-network, which only contains the mentioned two types of nodes and corresponding relations (paper–author and author–author). Then in order to preserve rich inherent structural information in heterogeneous bibliographic network, the task-guided embedding method called TaskGE is presented to learn the low-dimensional representations of nodes, which can be further used to determine the weights of edges in the constructed network. For the sake of solving the second challenge, we introduce the concept of quasi-clique in dense subgraph and convert the optimal author set identification into the quasi-clique discovery in the weighted paper-ego-network. Specifically, we design the local-search heuristic method under the guidance of a novel density function to find the optimal quasi-clique (author set). Meanwhile, we regard the anonymous paper as a constraint and claim the discovered set of closely connected authors must be related to the anonymous paper.

7.4.2 The ASI Model

In this section, we study the novel problem of author set identification in bibliographic network, which can be defined as follows.

Definition 2 (Author Set Identification Problem) Given a bibliographic network $G = (V, E)$, which includes a set of papers and papers' relevant information (i.e., authors, venues, terms, and year), the goal is to design a method to acquire an author set S'_A from C_A for a new anonymous paper p, such that S'_A is the optimal set to collaborate on the paper p among all subsets of C_A, where $C_A = \{a_1, a_2, \cdots, a_m\}$ denotes the set of all candidate authors.

In order to find the optimal author set, we present the proposed method that leverages quasi-clique for Author Set Identification, called **ASI**. In order to find the optimal author set, we introduce the concept of quasi-clique, which can be defined as follows.

Definition 3 (Quasi-Clique [33]) A set of nodes S is an α-quasi-clique if $e[S] \geq \alpha\binom{|S|}{2}$, i.e., if the edge density of the subgraph induced by S exceeds a threshold parameter $\alpha \in (0, 1)$. The edge density is defined as $e[S]/\binom{|S|}{2}$, where $e[S]$ is the size of edges in the subgraph induced by S.

7.4.2.1 Model Framework

The overall architecture of ASI is shown in Fig. 7.6. Given a heterogeneous bibliographic network and an anonymous paper (Fig. 7.6a), we first construct a weighted paper-ego-network for each anonymous paper (Fig. 7.6b) and then find the optimal quasi-clique with constraint (OQCC) in the weighted paper-ego-network (Fig. 7.6c). In the following, we will clarify the basic idea and specific details about these two phases. We aim to find a set of closely connected authors that are related to the anonymous paper. However, it is challenging to incorporate interactions between anonymous paper and authors, meanwhile preserving rich inherent structural infor-

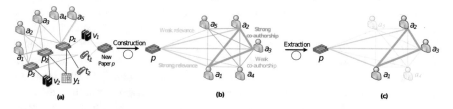

Fig. 7.6 The overall architecture of proposed method ASI (weighted paper-ego-network construction and optimal quasi-clique with constraint extraction). (**a**) Input. (**b**) Construct a weighted paper-ego-network for each new paper p. (**c**) Extract the optimal quasi-clique with constraint in the constructed network

mation among authors in heterogeneous bibliographic network. We consider to construct a weighted paper-ego-network only containing the anonymous paper and authors. Because there are no direct links between anonymous paper and authors, as well as between authors in bibliographic network, we need to devise an approach to determine the weight of these two kinds of edges. Hence, we propose the task-guided embedding method to learn vector representations of nodes, which can be further used to determine the weights of edges through proper distance function for constructing the weighted paper-ego-network. Then we transform the author set identification into the problem of quasi-clique extraction with constraint. The constraint condition means that the discovered optimal quasi-clique should contain the node of anonymous paper, which also implies the close relationship between author set and anonymous paper. Finally, we propose an approach of local-search heuristic under the guidance of designed novel density function, so as to discover the optimal quasi-clique in the constructed network.

7.4.2.2 Weighted Paper-Ego-Network Construction

Because we aim to find a set of closely connected authors that are related to the anonymous paper, so we just need to focus on two kinds of relationships, including that between the anonymous paper and author, as well as that between authors. Therefore, we consider to construct a weighted paper-ego-network, which naturally should only contain nodes of anonymous paper p and candidate authors except for the other types of redundant nodes (V, T, and Y). The key of constructing the network is to determine the weights of edges between anonymous paper and authors, and that between authors. For edges between authors, we propose the task-guided embedding (TaskGE) to learn the low-dimensional representations of nodes. Since the feature representation for the anonymous paper is unknown, we first employ the weighted combination of feature vectors of its observed neighbors in the network to calculate its vector. Then we can easily determine the edges between anonymous paper and authors based on the computed representation of the anonymous paper and the vectors of authors obtained in TaskGE.

Specifically, for each anonymous paper p, we denote the constructed weighted paper-ego-network by $G_p = (V, E, W)$, where V is a set of nodes, E is a set of edges, and W is a set of the weight on each edge. V includes two types of nodes, that is, anonymous paper p and candidate authors. Correspondingly, E contains two types of edges, namely, the edge between paper p and any candidate author a, and the edge between any two candidate authors a_1, a_2. We denote the weights of these two types of edges by w_{pa} and $w_{a_1 a_2}$, respectively. Different from the existing general-purpose embedding, our embedding method is totally dependent of specific task. We exploit two unique characteristics or significant aspects of author set identification task. One is the proximity between anonymous paper and authors, and we model it as paper–author-aware embedding. The other is the strong relationship between authors, and we model it as author–author-aware embedding.

Paper–Author–Aware Embedding Intuitively, for a given paper p, the relevance score of p and any one a of its true authors should be ξ larger than that of p and other author a' who is not the author of p. Otherwise, a loss penalty will incur. Here, we employ the hinge loss [41] to define a general function to model the relationship between paper and author as follows:

$$\mathcal{L}_{R_{P \curvearrowright A}} = \sum_{r \in \mathcal{P}_{P \curvearrowright A}} \mathbb{E}_{<p,a,a'>|r} [\xi + f(p,a) - f(p,a')]_+, \qquad (7.29)$$

where $[x]_+ = max(x,0)$ is the standard hinge loss, and ξ is the safety margin size [2]. $< p, a, a' >$ denotes the triples <paper, positive author, negative author>. r and $\mathcal{P}_{P \curvearrowright A}$ denote any meta-path and the set of meta-paths between paper and author, respectively. Generally, we can add any proper meta-paths between paper and author to $\mathcal{P}_{P \curvearrowright A}$ for leveraging multiple information. Actually, there exist multiple indirect relations besides the direct relation between paper and author. For example, $\mathcal{P}_{P \curvearrowright A} = \{PA, PTPA\}$ denotes we not only consider the direct author but also take the potential authors into account. Correspondingly, $\mathcal{P}_{P \curvearrowright A} = \{PA\}$ means we only consider the direct author of paper. $f(p,a)$ stands for the metric between paper p and author a. As demonstrated by CML [12], distance metric [37] satisfies better triangle inequality and transition property than inner product, we use the Euclidean distance to define the metric:

$$f(p,a) = \|\mathbf{X}_p - \mathbf{X}_a\|_2^2, \qquad (7.30)$$

where $\mathbf{X}_p$ and $\mathbf{X}_a$ are the embedding vectors of p and a, respectively. For a new anonymous paper, we adopt similar approach to Chen et al. [4] to calculate its vector representation. That is, the embedding of a paper is represented as the weighted combination of the vectors of observed different types of neighbors in the network as follows:

$$\mathbf{X}_p = \sum_{t=1}^{n} w_t \mathbf{X}_p^t, \qquad (7.31)$$

where n is the number of neighbors' types of paper p, $\mathbf{X}_p^t$ is the mean of vectors of the t-th node type, $\mathbf{X}_p^t = \sum_{i \in N_p^t} \frac{\mathbf{X}_i}{|N_p^{(t)}|}$, and $N_p^{(t)}$ denotes the set of nodes of the t-th type. In this section, we do not employ the reference type of nodes due to the lack of citation data.

Author–Author-Aware Embedding $\mathcal{L}_{R_{P \curvearrowright A}}$ models the relationship between paper and author; in this subsection, we will consider how to model the relationship between authors. It is reasonable that there should be strong relationships between co-authors. In other words, the relevance score between co-authors should be larger than that of authors who have never collaborated with each other. Correspondingly, there might exist some potential co-authorship between authors implicitly indicated by meta-paths like $APTPA$. Therefore, we also define a general function to formulate the triple relation $< a^*, a^+, a^- >$.

$$\mathcal{L}_{R_{A \curvearrowright A}} = \sum_{r \in \mathcal{P}_{A \curvearrowright A}} \mathbb{E}_{(a^*, a^+, a^-)|r} [\xi + f(a^*, a^+) - f(a^*, a^-)]_+, \qquad (7.32)$$

where a^+ means any co-author of a^*, and a^- denotes any author who has never cooperated with a^*. f is the metric function that has been introduced in Eq. (7.30). r denotes any meta-path between authors. $\mathcal{P}_{A \curvearrowright A}$ is the set of meta-paths between authors. $\mathcal{P}_{A \curvearrowright A} = \{APA\}$ means we only consider the existing co-authors.

Regularization Recently, Cogswell et al. [6] propose a new regularization technique called covariance regularization, which is initially used to reduce the correlation between activations in a deep neural network. Afterward, Hsieh et al. [12] find that it is useful in de-correlating the dimensions. As covariances can be seen as a measure of linear redundancy between dimensions, this loss of covariance regularization essentially tries to prevent each dimension from being redundant. Therefore, we employ loss of covariance regularization as follows:

$$\mathcal{L}_{reg} = \frac{1}{N} (\|C\|_f - \|diag(C)\|_2^2), \qquad (7.33)$$

where $\| \cdot \|_f$ is the Frobenius norm, C is the covariance matrix between all pairs of dimensions i and j, $C_{ij} = \frac{1}{N} \sum_{k=1}^N (\mathbf{X}_k^{(i)} - u_i)(\mathbf{X}_k^{(j)} - u_j)$, $u_i = \frac{1}{N} \sum_{k=1}^N \mathbf{X}_k^{(i)}$, and $\mathbf{X}_k^{(i)}$ denotes the i-th dimension of embedding vector of node k.

Finally, we combine three parts above to get the unified objective function for task-guided embedding as follows:

$$\mathcal{L} = \mathcal{L}_{R_{P \curvearrowright A}} + \gamma \mathcal{L}_{R_{A \curvearrowright A}} + \lambda \mathcal{L}_{reg}, \qquad (7.34)$$

where $\mathcal{L}_{reg}$ is the regularization term for avoiding over-fitting, λ controls penalty of regularization, and γ is a harmonic factor to balance two components. In this section, we only consider the direct relation PA in $\mathcal{L}_{R_{P \curvearrowright A}}$ and APA in $\mathcal{L}_{R_{A \curvearrowright A}}$.

To minimize $\mathcal{L}$, we design a sampling-based mini-batch Adam optimizer [14]. To get the training triples $< p, a, a' >$ and $< a^*, a^+, a^- >$, we draw positive samples according to the proportion of path instances of different meta-paths. This sampling strategy can avoid the problem of under-sampling for relations with a large number of links or over-sampling for those with a small number of links. For each sampled positive example $< p, a >$, we first fix vertex p and the corresponding relation. Then we randomly generate negative vertex a' that has not the same relation with p to construct training triples $< p, a, a' >$. Similarly, we can fix a^* and corresponding relation to acquire training triples $< a^*, a^+, a^- >$. Given the low-dimension representation learned above, we can easily calculate w_{pa} and $w_{a_1 a_2}$ using distance function such as cosine.

7.4.2.3 Optimal Quasi-Clique with Constraint Extraction in Weighted Paper-Ego-Network

For each new paper p, we construct a weighted paper-ego-network $G_p = (V, E, W)$. In order to find the optimal author set for the given paper p in G_p, we propose a new method called OQCCE that is an adaptation of the local-search heuristic by Tsourakakis et al. [33]. The algorithm selects p as initial set. Then under the guidance of designed novel density function, algorithm iterates two phases of adding or removing the designated nodes until the quasi-clique with maximum density function is discovered. What is more, the novel density function considers two kinds of heterogeneous relationships, including the close relationship between the anonymous paper and author, as well as that between authors.

In specific, we regard the node p as constraint, which means that the extracted subgraph must contain node p. In [33], there is only one type of edge. However, there exist two types of edges in weighted paper-ego-network. The simplest method is to assign equal significance to two types of edges. In fact, the importance may vary. Therefore, we introduce a variable β to adjust the importance of two types of edges. Meanwhile, we also adapt the density function to accommodate the weighted network. Accordingly, the proposed novel density function can be defined as follows:

$$
g_{\alpha,\beta}(S) = \beta \sum_{(i,j)\in D_{PA}} w_{ij} + \sum_{(k,l)\in D_{AA}} w_{kl} - \alpha \binom{|S|}{2}, \tag{7.35}
$$

where S represents a subset of vertices of network G_p having $S \subseteq V$, $|S|$ denotes the number of nodes in the subgraph induced by S, and w_{ij} is the weight of edge between nodes i and j in the subgraph induced by S. D_{PA} represents the set of edges between given paper p and candidate authors in the subgraph induced by S. Likewise, D_{AA} represents the set of edges between authors in the subgraph induced by S. β controls the importance of paper–author edge. α is a constant. The first two parts in Eq. (7.35) favor subgraphs with abundant edges, while the third part penalizes large subgraphs.

Based on the proposed density function above, next we will describe how to find the optimal quasi-clique with constraint in G_p. The algorithm firstly selects constrained node p as the initial set. Then it traverses all nodes one by one and adds u to S if $g_{\alpha,\beta}(S \cup \{u\})$ improves. Afterward, the algorithm traverses every vertex v in S and removes v if $g_{\alpha,\beta}(S \setminus \{v\})$ enhances. Note that we cannot remove constrained node p during the period of removal. The algorithm repeats these two phases of addition and removal until an optimum is reached or the number of iterations exceeds I_{max}.

7.4.3   *Experiments*

7.4.3.1   Experimental Settings

Datasets AMiner [31] is a classical academic network. Specifically, we extract two subsets with different scales, denoted by AMiner-I and AMiner-II. AMiner-I is a small subset data of some important venues in data mining area, which includes 5 venues, namely KDD, ICDM, SDM, CIKM, and PKDD. AMiner-II is a large subset of four areas, including Artificial Intelligence (AI), Data Mining (DM), Databases (DB), and Information System (IS). For each area, we choose some important venues[8] that have influential publications.

Baselines In order to examine the effectiveness of our approach, we compare against the following three kinds of representative methods. (1) Similarity measure. We design two kinds of similarity measure methods based on meta-paths $PTPA$ and $PCPA$, which can indirectly connect the new paper and candidate authors with term or venue. Then we rank candidate authors according to the similarity scores (i.e., the number of path instances) between candidate authors and the new paper. (2) Feature method. Following the work of Chen et al. [4], we extracted 17 features for each paper–author pair. We choose LR, SVM, and Bayes as learning algorithms. (3) HetNetE. HetNetE is recently proposed in [4] for author identification problem. It first learns the low-dimensional feature vectors of nodes to predict author of the given paper.

Parameter Settings For our method ASI, we set the embedding dimension d to 128, the size of negative samples to 2, the margin ξ to 2, the learning rate to 0.00001, the batch size to 200, the regularization penalty λ to 10, the trade-off factor γ to 1.0, α to 0.01, and β to 0.1. For HetNetE and feature method, we choose the optimal parameter. Three meta-paths APC, APW, APP are jointly used in HetNetE. In addition, for fairness comparison, we do not adopt the reference types of nodes when computing the embedding vectors of papers due to the lack of most citations in HetNetE and ASI.

Evaluation Metrics We adopt *Precision* (P), *Recall* (R), $F1$ score, *Jaccard* index (J), MAP (mean average precision), and $RMSE$ as evaluation metrics. (1) P. It reflects the accuracy of returned author set, which can be defined as the ratio of the true authors in the returned author set. $P = \frac{|S'_A \cap S_A|}{|S'_A|}$, where S'_A denotes the returned author set or the returned top-k author set in $P@k$. S_A means the true author set. (2) R. It shows the ratio of returned true authors in the whole true author set. It can be computed as follows: $R = \frac{|S'_A \cap S_A|}{|S_A|}$, where S'_A and S_A have the same meanings introduced above. (3) $F1$. It is the harmony average of P and R, which is defined as: $F1 = \frac{2*P*R}{P+R}$. (4) *Jaccard* index. It measures similarity between two sets and is

---

[8] AI: ICML, AAAI, IJCAI, NIPS. DM: KDD, WSDM, ICDM, PKDD. DB: SIGMOD, VLDB, ICDE. IS: SIGIR, CIKM.

formulated as: $J = \frac{|S'_A \cap S_A|}{|S'_A \cup S_A|}$, which means the ratio of the intersection and the union of two sets. (5) MAP. It is computed as mean of AP at different k for a paper. $AP = \frac{\sum_{i=1}^k p@i \times rel_i}{\text{# of correct author}}$, where rel_i equals 1 if the result at rank i is correct author and 0 otherwise. (6) $RMSE$. It is a measure of difference between the number of authors returned by model and the number of true authors. $RMSE = \sqrt{\frac{\sum(|S'_A| - |S_A|)^2}{|m|}}$, where m is the number of test papers, and S'_A and S_A are the number of returned author and true author, respectively.

7.4.3.2 Comparisons and Analysis

To evaluate the performance, we regard papers published before 2014 as training set and papers published in 2014 and 2015 as test set. Since it is time-consuming to rank all candidate authors for each anonymous paper in the evaluation procedure, following the strategy in [4], for each paper in the test set, we randomly sample some negative authors and obtain 100 candidate authors in all. Then, we rank the 100 candidate authors consisting of the positive and sampled negative authors for each paper. For our method ASI, we also select the same 100 candidate authors to construct the weighted paper-ego-network for each test paper. The final results are averaged over all the test papers for each evaluation metric.

We report the results of performance comparison in Tables 7.5 and 7.6. There are some observations and analysis. (1) Our method ASI achieves better performance than all baselines on all measures except R and MAP. It improves the performance by more than 15% on P, J, and $F1$ averagely. Although ASI does not achieve the best performance on R, it is also near the best value. (2) ASI can automatically confirm the appropriate number of authors for a given paper, which can be clearly demonstrated by the lowest value on metric $RMSE$. In a word, ASI not only can discover a set of authors with strong relationship but also can determine the proper number of authors for an anonymous paper. (3) To our surprise, the similarity measure method based on PTPA has very good performance, which indicates that the term has a significant role in finding author set for a given paper.

The more detailed method description and experiment validation can be seen in [43].

Table 7.5 Results of effectiveness experiments on AMiner-I. We use bold to mark the best performance for each comparison. ↑ indicates higher is better, and ↓ indicates lower is better. "Avg." means the average rank of different methods

Methods		Evaluation					
		P (↑)	R (↑)	J (↑)	F1 (↑)	MAP (↑)	RMSE (↓)
Top-5	Similarity measure PTPA	0.2716 (2)	0.5007 (7)	0.2310 (2)	0.3356 (2)	**0.6109** (1)	0.1714 (2)
	PCPA	0.2098 (7)	0.3937 (11)	0.1680 (7)	0.2614 (7)	0.4718 (9)	0.1714 (2)
	Feature method LR	0.2160 (5)	0.3915 (12)	0.1827 (6)	0.2657 (4)	0.4834 (7)	0.1714 (2)
	SVM	0.2493 (3)	0.4562 (9)	0.2154 (4)	0.3081 (3)	0.5451 (3)	0.1714 (2)
	Bayes	0.2209 (4)	0.4075 (10)	0.1888 (5)	0.2733 (5)	0.4951 (6)	0.1714 (2)
	HetNetE	0.2123 (6)	0.3870 (13)	0.1669 (8)	0.2616 (6)	0.4571 (11)	0.1714 (2)
Top-10	Similarity measure PTPA	0.1555 (9)	0.5779 (2)	0.1454 (10)	0.2365 (9)	0.5897 (2)	0.5023 (3)
	PCPA	0.1388 (11)	0.5066 (5)	0.1257 (13)	0.2110 (11)	0.4517 (12)	0.5023 (3)
	Feature method LR	0.1358 (13)	0.5005 (8)	0.1270 (12)	0.2059 (13)	0.4664 (10)	0.5023 (3)
	SVM	0.1629 (8)	**0.5988** (1)	0.1538 (9)	0.2477 (8)	0.5296 (4)	0.5023 (3)
	Bayes	0.1364 (12)	0.5010 (6)	0.1277 (11)	0.2069 (12)	0.4767 (8)	0.5023 (3)
	HetNetE	0.1506 (10)	0.5347 (3)	0.2269 (3)	0.2275 (10)	0.4435 (13)	0.5023 (3)
ASI		**0.4589** (1)	0.5284 (4)	**0.4009** (1)	**0.4712** (1)	0.5295 (5)	**0.1123** (1)

Table 7.6 Results of effectiveness experiments on AMiner-II. We use bold to mark the best performance for each comparison. ↑ indicates higher is better, and ↓ indicates lower is better. "Avg." means the average rank of different methods

Methods			Evaluation					
			P (↑)	R (↑)	J (↑)	F1 (↑)	MAP (↑)	RMSE (↓)
Top5	Similarity measure	PTPA	0.3391 (2)	0.5899 (6)	0.2886 (2)	0.4108 (2)	0.7165 (3)	0.2880 (2)
		PCPA	0.3287 (3)	0.5743 (8)	0.2776 (4)	0.3986 (3)	0.6595 (6)	0.2880 (2)
	Feature method	LR	0.3113 (4)	0.5400 (9)	0.2645 (5)	0.3769 (4)	0.6605 (5)	0.2880 (2)
		SVM	0.2202 (7)	0.4553 (12)	0.1674 (11)	0.2803 (9)	**0.9948** (1)	0.2880 (2)
		Bayes	0.2964 (5)	0.5144 (10)	0.2491 (6)	0.3587 (5)	0.6458 (8)	0.2880 (2)
	HetNetE		0.2645 (6)	0.4561 (11)	0.2078 (7)	0.3191 (6)	0.6021 (12)	0.2880 (2)
Top10	Similarity measure	PTPA	0.1927 (8)	**0.6624** (1)	0.1795 (8)	0.2884 (7)	0.6913 (4)	0.8536 (3)
		PCPA	0.1913 (9)	0.6531 (2)	0.1778 (9)	0.2860 (8)	0.6363 (10)	0.8536 (3)
	Feature method	LR	0.1857 (10)	0.5779 (7)	0.1729 (10)	0.2775 (10)	0.6382 (9)	0.8536 (3)
		SVM	0.1101 (13)	0.4553 (12)	0.0943 (13)	0.1702 (13)	0.9948 (1)	0.8536 (3)
		Bayes	0.1786 (11)	0.6157 (4)	0.1661 (12)	0.2673 (11)	0.6227 (11)	0.8536 (3)
	HetNetE		0.1720 (12)	0.6350 (3)	0.2858 (3)	0.2564 (12)	0.5602 (13)	0.8536 (3)
ASI			**0.5981** (1)	0.6019 (5)	**0.4943** (1)	**0.5720** (1)	0.6566 (7)	**0.2058** (1)

7.5 Conclusions

In recent years, to characterize the complex and heterogeneous auxiliary data in web services, HG representation techniques have become a very popular approach for recommender systems. In this chapter, we present three HG representation based recommendation systems, respectively, solving the unique challenges existing in diverse real-world scenarios. Particularly, we study the Top-N recommendation scenario and propose the MCRec framework, which is a three-way neural interaction model based HG representation method. In addition, we study the cold-start problem in the recommendation and propose a meta-learning based method for HG representation, named MetaHIN. At last, we study the author set identification problem in the bibliographic recommendation and propose ASI method to solve this problem. The experiments demonstrate the effectiveness of HG representation in each application.

In future work, we will consider how to combine more auxiliary multi-modal information to improve performance. In addition, we will extend the HG representation approach to other more challenging applications.

References

1. Berg, R.v.d., Kipf, T.N., Welling, M.: Graph convolutional matrix completion. arXiv preprint arXiv:1706.02263 (2017)
2. Bordes, A., Usunier, N., Garcia-Duran, A., Weston, J., Yakhnenko, O.: Translating embeddings for modeling multi-relational data. In: Advances in Neural Information Processing Systems (NeurIPS), pp. 2787–2795 (2013)
3. Chai, T., Draxler, R.R.: Root mean square error (RMSE) or mean absolute error (MAE)?– arguments against avoiding RMSE in the literature. Geosci. Model Dev. 7(3), 1247–1250 (2014)
4. Chen, T., Sun, Y.: Task-guided and path-augmented heterogeneous network embedding for author identification. In: Proceedings of the Tenth ACM International Conference on Web Search and Data Mining (WSDM), pp. 295–304. ACM, New York (2017)
5. Chen, T., Zhang, W., Lu, Q., Chen, K., Zheng, Z., Yu, Y.: SVDFeature: a toolkit for feature-based collaborative filtering. J. Mach. Learn. Res. 13, 3619–3622 (2012)
6. Cogswell, M., Ahmed, F., Girshick, R., Zitnick, L., Batra, D.: Reducing overfitting in deep networks by decorrelating representations. arXiv preprint arXiv:1511.06068 (2015)
7. Dong, Y., Chawla, N.V., Swami, A.: metapath2vec: Scalable representation learning for heterogeneous networks. In: Proceedings of the 23rd ACM SIGKDD International Conference on Knowledge Discovery and Data Mining (KDD), pp. 135–144 (2017)
8. Feng, W., Wang, J.: Incorporating heterogeneous information for personalized tag recommendation in social tagging systems. In: Proceedings of the 18th ACM SIGKDD International Conference on Knowledge Discovery and Data Mining (KDD), pp. 1276–1284 (2012)
9. Finn, C., Abbeel, P., Levine, S.: Model-agnostic meta-learning for fast adaptation of deep networks. In: International Conference on Machine Learning (ICML), pp. 1126–1135 (2017)
10. He, K., Zhang, X., Ren, S., Sun, J.: Deep residual learning for image recognition. In: Proceedings of the IEEE Conference on Computer Vision and Pattern Recognition (CVPR), pp. 770–778 (2016)

11. He, X., Liao, L., Zhang, H., Nie, L., Hu, X., Chua, T.S.: Neural collaborative filtering. In: Proceedings of the 26th International Conference on World Wide Web (WWW), pp. 173–182 (2017)

12. Hsieh, C.K., Yang, L., Cui, Y., Lin, T.Y., Belongie, S., Estrin, D.: Collaborative metric learning. In: Proceedings of the 26th International Conference on World Wide Web (WWW), pp. 193–201 (2017)

13. Hu, B., Shi, C., Zhao, W.X., Yu, P.S.: Leveraging meta-path based context for Top-N recommendation with a neural co-attention model. In: Proceedings of the 24th ACM SIGKDD International Conference on Knowledge Discovery and Data Mining (KDD), pp. 1531–1540 (2018)

14. Kingma, D.P., Ba, J.: Adam: A method for stochastic optimization. In: Third International Conference on Learning Representations (2015)

15. Koren, Y., Bell, R.: Advances in collaborative filtering. In: Recommender Systems Handbook, pp. 77–118 (2015)

16. Koren, Y., Bell, R., Volinsky, C.: Matrix factorization techniques for recommender systems. Computer **42**(8), 30–37 (2009)

17. Lee, H., Im, J., Jang, S., Cho, H., Chung, S.: MeLU: Meta-learned user preference estimator for cold-start recommendation. In: Proceedings of the 25th ACM SIGKDD International Conference on Knowledge Discovery and Data Mining (KDD), pp. 1073–1082 (2019)

18. Lu, Y., Fang, Y., Shi, C.: Meta-learning on heterogeneous information networks for cold-start recommendation. In: Proceedings of the 26th ACM SIGKDD International Conference on Knowledge Discovery and Data Mining (SIGKDD), pp. 1563–1573 (2020)

19. Pan, F., Li, S., Ao, X., Tang, P., He, Q.: Warm up cold-start advertisements: Improving CTR predictions via learning to learn ID embeddings. In: Proceedings of the 42nd International ACM SIGIR Conference on Research and Development in Information Retrieval (SIGIR), pp. 695–704 (2019)

20. Pham, T.A.N., Li, X., Cong, G., Zhang, Z.: A general recommendation model for heterogeneous networks. IEEE Trans. Knowl. Data Eng. **28**, 3140–3153 (2016)

21. Phan, M.C., Sun, A., Tay, Y., Han, J., Li, C.: NeuPL: Attention-based semantic matching and pair-linking for entity disambiguation. In: Proceedings of the 2017 ACM on Conference on Information and Knowledge Management (CIKM), pp. 1667–1676 (2017)

22. Rendle, S., Freudenthaler, C., Gantner, Z., Schmidt-Thieme, L.: BPR: Bayesian personalized ranking from implicit feedback. In: Proceedings of the Twenty-Fifth Conference on Uncertainty in Artificial Intelligence (UAI) (2009)

23. Rendle, S., Gantner, Z., Freudenthaler, C., Schmidt-Thieme, L.: Fast context-aware recommendations with factorization machines. In: Proceedings of the 34th International ACM SIGIR Conference on Research and Development in Information Retrieval (SIGIR), pp. 635–644 (2011)

24. Sarwar, B., Karypis, G., Konstan, J., Riedl, J.: Item-based collaborative filtering recommendation algorithms. In: Proceedings of the 10th International Conference on World Wide Web (WWW), pp. 285–295 (2001)

25. Shi, C., Zhang, Z., Luo, P., Yu, P.S., Yue, Y., Wu, B.: Semantic path based personalized recommendation on weighted heterogeneous information networks. In: Proceedings of the 24th ACM International on Conference on Information and Knowledge Management (CIKM), pp. 453–462 (2015)

26. Shi, C., Li, Y., Zhang, J., Sun, Y., Philip, S.Y.: A survey of heterogeneous information network analysis. IEEE Trans. Knowl. Data Eng. **29**, 17–37 (2017)

27. Shi, C., Hu, B., Zhao, W.X., Yu, P.S.: Heterogeneous information network embedding for recommendation. IEEE Trans. Knowl. Data Eng. **31**(2), 357–370 (2018)

28. Shi, C., Hu, B., Zhao, X., Yu, P.: Heterogeneous information network embedding for recommendation. IEEE Trans. Knowl. Data Eng. **31**(2), 357–370 (2018)

29. Sun, Y., Han, J.: Mining heterogeneous information networks: principles and methodologies. Synth. Lect. Data Mining Knowl. Discovery **3**(2), 1–159 (2012)

30. Sun, Y., Han, J., Yan, X., Yu, P.S., Wu, T.: PathSim: Meta path-based top-k similarity search in heterogeneous information networks. Very Large Data Base Endowment **4**, 992–1003 (2011)
31. Tang, J., Zhang, J., Yao, L., Li, J., Zhang, L., Su, Z.: ArnetMiner: extraction and mining of academic social networks. In: Proceedings of the 14th ACM SIGKDD International Conference on Knowledge Discovery and Data Mining (SIGKDD), pp. 990–998. ACM, New York (2008)
32. Tang, J., Qu, M., Wang, M., Zhang, M., Yan, J., Mei, Q.: Line: Large-scale information network embedding. In: Proceedings of the 24th International Conference on World Wide Web (WWW), pp. 1067–1077 (2015)
33. Tsourakakis, C., Bonchi, F., Gionis, A., Gullo, F., Tsiarli, M.: Denser than the densest subgraph: extracting optimal quasi-cliques with quality guarantees. In: Proceedings of the 19th ACM SIGKDD International Conference on Knowledge Discovery and Data Mining (SIGKDD), pp. 104–112. ACM, New York (2013)
34. Vartak, M., Thiagarajan, A., Miranda, C., Bratman, J., Larochelle, H.: A meta-learning perspective on cold-start recommendations for items. In: Advances in Neural Information Processing Systems (NeurIPS), pp. 6904–6914 (2017)
35. Volkovs, M., Yu, G., Poutanen, T.: DropoutNet: Addressing cold start in recommender systems. In: Advances in Neural Information Processing Systems (NeurIPS), pp. 4957–4966 (2017)
36. Wang, X., He, X., Wang, M., Feng, F., Chua, T.: Neural graph collaborative filtering. In: Proceedings of the 42nd international ACM SIGIR conference on Research and development in Information Retrieval (SIGIR), pp. 165–174 (2019)
37. Weinberger, K.Q., Saul, L.K.: Distance metric learning for large margin nearest neighbor classification. J. Mach. Learn. Res. **10**(Feb), 207–244 (2009)
38. Xu, K., Ba, J., Kiros, R., Cho, K., Courville, A., Salakhudinov, R., Zemel, R., Bengio, Y.: Show, attend and tell: Neural image caption generation with visual attention. In: International Conference on Machine Learning (ICML), pp. 2048–2057 (2015)
39. Yu, X., Ren, X., Sun, Y., Gu, Q., Sturt, B., Khandelwal, U., Norick, B., Han, J.: Personalized entity recommendation: a heterogeneous information network approach. In: Proceedings of the Tenth ACM International Conference on Web Search and Data Mining (WSDM), pp. 283–292 (2014)
40. Zhang, Y., Ai, Q., Chen, X., Croft, W.B.: Joint representation learning for top-n recommendation with heterogeneous information sources. In: Proceedings of the 2017 ACM on Conference on Information and Knowledge Management (CIKM), pp. 1449–1458 (2017)
41. Zhang, C., Huang, C., Yu, L., Zhang, X., Chawla, N.V.: Camel: Content-aware and meta-path augmented metric learning for author identification. In: Proceedings of the 2018 World Wide Web Conference (WWW), pp. 709–718 (2018)
42. Zhao, H., Yao, Q., Li, J., Song, Y., Lee, D.L.: Meta-graph based recommendation fusion over heterogeneous information networks. In: Proceedings of the 23rd ACM SIGKDD International Conference on Knowledge Discovery and Data Mining (KDD), pp. 635–644 (2017)
43. Zheng, Y., Shi, C., Kong, X., Ye, Y.: Author set identification via quasi-clique discovery. In: Proceedings of the 28th ACM International Conference on Information and Knowledge Management (CIKM), pp. 771–780 (2019)
44. Zhu, Y., Lin, J., He, S., Wang, B., Guan, Z., Liu, H., Cai, D.: Addressing the item cold-start problem by attribute-driven active learning. IEEE Trans. Knowl. Data Eng. **32**(4), 631–644 (2019)

Chapter 8
Heterogeneous Graph Representation for Text Mining

Abstract Heterogeneous graph representation techniques can be applied in many real-world applications. Even the natural languages that are usually modeled as sequential data can also be constructed as a heterogeneous graph by some techniques, so as to widely and accurately capture the complex interactions among the words, entities, topics, instances, and other components of the texts. In this chapter, we focus on summarizing the heterogeneous graph representation applications on text mining. Particularly, we introduce several heterogeneous graph based text mining methods, including HGAT for short text classification, GUND and GNewsRec for news recommendation. In the field of heterogeneous graph representation for text mining, methods mainly contain two key components: heterogeneous graph construction from texts and heterogeneous graph representation algorithm for tasks. We will roughly illustrate heterogeneous graph modeling for text mining tasks from these two points.

8.1 Introduction

With the rapid development of online social media and e-commerce, the text corpus on the Internet has grown tremendously, including short texts such as queries, reviews, tweets, etc. [30], and long texts such as news, articles, papers, etc. Therefore, there is a pressing need to successfully and accurately analyze them. For example, as the most basic task, text classification could categorize these text corpus into several groups, thus facilitating storage and rapid retrieval [1, 22]. News recommendation could keep users from information overloading and help them quickly to find their interests [5, 38, 39].

However, many of the text analysis tasks will face the problem of data sparsity [26, 42]. Fortunately, graphs, especially heterogeneous graphs, have powerful capabilities in integrating extra information and modeling interactions among objects. Hence, researchers explore to construct a suitable heterogeneous graph for these texts, containing different types of objects (e.g., words, entities, topics, instances, and other components of the texts) and one/multiple types of edges

© The Author(s), under exclusive license to Springer Nature Singapore Pte Ltd. 2022
C. Shi et al., *Heterogeneous Graph Representation Learning and Applications*,
Artificial Intelligence: Foundations, Theory, and Algorithms,
https://doi.org/10.1007/978-981-16-6166-2_8

connecting the objects together, which could be beneficial for overcoming data sparsity problem and improving many natural language processing tasks. Moreover, different tasks will encounter some of their own unique challenges, which can also be dealt with by the properly constructed heterogeneous graphs followed by a well-designed heterogeneous graph representation method.

In this chapter, we focus on the methods on the two aforementioned typical tasks: short text classification and news recommendation. Specifically, in the task of short text classification, in addition to the challenge of data sparsity, there are also problems such as ambiguity and a lack of labeled data. To address these issues, we introduce a novel **Heterogeneous Graph ATtention** network (named **HGAT**) [18] for semi-supervised short text classification in Sect. 8.2. In terms of the long texts, in the task of news recommendation, to tackle the problem that existing methods ignore considering the latent topic information and users' long-term and short-term interests, we introduce a novel **Graph neural News Rec**ommendation model (named **GNewsRec**) [11] in Sect. 8.3. Moreover, we introduce another news recommendation method, **Graph neural News** recommendation model with **Unsupervised preference Disentanglement** (named **GNUD**) [12], which further considers users' great diversity of preferences.

8.2 Short Text Classification

8.2.1 Overview

Short text classification can be widely applied in many domains, ranging from sentiment analysis to news tagging/categorization and query intent classification [1, 22]. Nevertheless, short text classification is non-trivial due to the following challenges. First, short texts are usually semantically sparse and ambiguous, lacking contexts [26]. Although some methods have been proposed to incorporate additional information such as entities [36, 37], they are unable to consider the relational data such as the semantic relations among entities. Second, the labeled training data is limited, which leads to traditional and neural supervised methods [15, 28, 29, 35, 50] ineffective. As such, how to make full use of the limited labeled data and a large number of unlabeled data has become a key problem for short text classification [1]. Finally, it needs to capture the importance of different information that is incorporated to address sparsity at multiple granularity levels and reduce the weights of noisy information to achieve more accurate classification results.

In this section, we introduce a novel heterogeneous graph neural network-based method for semi-supervised short text classification [18], which makes full use of both limited labeled data and large unlabeled data by allowing information propagation through the automatically constructed graph. Particularly, it first presents a flexible heterogeneous graph framework for modeling the short texts, which is able to incorporate any additional information (e.g., entities and topics) as well as capture the rich relations among the texts and the additional information. Then, it proposes **H**eterogeneous **G**raph **AT**tention networks (**HGAT**) to embed the HG for short text classification based on a new dual-level attention mechanism including node-level and type-level attentions. The HGAT method considers the heterogeneity of different node types. Additionally, the dual-level attention mechanism captures both the importance of different neighboring nodes (reducing the weights of noisy information) and the importance of different node (information) types to a current node. Finally, extensive experimental results demonstrate that the proposed HGAT model significantly outperforms seven state-of-the-art methods across six benchmark datasets.

8.2.2   HG Modeling for Short Texts

We first present the HG framework for modeling the short texts, which enables integration of any additional information and captures the rich relations among the texts and the added information. In this way, the sparsity of the short texts is alleviated.

Previous studies have exploited latent topics [49] and external knowledge (e.g., entities) from knowledge bases to enrich the semantics of the short texts [36, 37]. However, they fail to consider the semantic relation information, such as entity relations. This HG framework for short texts is flexible for integrating any additional information and modeling their rich relations.

Here, two types of additional information are considered, i.e., topics and entities. As shown in Fig. 8.1, the HG $\mathcal{G} = (\mathcal{V}, \mathcal{E})$ is constructed containing the short texts $D = \{d_1, \ldots, d_m\}$, topics $T = \{t_1, \ldots, t_K\}$, and entities $E = \{e_1, \ldots, e_n\}$ as nodes, i.e., $\mathcal{V} = D \cup T \cup E$. The set of edges $\mathcal{E}$ represent their relations. The details of constructing the network are described as follows.

First, the latent topics T are mined to enrich the semantics of short texts using LDA [3]. Each topic $t_i = (\theta_1, \ldots, \theta_w)$ (w denotes the vocabulary size) is represented by a probability distribution over the words. Each document is assigned to the top P topics with the largest probabilities. Thus, the edge between a document and a topic is built if the document is assigned to the topic.

Second, recognize the entities E in the documents D and map them to Wikipedia with the entity linking tool TAGME.[1] The edge between a document and an entity is built if the document contains the entity. An entity is taken as a whole word, and learn the entity embeddings using word2vec[2] based on the Wikipedia corpus. To further enrich the semantics of short texts and advance the information propagation, HGAT considers the relations between entities. Particularly, if the similarity score

---

[1] https://sobigdata.d4science.org/group/tagme/.

[2] https://code.google.com/archive/p/word2vec/.

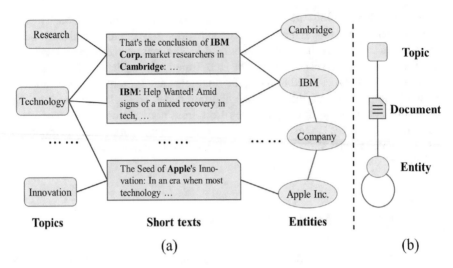

Fig. 8.1 An example of HG for short texts on AGNews. (Note that HIN, heterogeneous information network, in the above figure can be seen as an alias of heterogeneous graph. Here we use the original figure of [18])

(cosine similarity) between two entities, computed based on their embeddings, is above a pre-defined threshold δ, an edge is built between them.

By incorporating the topics, entities, and the relations, the semantics of the short texts are enriched and thus greatly benefit the following classification task. For example, as shown in Fig. 8.1, the short text "the seed of Apple's Innovation: In an era when most technology..." is semantically enriched by the relations with the entities "Apple Inc." and "company," as well as the topic "technology." Thus, it can be correctly classified into the category of "business" with high confidence.

8.2.3   The HGAT Model

We then introduce the HGAT model (shown in Fig. 8.2) to embed the HG for short text classification based on a new dual-level attention mechanism including node level and type level. HGAT considers the heterogeneity of different types of information with heterogeneous graph convolution. In addition, the dual-level attention mechanism captures the importance of different neighboring nodes (reducing the weights of noisy information) and the importance of different node (information) types to a specific node. Finally, it predicts the labels of documents through a softmax layer.

We first describe the heterogeneous graph convolution in HGAT, considering the heterogeneous types of nodes (information).

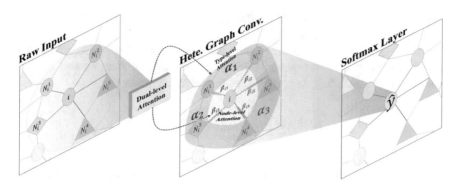

Fig. 8.2 Illustration of the model HGAT

As known, GCN [16] is a multi-layer neural network that operates directly on a homogeneous graph and induces the embedding vectors of nodes based on the properties of their neighborhoods. Formally, consider a graph $\mathcal{G} = (\mathcal{V}, \mathcal{E})$ where $\mathcal{V}$ and $\mathcal{E}$ represent the set of nodes and edges, respectively. Let $X \in \mathbb{R}^{|\mathcal{V}| \times q}$ be a matrix containing the nodes with their features $x_v \in \mathbb{R}^q$ (each row x_v is a feature vector for a node v). For the graph $\mathcal{G}$, we introduce its adjacency matrix $A' = A + I$ with added self-connections and degree matrix M, where $M_{ii} = \sum_j A'_{ij}$. Then the layer-wise propagation rule is as follows:

$$H^{(l+1)} = \sigma(\tilde{A} \cdot H^{(l)} \cdot W^{(l)}). \tag{8.1}$$

Here, $\tilde{A} = M^{-\frac{1}{2}} A' M^{-\frac{1}{2}}$ represents the symmetric normalized adjacency matrix. $W^{(l)}$ is a layer-specific trainable transformation matrix. $\sigma(\cdot)$ denotes an activation function such as ReLU. $H^{(l)} \in \mathbb{R}^{|\mathcal{V}| \times q}$ denotes the hidden representations of nodes in the lth layer. Initially, $H^{(0)} = X$.

Unfortunately, GCN cannot be directly applied to the HG for short texts due to the node heterogeneity issue. Specifically, in the HG, there are three types of nodes: documents, topics, and entities with different feature spaces. For a document $d \in D$, the TF-IDF vector is used as its feature vector x_d. For the topic $t \in T$, the word distribution is used to represent the topic $x_t = \{\theta_i\}_{i=[1,w]}$. For each entity, to make full use of relevant information, the entity x_v is represented by concatenating its embedding and TF-IDF vector of its Wikipedia description text.

A straightforward way to adapt GCN for the HG containing different types of nodes $\mathcal{T} = \{\tau_1, \tau_2, \tau_3\}$ is to construct a new large feature space by concatenating together the feature spaces of different types of nodes. For example, each node is denoted as a feature vector with 0 values for the irrelevant dimensions for other types. We name this basic method for adapting GCN to HG as *GCN-HIN*.[3]

---

[3] Here we follow the original naming in [18].

However, the performance of GCN-HIN is limited since it ignores the heterogeneity of different information types.

To address the issue, the heterogeneous graph convolution is proposed to consider the difference of various types of information and project them into an implicit common space with their respective transformation matrices.

$$H^{(l+1)} = \sigma \left(\sum_{\tau \in \mathcal{T}} \tilde{A}_\tau \cdot H_\tau^{(l)} \cdot W_\tau^{(l)} \right), \tag{8.2}$$

where $\tilde{A}_\tau \in \mathbb{R}^{|\mathcal{V}| \times |\mathcal{V}_\tau|}$ is the submatrix of $\tilde{A}$, whose rows represent all the nodes and columns represent their neighboring nodes with the type τ. The representation of the nodes $H^{(l+1)}$ is obtained by aggregating information from the features of their neighboring nodes $H_\tau^{(l)}$ with different types τ using different transformation matrices $W_\tau^{(l)} \in \mathbb{R}^{q^{(l)} \times q^{(l+1)}}$. The transformation matrix $W_\tau^{(l)}$ considers the heterogeneity of different feature spaces and projects them into an implicit common space $\mathbb{R}^{q^{(l+1)}}$. Initially, $H_\tau^{(0)} = X_\tau$.

Then, we present the dual-level attention mechanism. Typically, given a specific node, different types of neighboring nodes may have different impacts on it. For example, the neighboring nodes of the same type may carry more useful information. Additionally, different neighboring nodes of the same type could also have different importance. To capture both the different importance at both node level and type level, a new dual-level attention mechanism is designed as follows.

Type-Level Attention Given a specific node v, the type-level attention learns the weights of different types of neighboring nodes. Specifically, first represent the embedding of the type τ as $h_\tau = \sum_{v'} \tilde{A}_{vv'} h_{v'}$, which is the sum of the neighboring node features $h_{v'}$ where the nodes $v' \in \mathcal{N}_v$ are with the type τ. Then, calculate the type-level attention scores based on the current node embedding h_v and the type embedding h_τ:

$$a_\tau = \sigma(\mu_\tau^T \cdot [h_v || h_\tau]), \tag{8.3}$$

where μ_τ is the attention vector for the type τ, $||$ means "concatenate," and $\sigma(\cdot)$ denotes the activation function, such as Leaky ReLU.

Then the type-level attention weights can be obtained by normalizing the attention scores across all the types with the softmax function:

$$\alpha_\tau = \frac{\exp(a_\tau)}{\sum_{\tau' \in \mathcal{T}} \exp(a_{\tau'})}. \tag{8.4}$$

Node-Level Attention The node-level attention is designed to capture the importance of different neighboring nodes and reduce the weights of noisy nodes. Formally, given a specific node v with the type τ and its neighboring node $v' \in \mathcal{N}_v$ with the type τ', compute the node-level attention scores based on the node embeddings h_v and $h_{v'}$ with the type-level attention weight $\alpha_{\tau'}$ for the node v':

$$b_{vv'} = \sigma(v^T \cdot \alpha_{\tau'}[h_v || h_{v'}]), \tag{8.5}$$

where v is the attention vector. Then normalize the node-level attention scores with the softmax function:

$$\beta_{vv'} = \frac{\exp(b_{vv'})}{\sum_{i \in \mathcal{N}_v} \exp(b_{vi})}. \tag{8.6}$$

Finally, the dual-level attention mechanism including type-level and node-level attentions is incorporated into the heterogeneous graph convolution by replacing Eq. (8.2) with the following layer-wise propagation rule:

$$H^{(l+1)} = \sigma\left(\sum_{\tau \in \mathcal{T}} \mathcal{B}_\tau \cdot H_\tau^{(l)} \cdot W_\tau^{(l)}\right). \tag{8.7}$$

Here, $\mathcal{B}_\tau$ represents the attention matrix, whose element in the vth row v'th column is $\beta_{vv'}$ in Eq. (8.6).

After going through an L-layer HGAT, the embeddings of nodes (including short texts) in the HG can be obtained. The short text embeddings $H^{(L)}$ are then fed to a softmax layer for classification. Formally,

$$Z = \text{softmax}(H^{(L)}). \tag{8.8}$$

During model training, the cross-entropy loss is exploited over training data with the L2-norm. Formally,

$$\mathcal{L} = -\sum_{i \in D_{\text{train}}} \sum_{j=1}^{C} Y_{ij} \cdot \log Z_{ij} + \eta \|\Theta\|_2, \tag{8.9}$$

where C is the number of classes, D_{train} is the set of short text indices for training, Y is the corresponding label indicator matrix, Θ is the model parameters, and η is the regularization factor. For model optimization, HGAT adopts the gradient descent algorithm.

Table 8.1 Statistics of the datasets

	#docs	#tokens	#entities	#classes
AGNews	6000	18.4	0.9 (72%)	4
Snippets	12, 340	14.5	4.4 (94%)	8
Ohsumed	7400	6.8	3.1 (96%)	23
TagMyNews	32, 549	5.1	1.9 (86%)	7
MR	10, 662	7.6	1.8 (76%)	2
Twitter	10, 000	3.5	1.1 (63%)	2

8.2.4 Experiments

8.2.4.1 Experimental Settings

Datasets Extensive experiments are conducted on 6 benchmark short text datasets: **AGNews** [50], we randomly select 6000 pieces of news from AGNews, evenly distributed into 4 classes; **Snippets** [26], it is composed of the snippets returned by a web-search engine; **Ohsumed**[4] [48], it is a benchmark bibliographic classification dataset where the documents with multiple labels are removed. Here use the titles for short text classification; **TagMyNews** [33], it contains English news from really simple syndication (RSS) feeds. Here use the news titles as instances; **MR** [25], it is a movie review dataset, in which each review only contains one sentence annotated with positive or negative for binary sentiment classification; **Twitter**, provided by a library of Python, NLTK,[5] it is also a binary sentiment classification dataset. For each dataset, we randomly select 40 labeled documents per class, half of which for training and the other half for validation. Following [16], all the left documents are for testing, which are also used as unlabeled documents during training. All the datasets are preprocessed as follows. We remove non-English characters, the stop words, and low-frequency words appearing less than 5 times. Table 8.1 shows the statistics of the datasets, including the number of documents, the number of average tokens and entities, the number of classes, and the proportion of texts containing entities in parentheses. In these datasets, most of the texts (around 80%) contain entities.

Baselines To comprehensively evaluate the proposed method for semi-supervised short text classification, we compare it with the following nine state-of-the-art methods: **SVM**: SVM classifiers using TF-IDF features and LDA features [3] are denoted as SVM+TFIDF and SVM+LDA, respectively. **CNN**: CNN [15] with 2 variants: (1) CNN-rand, whose word embeddings are randomly initialized, and (2) CNN-pretrain, whose word embeddings are pre-trained with Wikipedia Corpus. **LSTM**: LSTM [19] with and without pre-trained word embeddings, named LSTM-rand and LSTM-pretrain, respectively. **PTE**: A semi-supervised representation learning

---

[4] http://disi.unitn.it/moschitti/corpora.htm.

[5] https://www.nltk.org/.

method for text data [31]. It first learns word embedding based on the heterogeneous text networks containing three bipartite networks of words, documents, and labels and then averages word embeddings as document embeddings for text classification. **TextGCN**: Text GCN [48] models the text corpus as a graph containing documents and words as nodes and applies GCN for text classification. **HAN**: HAN [43] embeds HGs by first converting an HG into several homogeneous sub-networks through pre-defined meta-paths and then applying graph attention networks. For fair comparison, all of the above baselines, such as SVMs, CNN, and LSTM, have used entity information.

Parameter Settings The parameter values of K, T, and δ are chosen according to the best results on the validation set. To construct HG for short texts, set the number of topics $K = 15$ in LDA for the datasets AGNews, TagMyNews, MR, and Twitter, and $K = 20$ for Snippets and $K = 40$ for Ohsumed. For all the datasets, each document is assigned to top $P = 2$ topics with the largest probabilities. The similarity threshold δ between entities is set $\delta = 0.5$. Following previous studies [32], we set the hidden dimension of the model HGAT and other neural models to $d = 512$ and the dimension of pre-trained word embeddings to 100, and the layer number L of HGAT, GCN-HIN, and TextGCN as 2. For model training, the learning rate is set to 0.005, dropout rate 0.8, and the regularization factor $\eta = 5e\text{-}6$. Early stopping is applied to avoid over-fitting.

8.2.4.2 Main Results

Table 8.2 shows the classification accuracy of different methods on 6 benchmark datasets. One can see that HGAT method significantly outperforms all the baselines by a large margin, which shows the effectiveness of the proposed method on semi-supervised short text classification.

The traditional method SVMs based on the human-designed features achieve better performance than the deep models with random initialization, i.e., CNN-rand

Table 8.2 Test accuracy (%) of different models on six standard datasets. The second best results are underlined. The note * means the proposed model significantly outperforms the baselines based on t-test (p < 0.01)

Dataset	SVM +TFIDF	SVM +LDA	CNN -rand	CNN -pretrain	LSTM -rand	LSTM -pretrain	PTE	TextGCN	HAN	HGAT
AGNews	57.73	65.16	32.65	67.24	31.24	66.28	36.00	67.61	62.64	**72.10***
Snippets	63.85	63.91	48.34	77.09	26.38	75.89	63.10	77.82	58.38	**82.36***
Ohsumed	41.47	31.26	35.25	32.92	19.87	28.70	36.63	41.56	36.97	**42.68***
TagMyNews	42.90	21.88	28.76	57.12	25.52	57.32	40.32	54.28	42.18	**61.72***
MR	56.67	54.69	54.85	58.32	52.62	60.89	54.74	59.12	57.11	**62.75***
Twitter	54.39	50.42	52.58	56.34	54.80	60.28	54.24	60.15	53.75	**63.21***

and LSTM-rand in most cases, while CNN-pretrain and LSTM-pretrain using the pre-trained vectors achieve significant improvements and outperform SVMs. The graph based model PTE achieves inferior performance compared to CNN-pretrain and LSTM-pretrain. The reason may be that PTE learns text embeddings based on word co-occurrences, which, however, are sparse in short text classification. Graph neural network-based models TextGCN and HAN achieve comparable results with the deep models CNN-pretrain and LSTM-pretrain. The proposed model HGAT consistently outperforms all the state-of-the-art models by a large margin, which shows the effectiveness of the proposed method. The reasons include that (1) HGAT constructs a flexible HG framework for modeling the short texts, enabling integration of additional information to enrich the semantics and (2) we propose a novel model HGAT to embed the HG for short text classification based on a new dual-level attention mechanism. The attention mechanism not only captures the importance of different neighboring nodes (reducing the weights of noisy information) but also the importance of different types of nodes.

8.2.4.3   Comparison of Variants of HGAT

We also compare the model HGAT with four variants to validate the effectiveness of the HGAT model. As shown in Table 8.3, the basic model GCN-HIN directly applies GCN on the constructed HG for short texts by concatenating the feature spaces of different types of information. It does not consider the heterogeneity of various information types. HGAT w/o ATT considers the heterogeneity through the proposed heterogeneous graph convolution, which projects different types of information to an implicit common space with respective transformation matrices. HGAT-Type and HGAT-Node, respectively, consider only type-level attention and node-level attention.

One can see from Table 8.2, HGAT w/o ATT consistently outperforms GCN-HIN on all datasets, demonstrating the effectiveness of the proposed heterogeneous graph convolution that considers the heterogeneity of various information types. HGAT-Type and HGAT-Node further improve HGAT w/o ATT by capturing the importance of different information (reducing the weights of noisy information). HGAT-Node achieves better performance than HGAT-Type, indicating that node-

Table 8.3 Test accuracy (%) of HGAT variants

Dataset	GCN -HIN	HGAT w/o ATT	HGAT -Type	HGAT -Node	HGAT
AGNews	70.87	70.97	71.54	71.76	**72.10***
Snippets	76.69	80.42	81.68	81.93	**82.36***
Ohsumed	40.25	41.31	41.95	42.17	**42.68***
TagMyNews	56.33	59.41	60.78	61.29	**61.72***
MR	60.81	62.13	62.27	62.31	**62.75***
Twitter	61.59	62.35	62.95	62.45	**63.21***

level attention is more important. Finally, HGAT significantly outperforms all the variants by considering the heterogeneity and applying dual-level attention mechanism including node-level and type-level attentions.

8.2.4.4 Case Study

As Fig. 8.3 shows, a short text from AGNews is taken as an example (which is classified to the class of sports correctly) to illustrate the dual-level attention of HGAT. The type-level attention assigns high weight (0.7) to the short text itself, while lower weights (0.2 and 0.1) to entities and topics. It means that the text itself contributes more for classification than the entities and topics. The node-level attention assigns different weights to neighboring nodes. The node-level weights of nodes belonging to a same type sum to 1. As one can see, the entities e_3 (Atlanta Braves, a baseball team), e_4 (Dodger Stadium, a baseball gym), e_1 (Shawn Green, a baseball player) have higher weights than e_2 (Los Angeles, referring to a city at most time). The topics t_1 (game) and t_2 (win) have almost the same importance for classifying the text to the class of sports. The case study shows that the proposed dual-level attention can capture key information at multiple granularities for classification and reduce the weights of noisy information.

The more detailed method description and experiment validation can be seen in [18, 47].

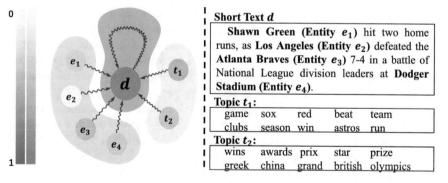

Fig. 8.3 Visualization of the dual-level attention including node-level attention (shown in red) and type-level attention (shown in blue). Each topic t is represented by top 10 words with highest probabilities

8.3 News Recommendation with Long/Short-Term Interest Modeling

8.3.1 Overview

News recommender systems that automatically recommend a small set of news articles for satisfying user's preferences have growingly attracted attentions in both industry and academics [5, 38, 39]. However, the most existing methods [5, 6, 13, 14, 17, 34, 39, 51] suffer from the data sparsity problem, since they fail to extensively exploit high-order structure information (e.g., the user1-news1-user2 relationship indicates the behavior similarity between user1 and user2). In addition, most of them ignore the latent topic information that would help indicate a user's interest and alleviate the sparse user–item interactions. The intuition is that news items with few user clicks can aggregate more information with the bridge of topics. What is more, the existing methods on news recommendation do not consider the user's long-term and short-term interests. A user usually has relatively stable long-term interests and may also be temporally attracted to certain things, i.e., short-term interests, which should be considered in news recommendation. For example, a user may continuously concern about political events, which is a long-term interest. In contrast, certain breaking news events such as attacks usually attract temporary interests.

To address the above issues, a novel **Graph Neural News Rec**ommendation model (**GNewsRec**) is proposed with long-term and short-term user interest modeling. First, a heterogeneous user–news–topic graph is constructed to explicitly model the interactions among users, news, and topics with complete historic user clicks. The topic information can help better reflect a user's interest and alleviate the sparsity issue of user–item interactions. To encode the high-order relationships among users, news items, and topics, we take advantage of graph neural networks (GNN) to learn user and news representations by propagating embeddings over the graph. The learned user embeddings with complete historic user clicks are supposed to encode a user's long-term interest. GNewsRec also models a user's short-term interest using recent user reading history with an attention-based LSTM [10, 20] model. It combines both long-term and short-term interests for user modeling, which are then compared to the candidate news representation for prediction.

8.3.2 Problem Formulation

The news recommendation problem in this chapter can be illustrated as follows. Given the click histories for K users $U = \{u_1, u_2, \cdots, u_K\}$ over M news items $I = \{d_1, d_2, \cdots, d_M\}$, the user–item interaction matrix $Y \in \mathbb{R}^{K \times M}$ is defined according to users' implicit feedback, where $y_{u,d} = 1$ indicates the user u clicked the news d, otherwise $y_{u,d} = 0$. Additionally, from the click history with timestamps, the recent click sequence $s_u = \{d_{u,1}, d_{u,2}, \cdots, d_{u,n}\}$ is obtained for a specific user u, where $d_{u,j} \in I$ is the j-th news the user u clicked.

Given the user–item interaction matrix Y as well as the users' recent click sequences S, the news recommendation problem aims to predict whether a user u has potential interest in a news item d that he/she has not seen before. This work considers the title and profile (a given set of entities E and their entity types C from the news page content) of news as features. Each news title T contains a sequence of words $T = \{w_1, w_2, \cdots, w_m\}$. The profile contains a sequence of entities $E = \{e_1, e_2, \cdots, e_n\}$ as well as its type set $C = \{c_1, c_2, \cdots, c_n\}$, where c_j is the type of the j-th entity e_j.

8.3.3 The GNewsRec Model

In this subsection, we present our graph neural news recommendation model GNewsRec with long-term and short-term interest modeling. Our model takes full advantage of the high-order structure information between users and news items by first constructing a heterogeneous graph modeling the interactions and then applying GNN to propagate the embeddings. As illustrated in Fig. 8.4, GNewsRec contains

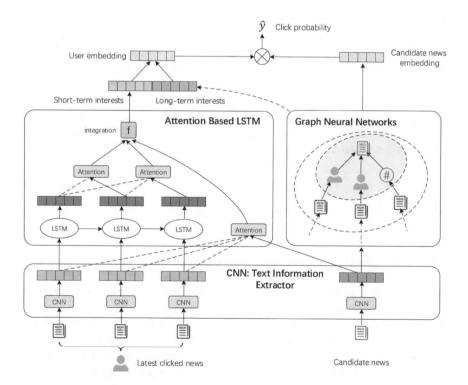

Fig. 8.4 The framework of GNewsRec

three main parts: CNN for text information extraction, GNN for long-term user interest modeling and news modeling, and attention-based LSTM model for short-term user interest modeling. The first part extracts the news feature from the news title and profile through CNN. The second part constructs a heterogeneous user–news–topic graph with complete historic user clicks and applies GNN to encode high-order structure information for recommendation. The incorporated latent topic information can alleviate the user–item sparsity since news items with few user clicks can aggregate more information with the bridge of topics. The learned user embeddings with complete historic user clicks are supposed to encode the relatively stable long-term user interest. GNewsRec also models the user's short-term interest with recent reading history through an attention-based LSTM in the third part. Finally, it combines a user's long-term and short-term interests for user representation and then compares and matches it to candidate news representation for recommendation. The three parts are detailed as follows.

8.3.3.1   Text Information Extractor

GNewsRec uses two parallel CNNs as the news text information extractor, which, respectively, take the title and profile of news as inputs and learn the title-level and profile-level representations of news. The concatenation of such two representations is regarded as the final text feature representation of news.

Specifically, the title is represented as $\mathbf{T} = [\mathbf{w}_1, \cdots, \mathbf{w}_m]^T$ and the profile as $\mathbf{P} = [\mathbf{e}_1, f(\mathbf{c}_1), \mathbf{e}_2, f(\mathbf{c}_2), \cdots, \mathbf{e}_n, f(\mathbf{c}_n)]^T$, where $\mathbf{P} \in \mathbb{R}^{2n \times k_1}$ and k_1 is the dimension of entity embedding. $f(\mathbf{c}) = \mathbf{W}_c \, \mathbf{c}$ is the transformation function. $\mathbf{W}_c \in \mathbb{R}^{k_1 \times k_2}$ (k_2 is the dimension of entity type embedding) is the trainable transformation matrix.

The title $\mathbf{T}$ and profile $\mathbf{P}$ are, respectively, fed into two parallel CNNs that have separate weight parameters. Hence, their feature representations are separately obtained as $\widetilde{\mathbf{T}}$ and $\widetilde{\mathbf{P}}$ through two parallel CNNs. Finally $\widetilde{\mathbf{T}}$ and $\widetilde{\mathbf{P}}$ are concatenated as the final news text feature representation:

$$\mathbf{d} = f_c([\widetilde{\mathbf{T}}; \widetilde{\mathbf{P}}]), \tag{8.10}$$

where $\mathbf{d} \in \mathbb{R}^D$ and f_c is a densely connected layer.

8.3.3.2   Long-Term User Interest Modeling and News Modeling

To model long-term user interest and news, GNewsRec first constructs a heterogeneous user–news–topic graph with users' complete historic clicks. The incorporated topic information can help better indicate a user's interest and alleviate the sparsity of user–item interactions. Then it applies graph convolutional networks for learning embeddings of users and news items, which encodes the high-order information between users and items through propagating embeddings over the graph.

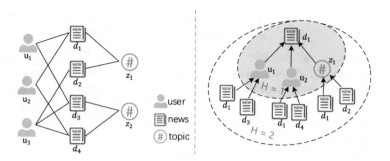

Fig. 8.5 Heterogeneous user–news–topic graph (left) and two-layer GNN (right)

Heterogeneous User–News–Topic Graph GNewsRec incorporates the latent topic information in news articles to better indicate the user's interest and alleviate the user–item sparsity issue. Hence, a heterogeneous undirected graph $G = (V, R)$ is constructed as illustrated in the left part of Fig. 8.5, where V and R are, respectively, the sets of nodes and edges. Our graph contains three types of nodes: users U, news items I, and topics Z. The topics Z can be mined through the topic model LDA [3].

The user–item edges are built if the user u clicked a news item d, i.e., $y_{u,d} = 1$. For each news document d, its topic distribution is obtained as $\theta_d = \{\theta_{d,i}\}_{i=1,\cdots,\mathcal{K}}$, $\sum_{i=1}^{\mathcal{K}} \theta_i = 1$ through LDA. Next, the connection of the news document d and the topic z is built with the largest probability.

Note that for testing, the estimated LDA model can infer the topics of new documents [23]. In this way, the new documents that do not exist in the graph can be connected with the constructed graph and update their embeddings through graph convolution. Hence, the topic information can alleviate the cold-start problem as well as the sparsity issue of user–item interactions.

GNN for Heterogeneous User–News–Topic Graph With the constructed heterogeneous user–news–topic graph, GNewsRec then applies GNN [9, 40, 41] to capture high-order relationships between users and news by propagating the embeddings through it. Following are the general form of computing a certain node embedding of a single GNN layer:

$$\mathbf{h}_{\mathcal{N}_v} = \text{AGGREGATE}(\{\mathbf{W}^t \mathbf{h}_u^t, \forall u \in \mathcal{N}_v\}), \tag{8.11}$$

$$\mathbf{h}_v = \sigma(\mathbf{W} \cdot \mathbf{h}_{\mathcal{N}_v} + \mathbf{b}), \tag{8.12}$$

where AGGREGATE is the aggregator function to aggregate information from neighboring nodes. Here GNewsRec uses the mean aggregator that simply takes the element-wise mean of the vectors of the neighbors. $\mathcal{N}_v$ denotes the neighborhood of a certain node v and $\mathbf{W}^t$ is the trainable transformation matrix for transforming different types of nodes h_u^t into the same space. $\mathbf{W}$ and $\mathbf{b}$ are the weight matrices and bias of one GNN layer to update the center node embedding $\mathbf{h}_v$.

In particular, considering the candidate pair of user u and news d, GNewsRec uses $U(d)$ and $Z(d)$[6] to, respectively, denote the set of users and topics directly connected to the news document d. In real applications, the size of $U(d)$ may vary significantly over all news documents. To keep the computational pattern of each batch fixed and more efficient, GNewsRec uniformly samples a set of neighbors $S(d)$ with fixed size for each news d instead of using all its neighbors, where the size $|S(d)| = L_u$.[7] Following Eqs. (8.11) and (8.12), to characterize the topological proximity structure of news d, first, we compute the linear average combination of all its sampled neighbors as follows:

$$\mathbf{d}_{\mathcal{N}} = \frac{1}{|S(d)|} \sum_{u \in S(d)} \mathbf{W}_u \mathbf{u} + \frac{1}{|Z(d)|} \sum_{z \in Z(d)} \mathbf{W}_z \mathbf{z}, \tag{8.13}$$

where $\mathbf{u} \in \mathbb{R}^D$ and $\mathbf{z} \in \mathbb{R}^D$ are the representations of the neighboring user and topic of news d. Note that $\mathbf{u}$ and $\mathbf{z}$ are initialized randomly, while $\mathbf{d}$ is initialized with the text feature embedding obtained from text information extractor (Sect. 4.1). $\mathbf{W}_u \in \mathbb{R}^{D \times D}$ and $\mathbf{W}_z \in \mathbb{R}^{D \times D}$ are, respectively, the trainable transformation matrices for users and topics, which map them from the different spaces to the same space of news embeddings.

Then the candidate news embedding is updated with the neighborhood representation $\mathbf{d}_{\mathcal{N}}$ by

$$\widetilde{\mathbf{d}} = \sigma(\mathbf{W}^1 \cdot \mathbf{d}_{\mathcal{N}} + \mathbf{b}^1), \tag{8.14}$$

where σ is the non-linear function $ReLU$, and $\mathbf{W}^1 \in \mathbb{R}^{D \times D}$ and $\mathbf{b}^1 \in \mathbb{R}^D$ are transformation weight and bias of the first layer of GNN, respectively.

This is a single-layer GNN, where the final embedding of the candidate news is only dependent on its immediate neighbors. In order to capture high-order relationships between users and news, GNewsRec can extend the GNN from one layer to multiple layers, propagating the embeddings in a broader and deeper way. As shown in Fig. 8.5, 2-order news embeddings can be obtained as follows. We first get its 1-hop neighboring user embeddings $\mathbf{u}_l$ and topic embeddings $\mathbf{z}$ by aggregating their neighboring news embeddings using Eqs. (8.11) and (8.12). Then their embeddings $\mathbf{u}_l$ and $\mathbf{z}$ are aggregated to get 2-order news embeddings $\widetilde{\mathbf{d}}$. Generally speaking, the H-order representation of an news is a mixture of initial representations of its neighbors up to H hops away.

Through the GNN, one can get the final user and news embeddings $\mathbf{u}_l$ and $\widetilde{\mathbf{d}}$ with high-order information encoded. The user embeddings learned with complete

---

[6] Here, we assume each news has only one topic, i.e., $|Z(d)| = 1$.

[7] $S(d)$ may contain duplicates if $|U(d)| < L_u$. If $U(d) = \emptyset$, then $S(d) = \emptyset$.

user click history are supposed to capture the relatively stable long-term user interests. However, we argue that a user could be temporally attracted to certain things, namely, a user has short-term interest, which should also be considered in personalized news recommendation.

8.3.3.3 Short-Term User Interest Modeling

This subsection presents how to model a user's short-term interest using his/her recent click history through an attention-based LSTM model. We pay attention to not only the news contents but also the sequential information.

Attention over Contents Given a user u with his/her latest l clicked news $\{\mathbf{d}_1, \mathbf{d}_2, \ldots, \mathbf{d}_l\}$,[8] GNewsRec uses an attention mechanism to model the different impacts of the user's recent clicked news on the candidate news d:

$$\mathbf{u}_j = tanh(\mathbf{W}'\mathbf{d}_j + \mathbf{b}'), \tag{8.15}$$

$$\mathbf{u} = tanh(\mathbf{W}\mathbf{d} + \mathbf{b}), \tag{8.16}$$

$$\alpha_j = \frac{exp(\mathbf{v}^T(\mathbf{u} + \mathbf{u}_j))}{\sum_j exp(\mathbf{v}^T(\mathbf{u} + \mathbf{u}_j))}, \tag{8.17}$$

$$\mathbf{u}_c = \sum_j \alpha_j \mathbf{d}_j, \tag{8.18}$$

where $\mathbf{u}_c$ is the user's current content-level interest embedding, α_j is the impact weight of clicked news $d_j(j = 1, \cdots, l)$ on candidate news d, $\mathbf{W}', \mathbf{W} \in \mathbb{R}^{D \times D}, \mathbf{d}_j$, $\mathbf{b}_w, \mathbf{b}_t, \mathbf{v}^T \in \mathbb{R}^D$, and D is the dimension of news embedding.

Attention over Sequential Information Besides applying attention mechanism to model user current content-level interest, we also take attention of the sequential information of the latest clicked news, and thus an attention-based LSTM [10] is used to capture the sequential features.

As is shown in Fig. 8.4, LSTM takes user's clicked news embeddings as input and outputs the user's sequential feature representation. Since each user's current click is affected by previous clicked news, the attention mechanism described above (for content-level interest modeling) is applied on each hidden state $\mathbf{h}_j$ and their previous hidden states $\{\mathbf{h}_1, \mathbf{h}_2, \cdots, \mathbf{h}_{j-1}\}$ ($\mathbf{h}_j = LSTM(\mathbf{h}_{j-1}, \mathbf{d}_j)$) of the LSTM to obtain richer sequential feature representation $\mathbf{s}_j(j = 1, \cdots, l)$ at different click times. These features $(\mathbf{s}_1, \cdots, \mathbf{s}_l)$ are integrated by a CNN to get the final sequential feature representation $\widetilde{\mathbf{s}}$ of user's latest l clicked history.

---

[8] If the click history sequence length is less than l, it will be padded with zero embeddings.

The concatenation of current content-level interest embedding and the sequence-level embedding is fed into a linear layer and gets the final user's short-term interest embedding:

$$\mathbf{u}_s = \mathbf{W}_s[\mathbf{u}_c; \widetilde{s}], \tag{8.19}$$

where $\mathbf{W}_s \in \mathbb{R}^{D \times 2D}$ is the parameter matrix.

8.3.3.4   Prediction and Training

Finally, the user embedding $\mathbf{u}$ is computed by taking linear transformation over the concatenation of the long-term and short-term embedding vectors:

$$\mathbf{u} = \mathbf{W}[\mathbf{u}_l; \mathbf{u}_s], \tag{8.20}$$

where $\mathbf{W} \in \mathbb{R}^{d \times 2d}$ is a parameter matrix to fuse into the final user embedding.

Then we compare the final user embedding $\mathbf{u}$ to the candidate news embedding $\widetilde{\mathbf{d}}$, and the probability of user u clicking news d is predicted by a DNN:

$$\hat{y} = DNN(\mathbf{u}, \widetilde{\mathbf{d}}). \tag{8.21}$$

To train our proposed model GNewsRec, positive samples are selected from the existing observed clicked reading history and equal amount of negative samples from unobserved reading. A training sample is denoted as $X = (u, x, y)$, where x is the candidate news to predict whether click or not. For each positive input sample, $y = 1$, otherwise $y = 0$. After our model, each input sample has a respective estimated probability $\hat{y} \in [0, 1]$ of the user whether will click the candidate news x. The cross-entropy loss is used as lost function:

$$\mathcal{L} = -\left\{ \sum_{X \in \Delta^+} y \, log \, \hat{y} + \sum_{X \in \Delta^-} (1 - y) \, log(1 - \hat{y}) \right\} + \lambda \|\Theta\|_2, \tag{8.22}$$

where Δ^+ is the positive sample set and Δ^- is the negative sample set, $\lambda \|\Theta\|_2$ is the L2 regularization to all the trainable parameters, and λ is the penalty weight. Besides, dropout and early stopping are also applied to avoid over-fitting.

Table 8.4 Statistics of the dataset

Number	Adressa-1 week	Adressa-10 week
#users	537,627	590,673
#news	14,732	49,994
#events	2,527,571	23,168,411
#vocabulary	116,603	279,214
#entity-type	11	11
#average words per title	4.03	4.10
#average entity per news	22.11	21.29

8.3.4 Experiments

8.3.4.1 Experimental Settings

Datasets Experiments are conducted on a real-world online news dataset Adressa[9] [7], which is a click log dataset with approximately 20 million page visits from a Norwegian news portal as well as a sub-sample with 2.7 million clicks. We use the two light versions, named Adressa-1week, which collects news click logs as long as 1 week (from 1 January to 7 January 2017), and Adressa-10 week, which collects 10 weeks (from 1 January to 31 March 2017) dataset. Following DAN [51], for each event, we just select the (sessionStart, sessionStop),[10] user id, news id, timestamp, the title and profile of news for building our datasets. In terms of data splits, for the Adressa-1 week dataset, the data is split as: the first 5 days' history data for graph construction and the latest l news clicked in the 5 days for short-term interest modeling, the 6-th day's for generating training pairs <u, d>, 20% of the last day's for validation, and the left 80% for testing. Note that during testing, we reconstruct the graph with the previous 6 days' history data and use the latest l news clicked in the 6 days to model short-term user interest. Similarly, for the Adressa-10 week dataset, in training period, the previous 50 days' data is used for graph construction, the following 10 days for generating training pairs, 20% of the left 10 days for validation and 80% for testing. The statistics of the final datasets are shown in Table 8.4.

Baselines The following state-of-the-art methods are used as baselines: **DMF** [45], a deep matrix factorization model, uses multiple non-linear layers to process raw rating vectors of users and items but ignores the news contents and takes the implicit feedback as its input. **DeepWide** [4], a deep learning based model, combines the linear model (Wide) and feed-forward neural network (Deep) to model low- and high-level feature interactions simultaneously. **DeepFM** [8], a general deep model for recommendation, combines a component of factorization machines and

---

[9] http://reclab.idi.ntnu.no/dataset/.

[10] sessionStart and sessionStop determine the session boundaries.

a component of deep neural networks that share the input to model low- and high-level feature interactions. **DKN** [39], a deep content-based recommendation framework, fuses semantic-level and knowledge-level representations of news by a multi-channel CNN. **DAN** [51], a deep attention-based neural network for news recommendation, improves DKN [39] by considering the user's click sequence information. Note that all the baseline models are based on deep neural networks. DMF is a collaborative filtering based model, while the others are all content-based.

8.3.4.2   Comparisons of Different Models

In this subsection, experiments are conducted to compare our model with the state-of-the-art baseline models on two datasets, and the results are reported in Table 8.5 in terms of AUC and $F1$ metrics.

As one can see from Table 8.5, our model consistently outperforms all the baselines on both datasets by more than 10.67% on F1 and 2.37% on AUC. We attribute the significant superiority of our model to its three advantages: (1) Our model constructs a heterogeneous user–news–topic graph and learns better user and news embeddings with high-order information encoded by GNN. (2) Our model considers not only the long-term user interest but also the short-term interest. (3) The topic information incorporated in the heterogeneous graph can help better reflect a user's interest and alleviate the sparsity issue of user–item interactions. The news items with few user clicks can still aggregate neighboring information through the topics. One can also find that all content-based models achieve better performance than the CF-based model DMF. This is because CF-based methods cannot work well in news recommendation due to cold-start problem. Our model as a hybrid model can combine the advantages of content-based models and CF-based model. In addition, new arriving documents without user clicks can also be connected to the existing graph via topics and update their embeddings through GNN. Thus, our model can achieve better performance.

Table 8.5 Comparison of different models

Model	Adressa-1 week		Adressa-10 week	
	AUC(%)	F1(%)	AUC(%)	F1(%)
DMF	55.66	56.46	53.20	54.15
DeepWide	68.25	69.32	73.28	69.52
DeepFM	69.09	61.48	74.04	65.82
DKN	75.57	76.11	74.32	72.29
DAN	75.93	74.01	76.76	71.65
GNewsRec	**81.16**	**82.85**	**78.62**	**81.01**

Table 8.6 Comparison of GNewsRec variants

Model	Adressa-1 week		Adressa-10 week	
	AUC(%)	F1(%)	AUC(%)	F1(%)
GNewsRec without GNN	75.93	74.01	76.76	71.65
GNewsRec without short-term interest	79.00	80.53	77.03	80.21
GNewsRec without topic	79.27	80.73	77.21	80.32
GNewsRec	**81.16**	**82.85**	**78.62**	**81.01**

8.3.4.3 Comparisons of GNewsRec Variants

Further, we compare among the variants of GNewsRec to demonstrate the efficacy of the design of our model with respect to the following aspects: GNN for learning user and news embeddings with high-order structure information encoded, combining of long-term and short-term user interests, and the incorporation of topic information. We can draw the following conclusion from the results shown in Table 8.6.

First, as one can see from Table 8.6, there is a great decline in performance when the GNN module is removed while modeling long-term user interest and news, which encodes high-order relationships on the graph. This demonstrates the superiority of our model by constructing a heterogeneous graph and applying GNN to propagate the embeddings over the graph. Second, removing short-term interest modeling module will decrease the performance by around 2% in terms of both AUC and F1. It demonstrates that considering both long-term and short-term user interests is necessary. Third, compared to the variant model without topic information, GNewsRec achieves significant improvements on both metrics. This is because that the topic information can alleviate the user–item sparsity issue as well as the cold-start problem. New documents with few user clicks can still aggregate neighboring information through topics. GNewsRec without topic performs slightly better than GNewsRec without short-term interest modeling, which shows that considering short-term interest is important.

The more detailed method description and experiment validation can be seen in [11].

8.4 News Recommendation with Preference Disentanglement

8.4.1 Overview

One of the core problems in news recommendation is how to learn better representations of users and news. Existing deep learning based methods [2, 24, 39, 44, 51] usually only focus on news contents, and seldom consider the collaborative signal in the form of high-order connectivity underlying the user–news interactions.

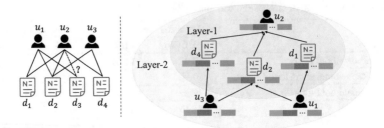

Fig. 8.6 An illustration of user–news interaction graph and high-order connectivity. The representations of user and news are disentangled with latent preference factors

Capturing high-order connectivity among users and news could deeply exploit structure characteristics and alleviate the sparsity, thus improving the recommendation performance [42]. For example, as shown in Fig. 8.6, the high-order relationship u_1–d_1–u_2 indicates the behavior similarity between u_1 and u_2 so that we may recommend d_3 to u_2 since u_1 clicked d_3, while d_1–u_2–d_4 implies d_1 and d_4 may have similar target users. Moreover, users may click different news due to their great diversity of preferences. The real-world user–news interactions arise from highly complex latent preference factors. For example, as shown in Fig. 8.6, u_2 might click d_1 under his/her preference to entertainment news, while chooses d_4 due to his/her interest in politics. When aggregating neighborhood information along the graph, different importance of neighbors under different latent preference factors should be considered. Learning representations that uncover and disentangle these latent preference factors can bring enhanced expressiveness and interpretability, which nevertheless remain largely unexplored by the existing literatures on news recommendation.

To address the above issues, the user–news interactions are modeled as a bipartite graph and propose a novel **G**raph **N**eural News Recommendation Model with **U**nsupervised preference **D**isentanglement (**GNUD**). The model is able to capture the high-order connectivities underlying the user–news interactions by propagating the user and news representations along the graph. Furthermore, the learned representations are disentangled by a neighborhood routing mechanism, which dynamically identifies the latent preference factors that may have caused the click between a user and news, and accordingly assigning the news to a subspace that extracts and convolutes features specific to that factor. To force each disentangled subspace to independently reflect an isolated preference, a novel preference regularizer is also designed to maximize the mutual information measuring dependency between two random variables in information theory to strengthen the relationship between the preference factors and the disentangled embeddings. It further improves the disentangled representations of users and news.

8.4.2 The GNUD Model

The news recommendation problem can be formalized as follows. Given the user–news historical interactions $\{(u, d)\}$, it aims to predict whether a user u_i will click a candidate news d_j that he/she has not seen before. In the following, we first introduce the news content information extractor that learns a news representation $\mathbf{h}_d$ from news content. Then we detail the proposed graph neural model GNUD with unsupervised preference disentanglement for news recommendation as shown in Fig. 8.7. The model not only exploits the high-order structure information underlying the user–news interaction graph but also considers the different latent preference factors causing the clicks between users and news. A novel preference regularizer is also introduced to force each disentangled subspace independently reflect an isolated preference factor.

8.4.2.1 News Content Information Extractor

For a news article d, the original paper considers the title T and profile P (a given set of entities E and their corresponding entity types C from the news content) as features. The entities E and their corresponding entity types C are already given in the datasets. Each news title T consists of a word sequence $T = \{w_1, w_2, \cdots, w_m\}$. Each profile P contains a sequence of entities defined as $E = \{e_1, e_2, \cdots, e_p\}$ and corresponding entity types $C = \{c_1, c_2, \cdots, c_p\}$. It is denoted that the title embedding as $\mathbf{T} = [\mathbf{w}_1, \mathbf{w}_2, \cdots, \mathbf{w}_m]^T \in R^{m \times n_1}$, entity set embedding as $\mathbf{E} = [\mathbf{e}_1, \mathbf{e}_2, \cdots, \mathbf{e}_p]^T \in R^{p \times n_1}$, and the entity-type set embedding as $\mathbf{C} = [\mathbf{c}_1, \mathbf{c}_2, \cdots, \mathbf{c}_p]^T \in R^{p \times n_2}$. $\mathbf{w}$, $\mathbf{e}$, and $\mathbf{c}$ are, respectively, the embedding

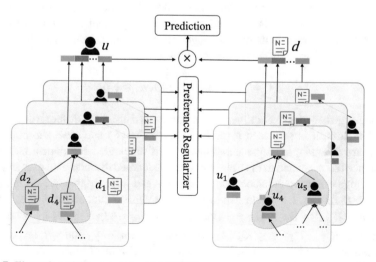

Fig. 8.7 Illustration of the proposed model GNUD

vectors of word w, entity e, and entity type c. n_1 and n_2 are the dimension of word (entity) and entity-type embeddings. These embeddings can be pre-trained from a large corpus or randomly initialized. Following [51], it defines the profile embedding $\mathbf{P} = [\mathbf{e}_1, g(\mathbf{c}_1), \mathbf{e}_2, g(\mathbf{c}_2), \cdots, \mathbf{e}_p, g(\mathbf{c}_p)]^T$ where $\mathbf{P} \in R^{2p \times n_1}$. $g(\mathbf{c})$ is the transformation function as $g(\mathbf{c}) = \mathbf{M}_c \mathbf{c}$, where $\mathbf{M}_c \in R^{n_1 \times n_2}$ is a trainable transformation matrix.

Following DAN [51], we also use two parallel convolutional neural networks (PCNN) taking the title $\mathbf{T}$ and profile $\mathbf{P}$ of news as input to learn the title-level and profile-level representation $\widehat{\mathbf{T}}$ and $\widehat{\mathbf{P}}$ for news. Finally we concatenate $\widehat{\mathbf{T}}$ and $\widehat{\mathbf{P}}$ and get the final news representation $\mathbf{h}_d$ through a fully connected layer f:

$$\mathbf{h}_d = f([\widehat{\mathbf{T}}; \widehat{\mathbf{P}}]). \tag{8.23}$$

8.4.2.2 Heterogeneous Graph Encoder

As illustrated in Fig. 8.7, to capture the high-order connectivity underlying the user–news interactions, the user–news interactions are modeled as a bipartite graph $\mathcal{G} = \{\mathcal{U}, \mathcal{D}, \mathcal{E}\}$, where $\mathcal{U}$ and $\mathcal{D}$ are the sets of users and news, $\mathcal{E}$ is the set of edges and each edge $e = (u, d) \in \mathcal{E}$ indicates that user u explicitly clicks news d. The model GNUD enables information propagation among users and news along the graph, thus capturing the high-order relationships among users and news. Additionally, GNUD learns disentangled embeddings that uncover the latent preference factors behind user–news interactions, enhancing expressiveness and interpretability. In the following, one single graph convolution layer with preference disentanglement is presented.

Graph Convolution Layer with Preference Disentanglement Given the user–news bipartite graph $\mathcal{G}$ where the user embedding $\mathbf{h}_u$ is randomly initialized and news embedding $\mathbf{h}_d$ is obtained with the news content information extractor (Sect. 8.4.2.1), a graph convolutional layer aims to learn the representation $\mathbf{y}_u$ of a node u by aggregating its neighbors' features:

$$\mathbf{y}_u = \text{Conv}(\mathbf{h}_u, \{\mathbf{h}_d : (u, d) \in \mathcal{E}\}). \tag{8.24}$$

Considering that users' click behaviors could be caused by different latent preference factors, it is proposed to derive a layer $\text{Conv}(\cdot)$ such that the outputs $\mathbf{y}_u$ and $\mathbf{y}_d$ are disentangled representations. Each disentangled component reflect one preference factor related to the user or news. The learned disentangled user and news embeddings can bring enhanced expressiveness and interpretability. Assuming that there are K factors, we would like to let $\mathbf{y}_u$ and $\mathbf{y}_d$ be composed of K independent components: $\mathbf{y}_u = [\mathbf{z}_{u,1}, \mathbf{z}_{u,2}, \cdots, \mathbf{z}_{u,K}]$, $\mathbf{y}_d = [\mathbf{z}_{d,1}, \mathbf{z}_{d,2}, \cdots, \mathbf{z}_{d,K}]$, where $\mathbf{z}_{u,k}$ and $\mathbf{z}_{d,k} \in R^{\frac{l_{out}}{K}}$ ($1 \leq k \leq K$) (l_{out} is the dimension of $\mathbf{y}_u$ and $\mathbf{y}_d$), respectively, characterizing the k-th aspect of user u and news d related to the k-th preference factor. Note that in the following, we can focus on user u and describe the learning process of its representation $\mathbf{y}_u$. The news d can be learned similarly, which is omitted.

Formally, given a u-related node $i \in \{u\} \bigcup \{d : (u, d) \in \mathcal{E}\}$, a subspace-specific projection matrix W_k is used to map the feature vector $\mathbf{h}_i \in R^{l_{in}}$ into the k-th preference related subspace:

$$\mathbf{s}_{i,k} = \frac{\text{ReLU}(\mathbf{W}_k^\top \mathbf{h}_i + \mathbf{b}_k)}{\| \text{ReLU}(\mathbf{W}_k^\top \mathbf{h}_i + \mathbf{b}_k) \|_2}, \tag{8.25}$$

where $\mathbf{W}_k \in R^{l_{in} \times \frac{l_{out}}{K}}$, and $\mathbf{b}_k \in R^{\frac{l_{out}}{K}}$. Note that $\mathbf{s}_{u,k}$ is not equal to the final representation of the k-th component of u: $\mathbf{z}_{u,k}$, since it has not mined any information from neighboring news yet. To construct $\mathbf{z}_{u,k}$, the information needs mining from both $\mathbf{s}_{u,k}$ and the neighborhood features $\{\mathbf{s}_{d,k} : (u, d) \in \mathcal{E}\}$.

The main intuition is that when constructing $\mathbf{z}_{u,k}$ characterizing the k-th aspect of u, it should only use the neighboring news articles d that connect with user u due to the preference factor k instead of all the neighbors. Therefore, a neighborhood routing algorithm [21] is applied to identify the subset of neighboring news that actually connect to u due to the preference factor k.

Neighborhood Routing Algorithm The neighborhood routing algorithm infers the latent preference factors behind user–news interactions by iteratively analyzing the potential subspace formed by a user and his/her clicked news. Formally, let $r_{d,k}$ be the probability that the user u clicks the news d due to the factor k. Then it is also the probability that we should use the news d to construct $\mathbf{z}_{u,k}$. $r_{d,k}$ is an unobserved latent variable that can be inferred in an iterative process. The motivation of the iterative process is as follows. Given $\mathbf{z}_{u,k}$, the value of the latent variables $\{r_{d,k} : 1 \leq k \leq K, (u, d) \in \mathcal{E}\}$ can be obtained by measuring the similarity between user u and his/her clicked news d under the k-th subspace, which is computed as Eq. (8.26). Initially, set $\mathbf{z}_{u,k} = \mathbf{s}_{u,k}$. On the other hand, after obtaining the latent variables $\{r_{d,k}\}$, one can find an estimate of $\mathbf{z}_{u,k}$ by aggregating information from the clicked news, which is computed as Eq. (8.27):

$$r_{d,k}^{(t)} = \frac{\exp(\mathbf{z}_{u,k}^{(t)\top} \mathbf{s}_{d,k})}{\sum_{k'=1}^{K} \exp(\mathbf{z}_{u,k}^{(t)\top} \mathbf{s}_{d,k})}, \tag{8.26}$$

$$\mathbf{z}_{u,k}^{(t+1)} = \frac{\mathbf{s}_{u,k} + \sum_{d:(u,d)\in\mathcal{G}} r_{d,k}^{(t)} \mathbf{s}_{d,k}}{\| \mathbf{s}_{u,k} + \sum_{d:(u,d)\in\mathcal{G}} r_{d,k}^{(t)} \mathbf{s}_{d,k} \|_2}, \tag{8.27}$$

where iteration $t = 0, \cdots, T - 1$. After T iterations, the output $\mathbf{z}_{u,k}^{(T)}$ is the final embedding of user u in the k-th latent subspace and we obtain $\mathbf{y}_u = [\mathbf{z}_{u,1}, \mathbf{z}_{u,2}, \cdots, \mathbf{z}_{u,K}]$.

The above shows a single graph convolutional layer with preference disentanglement, which aggregates information from the first-order neighbors. In order to capture information from high-order neighborhood and learn high-level features, multiple layers are stacked. Specially, L layers are used to get the final disentangled representation $\mathbf{y}_u^{(L)} \in R^{K\Delta n}$ ($K\Delta n = l_{out}$) for user u and $\mathbf{y}_d^{(L)}$ for news d, where Δn is the dimension of a disentangled subspace.

Preference Regularizer Naturally, we hope each disentangled subspace can reflect an isolated latent preference factor independently. Since there are no explicit labels indicating the user preferences in the training data, a novel preference regularizer is also designed to maximize the mutual information measuring dependency between two random variables in information theory to strengthen the relationship between the preference factors and the disentangled embeddings. According to [46], the mutual information maximization can be converted into the following form.

Given the representation of a user u in k-th ($1 \leq k \leq K$) latent subspace, the preference regularizer $P(k|\mathbf{z}_{u,k})$ estimates the probability of the k-th subspace (w.r.t. the k-th preference) that $\mathbf{z}_{u,k}$ belongs to

$$P(k|\mathbf{z}_{u,k}) = \text{softmax}(\mathbf{W}_p \cdot \mathbf{z}_{u,k} + \mathbf{b}_p), \tag{8.28}$$

where $\mathbf{W}_p \in R^{K \times \Delta n}$, and parameters in the regularizer $P(\cdot)$ are shared with all the users and news.

8.4.2.3 Model Training

For model training, a fully connected layer is added:

$$\mathbf{y}_u' = \mathbf{W}^{(L+1)\top} \mathbf{y}_u^{(L)} + \mathbf{b}^{(L+1)}, \tag{8.29}$$

where $\mathbf{W}^{(L+1)} \in R^{K\Delta n \times K\Delta n}$, $\mathbf{b}^{(L+1)} \in R^{K\Delta n}$. Then simple dot product is used to compute the news click probability score:

$$\hat{s}\langle u, d \rangle = \mathbf{y}_u'^\top \mathbf{y}_d'. \tag{8.30}$$

Once obtaining the click probability scores $\hat{s}\langle u, d \rangle$, we define the following base loss function for training sample (u, d) with the ground truth $y_{u,d}$:

$$\mathcal{L}_1 = -[y_{u,d} \ln(\hat{y}_{u,d}) + (1 - y_{u,d}) \ln(1 - \hat{y}_{u,d})], \tag{8.31}$$

where $\hat{y}_{u,d} = \sigma(\hat{s}\langle u, d \rangle)$. Then we add the preference regularization term of both u and d, which can be formulated as

$$\mathcal{L}_2 = -\frac{1}{K} \sum_{k=1}^{K} \sum_{i \in \{u,d\}} \ln P(k|\mathbf{z}_{i,k})[k]. \tag{8.32}$$

Finally, the overall training loss can be rewritten as

$$\mathcal{L} = \sum_{(u,d)\in\mathcal{T}_{\text{train}}} ((1-\lambda)\mathcal{L}_1 + \lambda\mathcal{L}_2) + \eta\|\Theta\|, \qquad (8.33)$$

where $\mathcal{T}_{\text{train}}$ is the training set. For each positive sample (u, d), a negative instance is sampled from unobserved reading history of u for training. λ is a balance coefficient. η is the regularization coefficient and Θ denotes all the trainable parameters.

Note that during training and testing, the news that have not been read by any users are taken as isolated nodes in the graph. Their representations are based on only content feature h_d without neighbor aggregation and can also be disentangled via Eq. (8.25).

8.4.3 Experiments

8.4.3.1 Experimental Settings

For datasets, we use the same settings of the dataset as Sect. 8.3. For comparisons, in addition to the aforementioned baselines in Sect. 8.3, the following state-of-the-art methods are included to be further compared with GNUD. **LibFM** [27], a feature-based matrix factorization method, concatenates the TF-IDF vectors of news title and profile as input. **CNN** [15] applies two parallel CNNs to word sequences in news titles and profiles, respectively, and concatenates them as news features. **DSSM** [13], a deep structured semantic model, models the user's clicked news as the query and the candidate news as the documents. **GNewsRec** [11], a graph neural network-based method, combines long-term and short-term interest modeling for news recommendation.

8.4.3.2 Comparison of Different Methods

The comparisons between different methods are summarized in Table 8.7. One can observe that the proposed model GNUD consistently outperforms all the state-of-the-art baseline methods on both datasets. GNUD improves the best deep neural models DKN and DAN more than 6.45% on AUC and 7.79% on F1 on both datasets. The main reason is that GNUD fully exploits the high-order structure information in the user–news interaction graph, learning better representations of users and news. Compared to the best-performed baseline method GNewsRec, GNUD achieves better performance on both datasets in terms of both AUC (+2.85% and +4.59% on the two datasets, respectively) and F1 (+1.05% and +0.08%, respectively). This is because that GNUD considers the latent preference factors that cause the user–news interactions and learns representations that uncover and disentangle these latent preference factors, which enhance expressiveness. From Table 8.7, one can

Table 8.7 The performance of different methods on news recommendation

Methods	Adressa-1 week		Adressa-10 week	
	AUC	F1	AUC	F1
LibFM	61.20±1.29	59.87±0.98	63.76±1.05	62.41±0.72
CNN	67.59±0.94	66.33±1.44	69.07±0.95	67.78±0.69
DSSM	68.61±1.02	69.92±1.13	70.11±1.35	70.96±1.56
DeepWide	68.25±1.12	69.32±1.28	73.28±1.26	69.52±0.83
DeepFM	69.09±1.45	61.48±1,31	74.04±1.69	65.82±1.18
DMF	55.66±0.84	56.46±0.97	53.20±0.89	54.15±0.47
DKN	75.57±1.13	76.11±0.74	74.32±0.94	72.29±0.41
DAN	75.93±1.25	74.01±0.83	76.76±1.06	71.65±0.57
GNewsRec	81.16±1.19	82.85±1.15	78.62±1.38	81.01±0.64
GNUD w/o Disen	78.33±1.29	79.09±1.22	78.24±0.13	80.58±0.45
GNUD w/o PR	83.12±1.53	81.67±1.56	80.61±1.07	80.92±0.31
GNUD	**84.01±1.16**	**83.90±0.58**	**83.21±1.91**	**81.09±0.23**

also see that all the content-based methods outperform the CF-based model DMF. This is because CF-based methods suffer a lot from cold-start problem since most news are new coming. Except for DMF, all the deep neural network-based baselines (e.g., CNN, DSSM, DeepWide, DeepFM, etc.) significantly outperform LibFM, which shows that deep neural models can capture more implicit but informative features for user and news representations. DKN and DAN further improve other deep neural models by incorporating external knowledge and applying a dynamic attention mechanism.

8.4.3.3   Comparison of GNUD Variants

To further demonstrate the efficacy of the design of the model GNUD, we compare among the variants of it. As one can see from the last three lines in Table 8.7, when the preference disentanglement is removed, the performance of the model GNUD w/o Disen (GNUD without preference disentanglement) drops largely by 5.68% and 4.97% in terms of AUC on the two datasets (4.81% and 0.51% on F1), respectively. This observation demonstrates the effectiveness and necessity of preference disentangled representations of users and news. Compared to GNUD w/o PR (GNUD without preference regularizer), one can see that introducing the preference regularizer, which enforces each disentangled embedding subspace independently reflect an isolated preference, can bring performance gains on both AUC (+0.89% and +2.6%, respectively) and F1 (+2.23% and +0.17%, respectively).

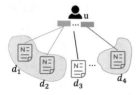

News	Keywords
d_1	norway oljebransjen (Norway oil industry), norskehavet (Norwegian sea), helgelandskysten (Helgeland coast), hygen (hygen), energy (energy), trondheim (a city)
d_2	Statkraft (State Power Corporation of Norway), trønderenergi (tronder energy), snillfjord (snill fjord), trondheimsfjorden (trondheim fjord), vindkraft (wind power), energy (energy)
d_3	Bolig (residence), hage (garden), hjemme (home), fossen (waterfall), hus (house), home (home)
d_4	health-and-fitness (health and fitness), mørk sjokolade (dark chocolate), vitaminrike (vitamin), olivenolje (olive oil), grønnsaker (vegetables), helse (health)

Fig. 8.8 Visualization of a user's clicked news that belong to different disentangled subspaces w.r.t. different preference factors. Here six keywords (translated into English) are used to illustrate a news

8.4.3.4 Case Study

To intuitively demonstrate the efficacy of GNUD, we randomly sample a user u and extract his/her logs from the test set. The representation of user u is disentangled into $K = 7$ subspaces and we randomly sample 2 subspaces. For each one, the top news is visualized that user u pay most attention to (with the probability $r_{d,k}$ larger than a threshold). As shown in Fig. 8.8, different subspaces reflect different preference factors. For example, one subspace (shown in blue) is related to "energy" as the top two news contain the keywords such as "oil industry," "hygen," and "wind power." The other subspace (shown in green) may indicate the latent preference factor about "healthy diet" as the related news contain the keywords such as "health," "vitamin," and "vegetables." The news d_3 about home has low probability in the both subspaces. It does not belong to any of the two preferences.

The more detailed method description and experiment validation can be seen in [12].

8.5 Conclusion

In recent years, heterogeneous graph based text mining has become a very popular research and industrial application direction. Considering the strong power of integrating additional information and modeling the relations between objects, heterogeneous graphs are widely explored to alleviate the data sparsity problem that is common in many tasks and applications. Therefore, it has gradually attracted attention from more researchers in the field of text mining that constructing a heterogeneous graph followed by a heterogeneous graph representation method. In this chapter, we have introduced three methods for text mining. HGAT, GNewsRec, and GUND, respectively, construct a heterogeneous graph to model the input short texts or long news. Hence the following designed heterogeneous graph neural network can make better use of the textual and auxiliary information and successfully outperforms.

In the future, the HG modeling can be explored for more other NLP tasks, such as relation extraction, question answering, etc. Moreover, it is also a valuable research direction to integrate graph-structured external knowledge, such as the knowledge graph, into the constructed heterogeneous graph for further improvement.

References

1. Aggarwal, C.C., Zhai, C.: A survey of text classification algorithms. In: Mining Text Data, pp. 163–222. Springer, Berlin (2012)
2. An, M., Wu, F., Wu, C., Zhang, K., Liu, Z., Xie, X.: Neural news recommendation with long- and short-term user representations. In: Proceedings of the 57th Annual Meeting of the Association for Computational Linguistics (ACL), pp. 336–345 (2019)
3. Blei, D.M., Ng, A.Y., Jordan, M.I.: Latent Dirichlet allocation. J. Mach. Learn. Res. **3**(Jan), 993–1022 (2003)
4. Cheng, H.T., Koc, L., Harmsen, J., Shaked, T., Chandra, T., Aradhye, H., Anderson, G., Corrado, G., Chai, W., Ispir, M., et al.: Wide and deep learning for recommender systems. In: Proceedings of the 1st Workshop on Deep Learning for Recommender Systems (DLRS@RecSys), pp. 7–10 (2016)
5. Das, A.S., Datar, M., Garg, A., Rajaram, S.: Google news personalization: scalable online collaborative filtering. In: Proceedings of the 16th International Conference on World Wide Web (WWW), pp. 271–280 (2007)
6. De Francisci Morales, G., Gionis, A., Lucchese, C.: From chatter to headlines: harnessing the real-time web for personalized news recommendation. In: Proceedings of the fifth ACM International Conference on Web Search and Data Mining (WSDM), pp. 153–162 (2012)
7. Gulla, J.A., Zhang, L., Liu, P., Özgöbek, Ö., Su, X.: The Adressa dataset for news recommendation. In: Proceedings of the International Conference on Web Intelligence (ICWI), pp. 1042–1048 (2017)
8. Guo, H., Tang, R., Ye, Y., Li, Z., He, X.: DeepFM: a factorization-machine based neural network for CTR prediction. In: Proceedings of the Twenty-Sixth International Joint Conference on Artificial Intelligence (IJCAI), pp. 1725–1731 (2017)
9. Hamilton, W., Ying, Z., Leskovec, J.: Inductive representation learning on large graphs. In: Proceedings of the 31st International Conference on Neural Information Processing Systems (NIPS), pp. 1024–1034 (2017)
10. Hochreiter, S., Schmidhuber, J.: Long short-term memory. Neural Comput. **9**(8), 1735–1780 (1997)
11. Hu, L., Li, C., Shi, C., Yang, C., Shao, C.: Graph neural news recommendation with long-term and short-term interest modeling. Inf. Process. Manage. **57**(2), 102142 (2020)
12. Hu, L., Xu, S., Li, C., Yang, C., Shi, C., Duan, N., Xie, X., Zhou, M.: Graph neural news recommendation with unsupervised preference disentanglement. In: Proceedings of the 58th Annual Meeting of the Association for Computational Linguistics (ACL), pp. 4255–4264 (2020)
13. Huang, P.S., He, X., Gao, J., Deng, L., Acero, A., Heck, L.: Learning deep structured semantic models for web search using clickthrough data. In: Proceedings of the 22nd ACM International Conference on Information and Knowledge Management (CIKM), pp. 2333–2338 (2013)
14. IJntema, W., Goossen, F., Frasincar, F., Hogenboom, F.: Ontology-based news recommendation. In: Proceedings of the 2010 EDBT/ICDT Workshops, p. 16 (2010)
15. Kim, Y.: Convolutional neural networks for sentence classification. In: Proceedings of the 2014 Conference on Empirical Methods in Natural Language Processing (EMNLP), pp. 1746–1751 (2014)
16. Kipf, T.N., Welling, M.: Semi-supervised classification with graph convolutional networks. In: Proceedings of the Conference ICLR (2017)

17. Li, L., Wang, D., Li, T., Knox, D., Padmanabhan, B.: Scene: a scalable two-stage personalized news recommendation system. In: Proceedings of the 34th International ACM SIGIR Conference on Research and Development in Information Retrieval (SIGIR), pp. 125–134 (2011)
18. Linmei, H., Yang, T., Shi, C., Ji, H., Li, X.: Heterogeneous graph attention networks for semi-supervised short text classification. In: Proceedings of the 2019 Conference on Empirical Methods in Natural Language Processing and the 9th International Joint Conference on Natural Language Processing (EMNLP-IJCNLP), pp. 4821–4830 (2019)
19. Liu, P., Qiu, X., Huang, X.: Recurrent neural network for text classification with multi-task learning. In: Proceedings of the Twenty-Fifth International Joint Conference on Artificial Intelligence (IJCAI), pp. 2873–2879 (2016)
20. Liu, M., Wang, X., Nie, L., Tian, Q., Chen, B., Chua, T.S.: Cross-modal moment localization in videos. In: Proceedings of the 26th ACM International Conference on Multimedia (MM), pp. 843–851 (2018)
21. Ma, J., Cui, P., Kuang, K., Wang, X., Zhu, W.: Disentangled graph convolutional networks. In: International Conference on Machine Learning (ICML), pp. 4212–4221 (2019)
22. Meng, Y., Shen, J., Zhang, C., Han, J.: Weakly-supervised neural text classification. In: Proceedings of the 27th ACM International Conference on Information and Knowledge Management (CIKM), pp. 983–992 (2018)
23. Newman, D., Smyth, P., Welling, M., Asuncion, A.U.: Distributed inference for latent Dirichlet allocation. In: Advances in Neural Information Processing Systems (NIPS), pp. 1081–1088 (2008)
24. Okura, S., Tagami, Y., Ono, S., Tajima, A.: Embedding-based news recommendation for millions of users. In: Proceedings of the 23rd ACM SIGKDD International Conference on Knowledge Discovery and Data Mining (KDD), pp. 1933–1942 (2017)
25. Pang, B., Lee, L.: Seeing stars: Exploiting class relationships for sentiment categorization with respect to rating scales. In: Proceedings of the 43rd Annual Meeting of the Association for Computational Linguistics (ACL), pp. 115–124 (2005)
26. Phan, X.H., Nguyen, L.M., Horiguchi, S.: Learning to classify short and sparse text and web with hidden topics from large-scale data collections. In: Proceedings of the 17th International Conference on World Wide Web (WWW), pp. 91–100 (2008)
27. Rendle, S.: Factorization machines with LIBFM. ACM Trans. Intell. Syst. Technol. 3(3), 57 (2012)
28. Shimura, K., Li, J., Fukumoto, F.: HFT-CNN: learning hierarchical category structure for multi-label short text categorization. In: Proceedings of the 2018 Conference on Empirical Methods in Natural Language Processing (EMNLP), pp. 811–816. Brussels, Belgium (2018)
29. Sinha, K., Dong, Y., Cheung, J.C.K., Ruths, D.: A hierarchical neural attention-based text classifier. In: Proceedings of the 2018 Conference on Empirical Methods in Natural Language Processing (EMNLP), pp. 817–823. Brussels, Belgium (2018)
30. Song, G., Ye, Y., Du, X., Huang, X., Bie, S.: Short text classification: A survey. J. Multimedia 9(5), 635 (2014)
31. Tang, J., Qu, M., Mei, Q.: PTE: Predictive text embedding through large-scale heterogeneous text networks. In: Proceedings of the 24th ACM SIGKDD International Conference on Knowledge Discovery and Data Mining (KDD), pp. 1165–1174 (2015)
32. Vaswani, A., Shazeer, N., Parmar, N., Uszkoreit, J., Jones, L., Gomez, A.N., Kaiser, L.u., Polosukhin, I.: Attention is all you need. In: Advances in Neural Information Processing Systems (NIPS), pp. 5998–6008 (2017)
33. Vitale, D., Ferragina, P., Scaiella, U.: Classification of short texts by deploying topical annotations. In: European Conference on Information Retrieval (ECIR), pp. 376–387 (2012)
34. Wang, C., Blei, D.M.: Collaborative topic modeling for recommending scientific articles. In: Proceedings of the 17th ACM SIGKDD International Conference on Knowledge Discovery and Data Mining (KDD), pp. 448–456 (2011)
35. Wang, S., Manning, C.D.: Baselines and bigrams: Simple, good sentiment and topic classification. In: Proceedings of the 50th Annual Meeting of the Association for Computational Linguistics (ACL), pp. 90–94 (2012)

36. Wang, X., Chen, R., Jia, Y., Zhou, B.: Short text classification using Wikipedia concept based document representation. In: Proceedings of the 2013 International Conference on Information Technology and Applications (ICITA), pp. 471–474 (2013)
37. Wang, J., Wang, Z., Zhang, D., Yan, J.: Combining knowledge with deep convolutional neural networks for short text classification. In: Proceedings of the Twenty-Sixth International Joint Conference on Artificial Intelligence (IJCAI), pp. 2915–2921 (2017)
38. Wang, X., Yu, L., Ren, K., Tao, G., Zhang, W., Yu, Y., Wang, J.: Dynamic attention deep model for article recommendation by learning human editors' demonstration. In: Proceedings of the 23rd ACM SIGKDD International Conference on Knowledge Discovery and Data Mining (KDD), pp. 2051–2059 (2017)
39. Wang, H., Zhang, F., Xie, X., Guo, M.: DKN: Deep knowledge-aware network for news recommendation. In: Proceedings of the 2018 World Wide Web Conference (WWW), pp. 1835–1844 (2018)
40. Wang, H., Zhao, M., Xie, X., Li, W., Guo, M.: Knowledge graph convolutional networks for recommender systems. In: Proceedings of the World Wide Web (WWW), pp. 3307–3313 (2019)
41. Wang, X., He, X., Cao, Y., Liu, M., Chua, T.S.: KGAT: Knowledge graph attention network for recommendation. In: Proceedings of the 25th ACM SIGKDD International Conference on Knowledge Discovery and Data Mining (KDD), pp. 950–958 (2019)
42. Wang, X., He, X., Wang, M., Feng, F., Chua, T.S.: Neural graph collaborative filtering. In: Proceedings of the 42nd international ACM SIGIR conference on Research and development in Information Retrieval (SIGIR), pp. 165–174 (2019)
43. Wang, X., Ji, H., Shi, C., Wang, B., Ye, Y., Cui, P., Yu, P.S.: Heterogeneous graph attention network. In: The World Wide Web Conference (WWW), pp. 2022–2032 (2019)
44. Wu, C., Wu, F., An, M., Huang, J., Huang, Y., Xie, X.: NPA: Neural news recommendation with personalized attention. In: Proceedings of the 25th ACM SIGKDD International Conference on Knowledge Discovery and Data Mining (KDD), pp. 2576–2584 (2019)
45. Xue, H.J., Dai, X., Zhang, J., Huang, S., Chen, J.: Deep matrix factorization models for recommender systems. In: Proceedings of the Twenty-Sixth International Joint Conference on Artificial Intelligence (IJCAI), pp. 3203–3209 (2017)
46. Yang, C., Sun, M., Yi, X., Li, W.: Stylistic Chinese poetry generation via unsupervised style disentanglement. In: Proceedings of the 2018 Conference on Empirical Methods in Natural Language Processing (EMNLP), pp. 3960–3969 (2018)
47. Yang, T., Hu, L., Shi, C., Ji, H., Li, X., Nie, L.: HGAT: Heterogeneous graph attention networks for semi-supervised short text classification. ACM Trans. Inf. Syst. **39**(3), 1–29 (2021)
48. Yao, L., Mao, C., Luo, Y.: Graph convolutional networks for text classification. In: Proceedings of the AAAI Conference on Artificial Intelligence (AAAI), pp. 7370–7377 (2019)
49. Zeng, J., Li, J., Song, Y., Gao, C., Lyu, M.R., King, I.: Topic memory networks for short text classification. In: Proceedings of the 2018 Conference on Empirical Methods in Natural Language Processing (EMNLP), pp. 3120–3131 (2018)
50. Zhang, X., Zhao, J., LeCun, Y.: Character-level convolutional networks for text classification. Adv. Neural Inf. Proces. Syst. **28**, 649–657 (2015)
51. Zhu, Q., Zhou, X., Song, Z., Tan, J., Guo, L.: Dan: Deep attention neural network for news recommendation. In: Proceedings of the AAAI Conference on Artificial Intelligence (AAAI), vol. 33, pp. 5973–5980 (2019)

Chapter 9
Heterogeneous Graph Representation for Industry Application

Abstract Heterogeneous graph (HG) representation is closely related with the real-world applications, as heterogeneous objects and interactions are ubiquitous in many practical systems. HG representation methods deployed in real-world system should consider capturing the complex interactions among objects as well as solving the unique challenges existing in real-world systems, such as large-scale, dynamics, and multi-source information. In this chapter, we focus on summarizing the industrial-level applications with HG representation. Particularly, we introduce several well deployed systems that have demonstrated the success of HG representation techniques in resolving real-world applications, including cash-out user detection, intent recommendation, share recommendation, and friend-enhanced recommendation. For industrial-level applications, we pay more attention on two key components: HG construction with industrial data and graph representation techniques on the HG.

9.1 Introduction

Heterogeneous objects and their relations are ubiquitous in many industrial-level applications. For example, in an e-commerce recommendation system, there are user, item, and shop objects, and the ternary interactions usually exist among these objects. However, the type information will be inevitably ignored if we utilize a homogeneous graph to model such data. Fortunately, the heterogeneous graph is a natural tool to model such complex data without information loss.

The existing methods applied in industrial applications can be roughly concluded into two categories. The first one focuses on performing subtle feature engineering from the historical user behavior data. However, this kind of method is labor-consuming. The other one is that the involved objects and their interaction are usually treated as a homogeneous graph and a homogeneous graph is adopted to learn node representation. Therefore, the heterogeneous information is largely ignored by this kind of method. But the heterogeneous information is very important for some scenarios.

In this chapter, we will introduce several successful cases that have applied HG representation methods on two categories of important industrial applications. The first category of task is the cash-out user detection that aims to predict whether a user will do cash-out transactions or not. And a novel **H**ierarchical **A**ttention mechanism based **C**ash-out **U**ser **D**etection model (named **HACUD**) is proposed to learn the users' features from the constructed HG. And the second category of task is the recommendation. We first study intent recommendation, where a novel **M**etapath guided **E**mbedding method for **I**ntent **Rec**ommendation (named **MEIRec**) is proposed to aggregate node information through multiple meta-paths in the triple-interacted HG. Moreover, share recommendation, aiming to predict whether a user will share an item with his friend, is first studied by the **H**eterogeneous **G**raph neural network-based **S**hare **Rec**ommendation model (named **HGSRec**) method. Finally, a novel friend-enhanced recommendation is studied, which multiplies the influence of friends in social recommendation. Unlike previous mentioned methods that need pre-defined meta-paths, we propose a novel **S**ocial **I**nfluence **A**ttentive **N**eural network (named **SIAN**), which does not require any manual selection of meta-paths. Next, we will introduce each case in detail.

9.2   Cash-Out User Detection

9.2.1   Overview

Cash-out frauds, which are to pursue cash gains with illegal or insincere means, have seriously influenced the security of credit payment services and have become major frauds on various kinds of credit payment services. The goal of cash-out user detection is to predict whether a user will do cash-out transactions or not in the future. Thus this problem can be formulated as a binary classification problem.

Conventional solutions first perform subtle feature engineering for each user, and then a classifier, such as tree-based model or neural network, is trained based on these features. However, these kinds of methods make prediction mainly based on the statistical features of a certain user but seldom fully exploit the interaction relations between users, which may be beneficial to the cash-out user detection problem. In fact, interactions between users are important to the cash-out user detection problem. Figure 9.1a demonstrates a general scenario of credit payment service, where there are three types of objects: users, merchants, and devices. Besides the attribute information, these objects also have rich interaction information, e.g., the fund transfer relation among users, the login relation between users and devices, and the transaction relation between users and merchants. The cash-out users not only have abnormal features but also behave abnormally in interaction relations.

In order to tackle these problems, we propose a novel **H**ierarchical **A**ttention mechanism based **C**ash-out **U**ser **D**etection model (named **HACUD**), an HG method

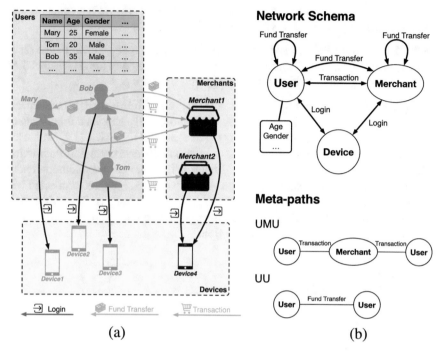

Fig. 9.1 The AHG of the scenario of credit payment service. (**a**) Scenario of credit payment service. (**b**) Network schema and meta-path examples

to predict whether a user will do cash-out transactions or not in the future. The basic idea of HACUD is to significantly enhance the feature representation of objects through fully exploiting interaction relations, i.e., with the help of meta-path based neighbors in Attributed Heterogeneous Graph (AHG). HACUD assumes that the feature representations of objects, besides intrinsic features, are also constituted by the features of their neighbors. We propose the concept of meta-path based neighbors to exploit rich structure information in AHG. Next we will explain HACUD specifically.

9.2.2   Preliminaries

Definition 1   Attributed Heterogeneous Graph (AHG). An AHG is denoted as $\mathcal{G} = \{\mathcal{V}, \mathcal{E}, \mathbf{X}\}$ consisting of an object set $\mathcal{V}$, a link set $\mathcal{E}$, and an attribute information matrix[1] $\mathbf{X} \in \mathbb{R}^{|\mathcal{V}| \times k}$. An AHG is also associated with a node type mapping function

---

[1] In this chapter, the original attributes are discretized to the same dimension.

$\phi : \mathcal{V} \to \mathcal{A}$ and a link type mapping function $\psi : \mathcal{E} \to \mathcal{R}$. $\mathcal{A}$ and $\mathcal{R}$ denote the sets of pre-defined object and link types, where $|\mathcal{A}| + |\mathcal{R}| > 2$.

Definition 2 Meta-Path Based Neighbors. Given a user u and a meta-path ρ (start form u) in an AHG, the meta-path based neighbors is defined as the set of all visited objects when the object u walks along the given meta-path ρ.

Example 1 As shown in Fig. 9.1a, we construct an AHG to model the scenario of credit payment service in which cash-out fraud usually happens. It consists of multiple types of objects (i.e., User (U), Merchant (M), Device (D)) with rich attributes and relations (i.e., fund transfer relation between users and transaction relation between users and merchants). Figure 9.1b is the corresponding network schema and meta-path example. In the AHG, two users can be connected via multiple meta-paths, "User-(fund transfer) -User" (UU) and "User-(transaction) -Merchant-(transaction)-User" (UMU). In addition, the meta-path based neighbor of Marry under meta-path UMU could be merchant and Bob.

9.2.3 The HACUD Model

9.2.3.1 Model Framework

We show the overall architecture of the model in Fig. 9.2. First, the model aggregates neighbors for each user based on different meta-paths to integrate multiple aspects of structure information in AHG and then transforms and fuses the original features for better representation learning. Considering that different features and meta-paths

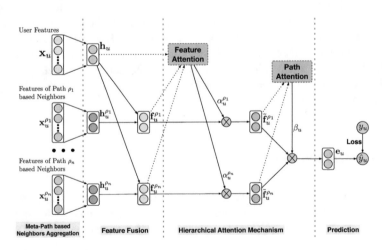

Fig. 9.2 The architecture of the proposed model

have different importances, a hierarchical attention mechanism is also used to model user preferences toward features and meta-paths.

9.2.3.2 Meta-Path Based Neighbors Aggregation

Similar to attributed network representation [16, 31], HACUD adopts to represent a node w.r.t. a certain meta-path via aggregating features of its neighbors rather than the one-hot representation of its neighbors. For each user u, the *aggregated features based on meta-path* ρ can be gotten from the formula as below:

$$\mathbf{x}_u^{\rho} = \sum_{j \in \mathcal{N}_u^{\rho}} w_{uj}^{\rho} * \mathbf{x}_j, \tag{9.1}$$

where $\mathcal{N}_u^{\rho}$ is the neighbors of node u based on meta-path ρ and $\mathbf{x}_j$ represents the attribute information vector associated with node j.

9.2.3.3 Feature Fusion

For each user u, we can obtain its own feature $\{\mathbf{x}_u\}$ as well as a set of its neighbor aggregation features based on multiple meta-paths $\{\mathbf{x}_u^{\rho}\}_{\rho \in \mathcal{P}}$. For better representation learning, a *feature fusion* part is used to transform and fuse the original features.

First, the original sparse features are projected to the low-dimensional dense representations in order to obtain the *latent representations* of user u and his/her neighbors based on different meta-paths (i.e., $\mathbf{h}_u$ and $\mathbf{h}_u^{\rho}$), respectively:

$$\mathbf{h}_u = \mathbf{W}\mathbf{x}_u + \mathbf{b}, \quad \mathbf{h}_u^{\rho} = \mathbf{W}^{\rho}\mathbf{x}_u^{\rho} + \mathbf{b}^{\rho}, \tag{9.2}$$

where $\mathbf{W}^* \in \mathbb{R}^{D \times d}$ and $\mathbf{b}^* \in \mathbb{R}^d$ are the weight matrix and bias vector, respectively. D is the dimension of the original feature space[2] and d is the dimension of latent representations. Next, the model fuses the latent representations of a user and his/her neighbors based on each meta-path and adds a fully connected layer for more complicated interaction. For a meta-path ρ, the above procedure to get *fusional representation* $\mathbf{f}_u^{\rho}$ w.r.t. meta-path ρ is formulated as below

$$\mathbf{f}_u^{\rho} = \text{ReLU}(\mathbf{W}_F^{\rho} g(\mathbf{h}_u, \mathbf{h}_u^{\rho}) + \mathbf{b}_F^{\rho}). \tag{9.3}$$

Here, $\mathbf{W}_F^{\rho} \in \mathbb{R}^{d \times 2d}$ and $\mathbf{b}_F^{\rho} \in \mathbb{R}^d$ represent the weight matrix and bias vector based on meta-path ρ, respectively.

---

[2] The original attributes are discretized to sparse D-dimensional feature as the model input.

9.2.3.4 Hierarchical Attention

Intuitively, different users are likely to have different preferences over the features based on different meta-paths as well as attribute information. Concretely, a user may place different importances to different aspect features based on meta-paths. Moreover, features also have different importances for the prediction task. Therefore, hierarchical attention mechanism is applied to capture user preferences toward features and meta-paths.

Feature Attention Since different features might not contribute to the prediction task equally, given the user latent representation $\mathbf{h}_u$ and latent representation of his/her neighbors $\mathbf{f}_u^\rho$ based on meta-path ρ, a two-layer neural network is adopted to implement the attention.

$$v_u^\rho = \text{ReLU}(\mathbf{W}_f^1[\mathbf{h}_u; \mathbf{f}_u^\rho] + \mathbf{b}_f^1), \tag{9.4}$$

$$\boldsymbol{\alpha}_u^\rho = \text{ReLU}(\mathbf{W}_f^2 v_u^\rho + \mathbf{b}_f^2), \tag{9.5}$$

where $\mathbf{W}_f^*$ and $\mathbf{b}_f^*$ denote the weight matrix and bias vector, respectively, and $[\cdot; \cdot]$ represents the concatenation of two vectors. The following is the standard setting of neural attention networks with the softmax function.

$$\hat{\alpha}_{u,i}^\rho = \frac{\exp(\alpha_{u,i}^\rho)}{\sum_{j=1}^K \exp(\alpha_{u,j}^\rho)}. \tag{9.6}$$

Then, the final representation of user u w.r.t. a meta-path ρ can be computed as follows:

$$\tilde{\mathbf{f}}_u^\rho = \hat{\boldsymbol{\alpha}}_u^\rho \odot \mathbf{f}_u^\rho, \tag{9.7}$$

where "$\odot$" denotes the element-wise product.

Path Attention Following [20], the attention weights over different meta-paths for collaboration can be learned. First, there are the attention weights of meta-path ρ for user u using a softmax unit as follows:

$$\beta_{u,\rho} = \frac{\exp(\mathbf{z}^{\rho\mathrm{T}} \cdot \tilde{\mathbf{f}}_u^C)}{\sum_{\rho' \in \mathcal{P}} \exp(\mathbf{z}^{\rho'\mathrm{T}} \cdot \tilde{\mathbf{f}}_u^C)}, \tag{9.8}$$

where $\mathbf{z}^\rho \in \mathbb{R}^{|\mathcal{P}|*d}$ is the attention vector for meta-path ρ and $\widetilde{\mathbf{f}}_u^C$ is the concatenation of user u's representations. After obtaining the path attention scores $\beta_{u,\rho}$, the final representation aggregating all meta-paths is given as follows:

$$\mathbf{e}_u = \sum_{\rho \in \mathcal{P}} \beta_{u,\rho} * \widetilde{\mathbf{f}}_u^\rho, \tag{9.9}$$

where $\widetilde{\mathbf{f}}_u^\rho$ is the representation of neighbors for user u based on meta-path ρ in Eq. 9.7.

9.2.3.5 Model Learning

In the end, the obtained final representation (i.e. $\mathbf{e}_u$) is fed into multiple fully connected neural networks as follows:

$$\mathbf{z}_u = \text{ReLU}(\mathbf{W}_L \cdots \text{ReLU}(\mathbf{W}_1 \mathbf{e}_u + \mathbf{b}_1) + \mathbf{b}_L), \tag{9.10}$$

where $\mathbf{W}_*$ and $\mathbf{b}_*$, respectively, denote the weight matrix and the bias vector for each layer. The predicted cash-out probability is obtained via a regression layer with a sigmoid unit:

$$p_u = \text{sigmoid}(\mathbf{w}_p^T \mathbf{z}_u + b_p). \tag{9.11}$$

Here $\mathbf{w}_p$ and b_p are the weight vector and the bias, respectively. The objective function is maximum likelihood estimation, which can be formulated as follows:

$$\mathcal{L}(\Theta) = \sum_{\langle u, y_u \rangle \in \mathcal{D}} (y_u \log(p_u) + (1 - y_u) \log(1 - p_u)) + \lambda ||\Theta||_2^2, \tag{9.12}$$

where y_u and p_u represent the ground truth and the predicted cash-out probability of user u, respectively. Θ is the parameter set of the proposed model and λ is the regularizer parameter.

9.2.4 Experiments

9.2.4.1 Experimental Settings

Dataset The datasets in this section are real-world data in Ant Credit Pay. We extract two sub-datasets for the evaluation, namely Ten Days Dataset and One Month Dataset. For both datasets, the model can predict the cash-out probability of users some day in the future. In the datasets, the positive samples are users who have involved in suspected cash-out transactions within one month, and the negative samples are users who have never involved in suspected cash-out transactions within one month. After preprocessing, an attributed HG based on the two datasets is constructed, consisting of 56.75 million users and 0.51 million merchants. In addition, the AHG contains 77.40 million fund transfer relations between users and 20.64 million transaction relations between users and merchants.

Metrics The metric is **AUC** (i.e., Area Under the ROC Curve), a widely used metric for the performance of cash-out user detection.

Implementation Details HACUD utilizes two hidden layers for prediction and randomly initializes the parameters with a Xavier initializer [9]. RMSProp [24] is used as the optimizer. The batch size is set to 256, the learning rate to 0.002, and set the regularizer parameter $\lambda = 0.01$ to prevent over-fitting.

9.2.4.2   Performance Comparison

We report the comparison results of the HACUD and baselines w.r.t. the dimension of latent representation d in Table 9.1. The major findings from the experimental results can be summarized as follows:

(1) HACUD outperforms all the baselines, which indicates that the model adopts a more principled way to leverage interaction relations and attribute information for improving prediction performance.
(2) Among these baselines, we can find that the overall performance order is as follows: (label + attribute + structure) based methods (i.e., GBDT$_{Struct}$, Structure2vec) > (attribute + structure) based methods (i.e. Node2vec + Feature, Metapath2vec + Feature) > structure or attribute only based method (i.e., Node2vec, Metapath2vec, GBDT). It indicates that the better performances can

Table 9.1 Results of effectiveness experiments on two datasets w.r.t. the dimension of latent representation d. A larger value indicates a better performance. The best results of all methods are indicated in bold

	AUC							
	Ten Days Dataset				One Month Dataset			
Algorithm	$d = 16$	$d = 32$	$d = 64$	$d = 128$	$d = 16$	$d = 32$	$d = 64$	$d = 128$
Node2vec [10]	0.5893	0.5913	0.5926	0.5930	0.5980	0.6063	0.6009	0.6021
Metapath2vec [6]	0.5914	0.5903	0.5917	0.5920	0.6005	0.5976	0.5995	0.5983
Node2vec + Feature	0.6455	0.6464	0.6510	0.6447	0.6541	0.6561	0.6607	0.6518
Metapath2vec + Feature	0.6456	0.6429	0.6469	0.6485	0.6550	0.6552	0.6523	0.6545
Structure2vec [5]	0.6537	0.6556	0.6598	0.6545	0.6641	0.6632	0.6657	0.6678
GBDT [8]	0.6389	0.6389	0.6389	0.6389	0.6467	0.6467	0.6467	0.6467
GBDT$_{Struct}$	0.6948	0.6948	0.6948	0.6948	0.6968	0.6968	0.6968	0.6968
HACUD	**0.7066**	**0.7115**	**0.7056**	**0.7049**	**0.7132**	**0.7160**	**0.7109**	**0.7154**

be achieved through fusing more information. In addition, structure information (i.e., interaction relations) is really helpful for performance improvement.

(3) Comparing the two variants of GBDT (i.e., traditional GBDT and GBDT$_{Struct}$), we can find that GBDT$_{Struct}$ significantly outperforms traditional GBDT and other baselines, which further demonstrates the contribution of structural features provided by meta-path based neighbors in AHG.

9.2.4.3   Effects of Hierarchical Attention

One of the major contributions of HACUD is hierarchical attention mechanism that learns the user preference toward features and meta-paths. In order to examine its effectiveness, we compare the model with its two variants, namely HACUD$_{\backslash PathAtt}$ (HACUD without path attention) and HACUD$_{\backslash PathAtt+FeaAtt}$ (HACUD without path and feature attention). For the performance comparison in Fig. 9.3, we can find that the overall performance order is as follows: HACUD > HACUD$_{\backslash PathAtt}$ > HACUD$_{\backslash PathAtt+FeaAtt}$. The results show that the hierarchical mechanism is able to better utilize the user feature and features generated by meta-paths in two aspects. First, different meta-paths have different contributions to cash-out user prediction, which cannot be treated equally (i.e., HACUD$_{\backslash PathAtt}$). Second, each user tends to place different importances to the various attributes for each meta-path. Ignoring such influence may not be able to achieve the promising performance for fully exploiting attribute and structure information (i.e., HACUD$_{\backslash PathAtt+FeaAtt}$).

9.2.4.4   Impact of Different Meta-Paths

Furthermore, there is another experiment about the performances based on single meta-path and the corresponding average attention value in Fig. 9.4. As we have observed, the performances of HACUD with different meta-paths and the corresponding attentions are positively correlated (i.e., important meta-paths tend to

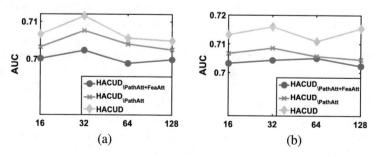

Fig. 9.3 Performance comparison of hierarchical attention w.r.t. the dimension of latent representation d. (**a**) Ten Days Dataset. (**b**) One Month Dataset

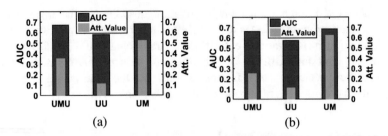

Fig. 9.4 Performances comparison on different meta-paths and corresponding attention values.
(**a**) Ten Days Dataset. (**b**) One Month Dataset

attract more attentions). In other words, the proposed HACUD model is potential
to let different users focus on the proper meta-paths.

The more detailed method description and experiment validation can be seen
in [11].

9.3   Intent Recommendation

9.3.1   Overview

With the development of mobile Internet, a novel recommendation service, named
intent recommendation, in many e-commerce Apps (e.g., Taobao and Amazon)
has emerged, which automatically recommends user intent (presented as several
words) in a search box according to users' historical behaviors when users open
an e-commerce App. Figure 9.5a illustrates an intent recommendation example on
the Taobao mobile App. According to user historic information, an intent (e.g.,
presented as "air jordan") will be automatically recommended in the search box
when a user opens the App. If the user clicks the search button, he/she will jump to
the corresponding item list page.

In this chapter, we define the intent recommendation as follows: automatically
recommend a personalized intent for a user according to his/her historical behaviors
without query input. Here, in our application scenario, intent is presented as a query,
consisting of several words or terms simply and directly reflecting user intent. The
existing methods for intent recommendation used in industry, such as Taobao and
Amazon, usually extract handcrafted features and then feed these features to a
classifier, e.g., GBDT [8] and XGBoost [4]. These methods heavily rely on domain
knowledge and need laboring feature engineering. They only utilize attribute and
statistic information of users and queries and fail to take full advantage of the rich
interaction information among objects. However, the interaction information is very
abundant in real systems, and it is really critical to capture user intent.

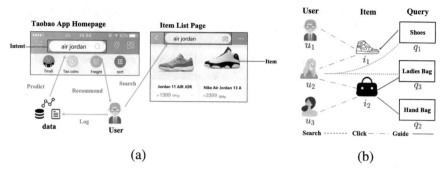

Fig. 9.5 Intent recommendation example on Taobao mobile application and the corresponding heterogeneous graph. (**a**) Intent recommendation. (**b**) Heterogeneous graph

As a general information modeling method, heterogeneous graph, consisting of multiple types of objects and links, has been widely applied to many data mining tasks [11, 21, 22]. In this chapter, we first propose to model the intent recommendation system with an HG, through which we can flexibly exploit its rich interaction information. As shown in Fig. 9.5b, obviously, HG clearly demonstrates objects in intent recommendation (e.g., users, items, and queries) and their interaction relations. Furthermore, we present a novel **M**eta-path guided **E**mbedding method for **I**ntent **Rec**ommendation (named **MEIRec**). In order to fully utilize rich interaction information in intent recommendation, we propose to learn structural feature representations of users and queries with Heterogeneous Graph Neural Network (HGNN). Concretely, we present the meta-path guided neighbors to aggregate rich neighbor information, where different aggregation functions are designed according to the characteristics of different types of neighboring information. In addition, a uniform term embedding mechanism is designed to significantly reduce the parameter space. With the static features used in the existing systems, as well as the embeddings of users and queries learned from interaction information, we build a prediction model for intent recommendation.

9.3.2 Problem Formulation

Definition 3 Intent Recommendation. Given a set $< \mathcal{U}, \mathcal{I}, \mathcal{Q}, \mathcal{W}, \mathcal{A}, \mathcal{B} >$, where $\mathcal{U} = \{u_1, \cdots, u_p\}$ denotes the set of p users, $\mathcal{I} = \{i_1, \cdots, i_q\}$ denotes the set of q items, $\mathcal{Q} = \{q_1, \cdots, q_r\}$ denotes the set of r queries, $\mathcal{W} = \{w_1, \cdots, w_n\}$ denotes the set of n terms, $\mathcal{A}$ denotes the attributes associated with objects, and $\mathcal{B}$ denotes the interaction behaviors between different types of objects. In our application, a query $q \in \mathcal{Q}$ or an item $i \in I$ is constituted by several terms $w \in \mathcal{W}$. The purpose of intent recommendation is to recommend the most related intent (i.e., query) $q \in \mathcal{Q}$ to a user $u \in \mathcal{U}$.

Example 2 Taking Fig. 9.5a for example, for a user $u \in \mathcal{U}$, when he refreshes the App, we can utilize information from $\mathcal{A}$ and $\mathcal{B}$ to calculate the preference score of u for a candidate query $q \in \mathcal{Q}$, and recommend the query with the highest score as user intent to the user u. It is worth noting that the recommended query reflects user intent through exploiting user historical interaction information.

9.3.3 The MEIRec Model

9.3.3.1 Model Framework

The basic idea of the proposed model MEIRec is to design a heterogeneous GNN for enriching the representations of users and queries. With the help of HG built from intent recommendation system, MEIRec leverages meta-paths to guide the selection of different step neighbors and designs a heterogeneous GNN to obtain the rich embeddings of users and queries. Moreover, we represent different types of objects with uniform term embedding for less parameters learning, since queries and titles of items are constituted by a small number of terms.

Figure 9.6 shows the overall framework of MEIRec. First, we use the triple-object HG containing $< user, item, query >$ as input. Second, we use the uniform term embedding to generate the initial embeddings of items and queries. Third, we aggregate the information of meta-path guided neighbors to learn the embeddings of users and queries via heterogeneous GNN. After that, we fuse the embeddings

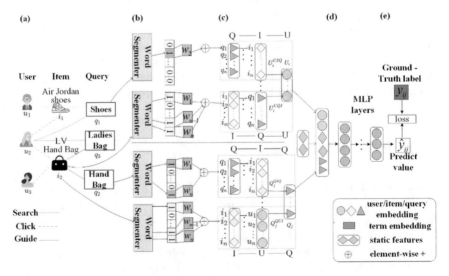

Fig. 9.6 The framework of MEIRec. (**a**) Input. (**b**) Term embedding layer. (**c**) Metapath-guided heterogeneous GNN layer. (**d**) Fusion layer. (**e**) Output layer

of users and queries based on different meta-paths, respectively. Finally, with the fused embeddings of users and queries, accompanying with static features of users and queries, we predict the probability that a user will search a specific query. We illustrate these steps in detail in the following sections.

9.3.3.2 Uniform Term Embedding

In the previous neural network-based recommendation, every user or query should have a unique embedding. In the intent recommendation scenario, there are billions of users and queries. If we employ traditional collaborative filtering or neural network-based methods to represent all users and queries, it will make the number of parameters tremendous. Note that queries and titles of items are constituted by terms and the number of terms is not many. So we propose to represent the queries and items with a small number of term embeddings. And thus we only need to learn the term embeddings, rather than all object embeddings. This method is able to significantly reduce the number of parameters.

Specifically, we extract terms from the queries and items' titles[3] and build a term lexicon $W = \{w_1, w_2, \cdots, w_{n-1}, w_n\}$. Note that queries and items (i.e., their titles) are the combination of several terms. For example, as shown in Fig. 9.6a and b, query "Hand Bag" is constituted by terms "Hand" and "Bag," and item "LV Hand Bag" is constituted by terms "LV," "Hand," and "Bag." Since the number of the lexicons W is far less than the number of the queries and users, the uniform term embedding can significantly reduce the number of learned parameters. More importantly, the new queries that have never been searched before can be represented by these terms.

9.3.3.3 Meta-Path Guided Heterogeneous Graph Neural Network

Inspired by the basic idea of the GCNs that generates node embeddings based on local neighbors [15, 28], we first propose a meta-path guided heterogeneous GNN. That is, we leverage meta-paths to obtain different step neighbors of an object, and the embeddings of users and queries are the aggregation of their neighbors under different meta-paths.

We present a toy example in Fig. 9.7 to illustrate this process. Here we describe how to obtain the embedding U_2 of user u_2 based on multiple meta-paths, such as UIQ and UQI. We first illustrate how we aggregate neighbor information along path UIQ. We use the uniform term embedding to obtain the initial embeddings of queries. And then we aggregate the meta-path guided neighbors to get the meta-path guided embedding of user u_2. According to the network structure in Fig. 9.5b, we get the 1-st step neighbors set of u_2, $\mathcal{N}_{\text{UIQ}}^1(u_2) = \{i_1, i_2\}$. For each node i_k

---

[3] Terms are important words or phrases. We use the AliWS (Alibaba Word Segmenter) to segment the queries and items' titles and select important words or phrases that contain rich meanings.

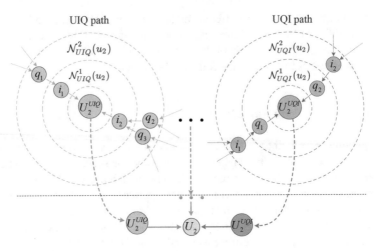

Fig. 9.7 A toy example of meta-path guided information aggregation

in the neighbor set $\mathcal{N}_{\text{UIQ}}^1(u_2)$, we extract the 2-nd step neighbor set $\mathcal{N}_{\text{UIQ}}^2(u_2) = \{q_1, q_2, q_3\}$. After we obtain the 1-st step and 2-nd step neighbor sets of u_2, we aggregate the embeddings of 2-nd step neighbors to obtain the 1-st step neighbors' embeddings. Finally, we aggregate the embeddings of 1-st step neighbors $\{i_1, i_2\}$ to obtain embedding U_2^{UIQ} of user u_2. Following this process, we can get different meta-path guided embeddings of u_2, such as U_2^{UQI}. Then we aggregate all the meta-path guided embeddings to get final embedding of u_2 (i.e., U_2).

9.3.3.4 User Modeling

In our model, we aggregate the information of different step neighbors to obtain the representation U_i of user u_i via meta-path guided heterogeneous GNN. In this section, we show how MEIRec models user embedding in detail.

As shown in the upper box in Fig. 9.6c, in order to get the embedding U_i of user u_i, we select meta-paths starting from target user. We first search different step neighbors along the meta-path and then aggregate the embeddings of neighbors step by step. Taking the meta-path UIQ (meaning user clicks the items that had been guided by queries) for example, we can obtain different step neighbors of a user u_i. After we get the 1-st step and 2-nd step neighbor sets, we aggregate the embeddings of 2-nd step neighbors (query) to obtain the 1-st step neighbors' (item) embeddings, and the embedding I_j^{UIQ} of item i_j in $\mathcal{N}_{\text{UIQ}}^1(u_i)$ based on the meta-path UIQ is

$$I_j^{\text{UIQ}} = g(E_{q_1}, E_{q_2}, \cdots), \tag{9.13}$$

where $g(\cdot)$ is the average aggregation function. And the queries $\{q_1, q_2, \cdots\}$ are the neighbors of item i_j.

Next, we aggregate 1-st step neighbors' (item) embeddings to obtain the embedding U_i^{UIQ} of user u_i:

$$U_i^{\text{UIQ}} = g(I_1^{\text{UIQ}}, I_2^{\text{UIQ}}, \cdots), \qquad (9.14)$$

where the items $\{i_1, i_2, \cdots\}$ are the neighbors of user u_i. Since users click queries or items with timestamp, we model the neighbors of users (i.e., items or queries) as a sequence data and utilize LSTM [2] to aggregate them.

Then we obtain the fused user embedding by aggregating embeddings based on different meta-paths $\{\rho_1, \rho_2, \cdots, \rho_k\}$:

$$U_i = g(U_i^{\rho_1}, U_i^{\rho_2}, \cdots, U_i^{\rho_k}), \qquad (9.15)$$

where the ρ is meta-path starting from user.

9.3.3.5 Query Modeling

Similar to user information aggregation, we also obtain the fused query embedding Q_i based on meta-paths $\{\rho_1, \rho_2, \cdots, \rho_k\}$:

$$Q_i = g(Q_i^{\rho_1}, Q_i^{\rho_2}, \cdots, Q_i^{\rho_k}), \qquad (9.16)$$

where the ρ is the meta-path starting from query.

9.3.3.6 Optimization Objective

In our model, we predict the probability $\hat{y}_{ij}$ of user u_i search the query q_j that is in the range of [0,1] to ensure that the output value is a probability. Through aggregating the neighbors of user and query, we obtain the fused user embedding U_i for user u_i and the fused query embedding Q_j for query q_j. In addition, there are raw static features used in traditional methods, including attributes of users (queries) and static features from interaction information. We feed these static features to a Multi-Layer Perceptron for obtaining the representation of the static features S_{ij}. Then, we concatenate the embeddings of user, query, and static features to fuse them. Finally, we feed the fused embeddings into MLP layers to get the predict score $\hat{y}_{ij}$. Then we have

$$\hat{y}_{ij} = sigmoid(f(U_i \oplus Q_j \oplus S_{ij})), \qquad (9.17)$$

where the $f(\cdot)$ is the MLP layers with only one output, $sigmoid(\cdot)$ is the sigmoid layer, and $\oplus$ is the embedding concatenate operation.

The loss function of our model is a point-wise loss function in Eq. 9.18.

$$J = \sum_{i,j \in \mathcal{Y} \cup \mathcal{Y}^-} \left(y_{ij} log \hat{y}_{ij} + (1 - y_{ij}) log(1 - \hat{y}_{ij}) \right), \qquad (9.18)$$

where y_{ij} is the label of the instance (i.e., 1 or 0) and the $\mathcal{Y}$ and the $\mathcal{Y}^-$ are the positive and negative instances set, respectively.

9.3.4 Experiments

9.3.4.1 Experimental Settings

Dataset We collect a real-world large-scale dataset from Taobao mobile application from Android and IOS online. We first extract static features for user and query. And we construct an HG based on interaction data collected during 10 days. For offline experiments, we utilize the interaction data during 5 days. Specifically, each raw interaction record in the collected dataset contains $<$ $user, query, timestamp, label >$ representing that the recommended query has been shown to user at timestamp. And the label indicates whether the user clicks the recommended query. Moreover, we use training data for different time periods (from 1 to 5 days) to predict the next one-day. Therefore, we have three datasets with different scales marked as 1-day, 3-day, and 5-day. The detailed statistics of the data are shown in Table 9.2.

Baselines and Evaluation Metrics To validate the effectiveness of our proposed model, we use the popular models used in industry (i.e., LR, DNN, and GBDT) with different feature settings and a popular neural network-based model NeuMF. In particular, LR/DNN/GBDT + DW/MP means that we feed the static features of users and queries, as well as the pre-training embeddings learned by

Table 9.2 The statistics of the datasets

Dataset	1-day	3-day	5-day
Training size (positive)	2,000,000	6,000,000	9,999,999
Training size (all)	8,000,000	23,999,998	39,999,997
Validation size (positive)	2,000,000	2,000,000	1,949,143
Validation size (all)	7,999,997	8,000,000	7,949,142
Train users	4,792,621	11,489,531	16,419,735
Train queries	871,133	1,653,865	2,163,574
Validation users	4,819,489	4,809,497	4,790,912
Validation queries	876,636	859,488	787,672
New users in validation set	3,666,692	2,613,695	2,064,564
Density	4.8×10^{-7}	3.1×10^{-7}	2.8×10^{-7}

DeepWalk (DW) [19]/MetaPath2vec (MP) [6] from structural information, into LR/DNN/GBDT model. In our experiments, we use AUC [17] to evaluate the performance of different models for comparison. The large AUC value means better performance.

9.3.4.2 Offline Performance Evaluation

The performances of MEIRec and the baselines are reported in Table 9.3. The major findings from the experimental results can be summarized as follows:

(1) MEIRec significantly outperforms all the compared baselines. Compared to the best performance of baselines (i.e., GBDT + MP or GBDT + DW, indicated with "*" at Table 9.3), MEIRec offers an improvement of 2.1%~4.3% in the three datasets. The results show that MEIRec achieves best results by using both static and structural features. It indicates that our model adopts a more comprehensive way to leverage static features and interaction relations for improving prediction performance.

(2) Among these baselines, we find that the order of overall performances is as follows: at the method level, GBDT > DNN > LR > NeuMF. Due to that NeuMF cannot learn the embeddings of new users and new queries appeared in the validation set, new objects' embeddings will be random variables, which makes the worst performances of NeuMF. And at the feature level, (static features + heterogeneous embeddings) based methods > (static features + homogeneous embeddings) based methods > static features based methods. This ranking indicates that fusing more information could usually get better performances. At both levels, we conclude that choosing a model plays a key role in intent recommendation, and adopting appropriate methods to fuse more information could significantly improve the performance. As a consequence, the MEIRec achieves best performances, due to the heterogeneous GNN model and utilization of rich heterogeneous interactions.

(3) As the scale of data increasing, our model outperforms the best baselines with an increased margin (from 2.1% to 4.3%). The result further confirms that our model is more scalable for large-scale datasets.

9.3.4.3 Online Experiments

To further evaluate the proposed model, we conduct online experiments in Taobao mobile App. We conduct a bucket testing (i.e., A/B testing) online to test the users' response to our model against baseline. We select one bucket for baseline, and another bucket for our model. And we select the GBDT model for comparison

Table 9.3 The AUC comparisons of different methods. The * indicates the best performance of the baselines. Best results of all methods are indicated in bold. The last row indicates the percentage of improvements gained by the proposed method compared to the best baseline

Method	1-day				3-day				5-day			
	40%	60%	80%	100%	40%	60%	80%	100%	40%	60%	80%	100%
NeuMF	0.6014	0.6066	0.6136	0.6143	0.6168	0.6218	0.6249	0.6291	0.6172	0.6224	0.6246	0.6295
LR	0.6854	0.6838	0.6884	0.6889	0.6844	0.6863	0.6857	0.6865	0.6817	0.6831	0.6827	0.6836
LR + DW	0.6878	0.6904	0.6898	0.6930	0.6888	0.6896	0.6898	0.6900	0.6838	0.6842	0.6863	0.6867
LR + MP	0.6918	0.6936	0.6950	0.6969	0.6919	0.6930	0.6933	0.6933	0.6874	0.6890	0.6898	0.6899
DNN	0.6939	0.6981	0.6991	0.6997	0.6966	0.6985	0.6999	0.7008	0.6996	0.7011	0.7017	0.7029
DNN + DW	0.6962	0.6980	0.7003	0.7024	0.7005	0.7017	0.7024	0.7030	0.7017	0.7029	0.7040	0.7047
DNN + MP	0.6984	0.6992	0.7024	0.7057	0.7025	0.7040	0.7051	0.7057	0.7017	0.7044	0.7060	0.7069
GBDT	0.7071	0.7071	0.7067	0.7073	0.7070	0.7071	0.7072	0.7071	0.7067	0.7068	0.7072	0.7066
GBDT + DW	0.7114	0.7119	0.7112*	0.7118*	0.7109	0.7106	0.7106	0.7104	0.7109	0.7112	0.7109	0.7114
GBDT + MP	0.7122*	0.7127*	0.7110	0.7111	0.7123*	0.7122*	0.7122*	0.7124*	0.7118*	0.7114*	0.7114*	0.7120*
MEIRec	**0.7273**	**0.7302**	**0.7339**	**0.7346**	**0.7352**	**0.7369**	**0.7380**	**0.7390**	**0.7372**	**0.7401**	**0.7409**	**0.7425**
Improvement	2.1%	2.5%	3.2%	3.2%	3.2%	3.5%	3.6%	3.7%	3.6%	4.0%	4.1%	4.3%

Table 9.4 Online A/B testing experiment results

Data	Methods	CTR	Unique click	UCTR
Android	GBDT	1.746%	256,116	13.939%
	MEIRec	1.758%	260,634	14.229%
	Improvement	0.70%	1.76%	2.07%
IOS	GBDT	0.7687%	62,462	5.2579%
	MEIRec	0.8056%	65,895	5.5436%
	Improvement	4.79%	5.50%	5.43%
Total	GBDT	1.4035%	318,578	10.5252%
	MEIRec	1.4252%	326,529	10.8052%
	Improvement	1.54%	2.50%	2.66%

for that GBDT is used in real system. We use the metric CTR, Unique Click,[4] and UCTR to evaluate the online performance, where CTR and UCTR = Unique Click/Unique Visitor indicate change of the click ratio and visit ratio.

The results are shown in Table 9.4. We can see that, compared to the GBDT, MEIRec achieves performance improvement in all metrics, which indicates that incorporating interaction information can better capture user latent intent. Our model gains the improvement of 0.70%, 4.79%, and 1.54% for Android, IOS, and Total, respectively, in CTR. Since the CTR is to measure the ratio of clicks against impressions, the improvement of CTR shows that our model can greatly improve the user's search experience. In addition, the metric UCTR indicates how many unique visitors click the recommended query, and it gains an improvement of 2.07%, 5.43%, and 2.66% for Android, IOS, and Total. The improvement of UCTR shows that our model has an advantage in attracting new users to search queries.

The more detailed method description and experiment validation can be seen in [7].

9.4 Share Recommendation

9.4.1 Overview

With the development of social e-commerce, a new recommendation paradigm, share recommendation, has sprung up recently. In particular, share recommendation aims to predict whether a user will share an item with his friend. Such recommendation demand is ubiquitous in social e-commerce. The share recommendation is significantly different from traditional recommendations, such as item recommendation [27] and friend recommendation [26]. As shown in Fig. 9.8, we can find

---

[4] The number of visitors who performed a click.

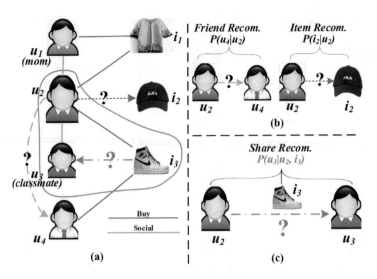

Fig. 9.8 Share recommendation vs. previous recommendations. (**a**) A toy example of recommendation. (**b**) Previous recommendation. (**c**) Share recommendation

that item recommendation aims to recommend an item to a user (i.e., essentially maximize the probability $P(i_2|u_2)$) and friend recommendation aims to recommend a friend to a user (i.e., maximize the probability $P(u_4|u_2)$). Significantly different from the above binary recommendations, the goal of share recommendation is to predict the ternary interactions among $\langle User, Item, Friend \rangle$, i.e., whether a user will share an item with his friend, maximizing the probability $P(u_3|u_2, i_3)$.

Deliberately considering the characteristics of share recommendation, we need to address the following challenges for modeling share recommendation. **Rich Heterogeneous Information.** Share recommendation usually contains complex heterogeneous information, including complex interactions among users and items, as well as rich feature information of users and items. **Complex ternary interaction.** Different from simple binary interaction in traditional recommendations, exemplified as $\langle u_2, i_2 \rangle$ interaction in the item recommendation and $\langle u_2, u_4 \rangle$ interaction in friend recommendation in Fig. 9.8, share recommendation faces complex ternary interaction (e.g., $\langle u_2, i_3, u_3 \rangle$ in Fig. 9.8). We need to consider the suitability of a share action, which evaluates the matching degree of three objects (e.g., u_2, i_3, u_3) in the share action. According to the characteristic of the recommended item, a user will recommend it to an appropriate friend, and thus how the item influences the user (or the friend) should be considered. **Asymmetric Share Action.** The share action is asymmetric and irreversible, which means the share action may not happen if we swap the roles of the user and the friend.

In this section, we first study the problem of share recommendation and propose a novel **H**eterogeneous **G**raph neural network-based **S**hare **Rec**ommendation model (**HGSRec**). We model the share recommendation system as an attributed heterogeneous graph to integrate rich heterogeneous information, and then we design HGSRec to learn the embeddings of u, i, v and predict the probability of share action $\langle u, i, v \rangle$ happening. Specifically, after initializing node embedding via encoding rich node features, a tripartite heterogeneous GNN is designed to learn the embeddings of u, i, v, respectively, via aggregating their meta-path based neighbors, which enables HGSRec to flexibly fuse different aspects of information. Furthermore, a dual co-attention mechanism is proposed to dynamically fuse the multiple embeddings of u (or v) under different meta-paths, considering the influence of item i to user u (or v), to improve the suitability of $\langle u, i, v \rangle$. Finally, a transitive triplet representation of $\langle u, i, v \rangle$ is employed to predict whether share action happens.

9.4.2 Problem Formulation

Definition 4 Share Recommendation. Given an attributed heterogeneous graph $\mathcal{G} = (\mathcal{V}, \mathcal{E}, \mathbf{X})$ representing a share recommendation system, share recommendation aims to predict a share action $\langle u, i, v \rangle$ (formulated with $\langle User, Item, Friend \rangle$, or abbreviated with $\langle U, I, V \rangle$). Specifically, the purpose of share recommendation is to recommend the most likely $Friend$ $v \in \mathcal{F}(u)$ to $User$ $u \in \mathcal{V}_U$ who would like to share the $Item$ $i \in \mathcal{V}_I$ ($\langle u, i \rangle \in \mathcal{E}_O$), i.e., $\arg\max_v P(v|u, i)$. The label $y_{u,i,v} \in \{0, 1\}$ indicates whether share action happens.

Example 3 Figure 9.9a shows the attributed heterogeneous graph of share recommendation. Here u_2 has two friends denoting as $\mathcal{F}(u_2) = \{u_1, u_3\}$. Meta-path [23], a composite relation connecting two nodes, is able to extract rich semantics. As shown in Fig. 9.9b, $User \xrightarrow{buy} Item \xrightarrow{buy} User$ (U-b-I-b-U for short) means the co-buying relations, $User \xrightarrow{social} User$ (U-s-U for short) means the social relations, $User \xrightarrow{buy} Item$ (U-b-I for short) means buy relations, and $User \xrightarrow{view} Item \xrightarrow{view} User$ (U-v-I-v-U for short) means the co-viewing relations. As shown in Fig. 9.8c, share recommendation will recommend a most likely friend, like $u_3 \in \mathcal{F}(u_2)$, to a user u_2 who would like to share the shoes i_3, which essentially maximizes the probability $P(u_3|u_2, i_3)$.

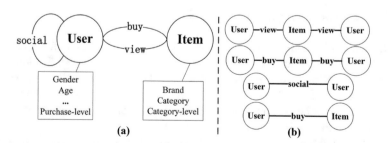

Fig. 9.9 A typical example for share recommendation. (**a**) Schema. (**b**) Meta-paths

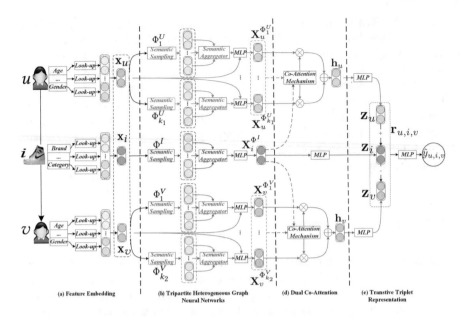

Fig. 9.10 The overall framework of the proposed HGSRec. (**a**) Initializing user and item embedding via feature embedding. (**b**) Updating node embedding via tripartite heterogeneous graph neural networks. (**c**) Fusing embedding dynamically via the dual co-attention mechanism. (**d**) Modeling asymmetric share action via transitive triplet representation

9.4.3   The HGSRec Model

9.4.3.1   Model Framework

The overall framework of HGSRec is shown in Fig. 9.10. Given a share action $<$ $u, i, v >$, the basic idea of HGSRec is to learn the embeddings of u, i, v to predict the probability of the action happening, with the help of delicate designs, such as tripartite heterogeneous GNNs, dual co-attention mechanism, and transitive triplet representation.

9.4.3.2   Initialization with Feature Embedding

First, we initialize node embedding via embedding their features. Different from ID embedding, feature embedding has two-fold benefits: (1) In real applications, there are numerous of newly coming nodes every day. The feature embedding effectively generates embeddings for previously unseen nodes by utilizing their features. (2) The number of features is much less than the number of nodes, which significantly reduces the number of learnable parameters.

For the k-th node feature $f_k \in \mathbb{R}^{|f_k|*1}$, we initialize a feature embedding matrix $\mathbf{M}^{f_k} \in \mathbb{R}^{d*|f_k|}$, where $|f_k|$ means the number of values of feature f_k and d is the dimension of feature embedding. The embedding of u's k-th feature is shown as follows:

$$\mathbf{e}_u^{f_k^U} = \mathbf{M}^{f_k^U} \cdot u^{f_k^U}. \tag{9.19}$$

Considering all the features of user u, we can get the initial user embedding $\mathbf{x}_u$, as follows:

$$\mathbf{x}_u = \sigma \left(\mathbf{W}_U \cdot \left(\overset{|f^U|}{\underset{k=1}{\|}} \mathbf{e}_u^{f_k^U} \right) + \mathbf{b}_U \right), \tag{9.20}$$

where $\|$ denotes the concatenation operation, and $\mathbf{W}_U$ and $\mathbf{b}_U$ denote the weight matrix and bias vector, respectively. The same process can be done for item/friend embedding.

9.4.3.3 Tripartite Heterogeneous Graph Neural Networks

Here we propose tripartite heterogeneous GNNs to learn embeddings of u, i, v via the corresponding heterogeneous GNN (i.e., $HeteGNN^U$, $HeteGNN^I$, and $HeteGNN^V$), respectively. Heterogeneous GNN usually follows a hierarchical manner: It first aggregates information from one kind of neighbors via one meta-path and learns the semantic-specific node embeddings in node-level. Then, it aggregates multiple semantics from different meta-paths and fuses a set of semantic-specific node embeddings in semantic-level.

Specifically, given one user u and k_1 user-related meta-paths $\{\Phi_1^U, \Phi_2^U, \cdots, \Phi_{k_1}^U\}$, $HeteGNN^U$ is able to get k_1 semantic-specific user embeddings $\{\mathbf{x}_u^{\Phi_1^U}, \mathbf{x}_u^{\Phi_2^U}, \cdots, \mathbf{x}_u^{\Phi_{k_1}^U}\}$.

$$\mathbf{x}_u^{\Phi_1^U}, \mathbf{x}_u^{\Phi_2^U}, \cdots, \mathbf{x}_u^{\Phi_{k_1}^U} = HeteGNN^U(u). \tag{9.21}$$

Note that the number of meta-path based neighbors of different nodes could be quite different, so we need to sample a fixed number of neighbors. Random sampling strategy causes heavy computation consumption and missing important nodes. Here we propose a top-N semantic sampling strategy: (1) If the number of meta-path based neighbors is more than fixed number N, we sample top-N meta-path based neighbors based on connection strength (e.g., how many times a user views an item). (2) Or else, we adopt resample to get N meta-path based neighbors.

Given a user u and the corresponding meta-path Φ^U, we propose a novel semantic aggregator $SemAgg_u^{\Phi^U}$ to aggregate sampled neighbors $\mathcal{N}_u^{\Phi^U}$ and obtain the meta-path based embedding $\mathbf{x}_u^{\mathcal{N}_u^{\Phi^U}}$, as follows:

$$\mathbf{x}_u^{\mathcal{N}_u^{\Phi^U}} = SemAgg_u^{\Phi^U}(\{\mathbf{x}_n | \forall n \in \mathcal{N}_u^{\Phi^U}\}). \tag{9.22}$$

Considering the time efficiency, we adopt $MeanPooling$ to accelerate aggregating processing for faster prediction. To emphasize the property of user u explicitly, we concatenate initial embedding $\mathbf{x}_u$ and meta-path based embedding $\mathbf{x}_u^{\mathcal{N}_u^{\Phi^U}}$ and get the semantic-specific user embedding $\mathbf{x}_u^{\Phi^U}$,

$$\mathbf{x}_u^{\Phi^U} = \sigma(\mathbf{W}^{\Phi^U} \cdot (\mathbf{x}_u || \mathbf{x}_u^{\mathcal{N}_u^{\Phi^U}}) + \mathbf{b}^{\Phi^U}), \tag{9.23}$$

where $\mathbf{W}^{\Phi^U}$ and $\mathbf{b}^{\Phi^U}$ denote the weight matrix and bias vector for meta-path Φ^U, respectively. Given a set of user-related meta-paths $\{\Phi_1^U, \Phi_2^U, \cdots, \Phi_{k_1}^U\}$, we can get k_1 semantic-specific user embeddings $\{\mathbf{x}_u^{\Phi_1^U}, \mathbf{x}_u^{\Phi_2^U}, \cdots, \mathbf{x}_u^{\Phi_{k_1}^U}\}$ that describe the characteristics of user u from different aspects. The same process can be done via $HeteGNN^V$ to learn multiple semantic-specific embeddings $\{\mathbf{x}_v^{\Phi_1^V}, \mathbf{x}_v^{\Phi_2^V}, \cdots, \mathbf{x}_v^{\Phi_{k_2}^V}\}$ of friend v. Since the characteristic of the item is much simple and stable than the user, we only adopt one meta-path Φ^I to get the embedding $\mathbf{x}_i^{\Phi^I}$ of item i via $HeteGNN^I$.

9.4.3.4   Dual Co-attention Mechanism

After obtaining a set of semantic-specific node embeddings (e.g., $\{\mathbf{x}_u^{\Phi_1^U}, \mathbf{x}_u^{\Phi_2^U}, \cdots, \mathbf{x}_u^{\Phi_{k_1}^U}\}$), we aim to fuse them properly based on the complex ternary interactions $\langle u, i, v \rangle$. So a dual co-attention mechanism is designed to dynamically fuse the embeddings of u (or v) under different meta-paths, considering the effect of item i, which consists of co-attention mechanism $CoAtt_{U,I}$ for $\langle U, I \rangle$ and co-attention mechanism $CoAtt_{V,I}$ for $\langle V, I \rangle$. Specifically, it learns the interaction-specific attention values of meta-paths for $\langle u, i, v \rangle$ and gets the most appropriate embedding of u, v, with the following benefits: (1) It reinforces the dependency of $\langle u, i, v \rangle$, making HGSRec more integrated. (2) It dynamically fuses the embeddings of u (or v), improving share suitabilities.

Taking $\langle U, I \rangle$ as an example, the co-attention mechanism $CoAtt_{U,I}$ aims to learn a set of interaction-specific co-attention weights $\{w_{u,i}^{\Phi_1^U}, w_{u,i}^{\Phi_2^U}, \cdots, w_{u,i}^{\Phi_{k_1}^U}\}$ for user u,

$$w_{u,i}^{\Phi_1^U}, w_{u,i}^{\Phi_2^U}, \cdots, w_{u,i}^{\Phi_{k_1}^U} = \text{CoAtt}_{U,I}(\mathbf{x}_u^{\Phi_1^U}, \cdots, \mathbf{x}_u^{\Phi_{k_1}^U}, \mathbf{x}_i^{\Phi^I}). \tag{9.24}$$

Specifically, we concatenate the semantic-specific embedding of u and i and project them into co-attention space. Then, we adopt a co-attention vector $\mathbf{q}_{U,I}$ to learn the importances of meta-paths for user u. The importance of meta-path Φ_m^U for u in the interaction $\langle u, i \rangle$, denoted as $\alpha_{u,i}^{\Phi_m^U}$,

$$\alpha_{u,i}^{\Phi_m^U} = \mathbf{q}_{U,I}^T \cdot \sigma(\mathbf{W}^{U,I} \cdot (\mathbf{x}_u^{\Phi_m^U} || \mathbf{x}_i^{\Phi^I}) + \mathbf{b}^{U,I}), \tag{9.25}$$

where $\mathbf{W}^{U,I}$ and $\mathbf{b}^{U,I}$ denote the weight matrix and bias vector, respectively. After obtaining the importances of meta-paths, we normalize them via softmax to get the co-attention weight $w_{u,i}^{\Phi_m^U}$ of meta-path Φ_m^U, shown as follows:

$$w_{u,i}^{\Phi_m^U} = \frac{\exp(\alpha_{u,i}^{\Phi_m^U})}{\sum_{m=1}^{k_1} \exp(\alpha_{u,i}^{\Phi_m^U})}, \tag{9.26}$$

where $w_{u,i}^{\Phi_m^U}$ reflects the contribution of meta-path Φ_m^U in improving share suitability. With the learned weights as coefficients, we can obtain the fused embedding $\mathbf{h}_u$ of u, shown as follows:

$$\mathbf{h}_u = \sum_{m=1}^{k_1} w_{u,i}^{\Phi_m^U} \cdot \mathbf{x}_u^{\Phi_m^U}. \tag{9.27}$$

Similar to $CoAtt_{U,I}$, $CoAtt_{V,I}$ learns a set of co-attention weights $\{w_{v,i}^{\Phi_1^V}, w_{v,i}^{\Phi_2^V}, \cdots, w_{v,i}^{\Phi_{k_2}^V}\}$ for friend v and gets the fused friend embeddings $\mathbf{h}_v$. Since we only select one meta-path for item, the fused embedding $\mathbf{h}_i$ of item i is actually $\mathbf{x}_i^{\Phi^I}$.

9.4.3.5 Transitive Triplet Representation

To predict the share action $\langle u, i, v \rangle$, we need to construct a triplet representation $\mathbf{r}_{u,i,v}$ based on $\mathbf{h}_u, \mathbf{h}_i, \mathbf{h}_v$. We first project all types of nodes in $\langle U, I, V \rangle$ into the same space via three type-specific MLPs, shown as follows:

$$\mathbf{z}_u = MLP^U(\mathbf{h}_u), \quad \mathbf{z}_i = MLP^I(\mathbf{h}_i), \quad \mathbf{z}_v = MLP^V(\mathbf{h}_v). \tag{9.28}$$

A simple way to construct the triplet representation $\mathbf{r}_{u,i,v}$ is to concatenate all node embeddings (a.k.a., $\mathbf{z}_u||\mathbf{z}_i||\mathbf{z}_v$). However, the simple concatenation cannot explicitly capture the remarkable characteristics of share action. Inspired by relational translation [1], we propose a transitive triplet representation $\mathbf{r}_{u,i,v}$ to explicitly model the characteristics of share action via item-translating, shown as follows:

$$\mathbf{r}_{u,i,v} = |\mathbf{z}_u + \mathbf{z}_i - \mathbf{z}_v|, \tag{9.29}$$

where $|\cdot|$ denotes the absolute operation. Then, we feed $\mathbf{r}_{u,i,v}$ into MLP and get the predict score $\hat{y}_{u,i,v}$, as follows:

$$\hat{y}_{u,i,v} = \sigma(\mathbf{W} \cdot \mathbf{r}_{u,i,v} + b), \tag{9.30}$$

where $\mathbf{W}$ and b denote the weight vector and bias scalar, respectively. Finally, we calculate the cross-entropy loss,

$$L = \sum_{u,i,v \in \mathcal{D}} (y_{u,i,v} \log \hat{y}_{u,i,v} + (1 - y_{u,i,v}) \log (1 - \hat{y}_{u,i,v})), \tag{9.31}$$

where $y_{u,i,v}$ is the label of the triplet and $\mathcal{D}$ denotes the dataset.

9.4.4 Experiments

9.4.4.1 Experimental Settings

Dataset We collect data from Taobao platform, ranging from 2019/10/09 to 2019/10/14, and construct an AHG (shown in Fig. 9.9). Each sample contains a share action $\langle u, i, v \rangle$ and the corresponding label $y_{u,i,v} \in \{0, 1\}$. We select four meta-paths including *U-s-U*, *U-b-I-b-U*, and *U-v-I-v-U* for the user and *U-b-I* for the item. In offline experiments, we use the last day (i.e., 2019/10/14) as validation set and the previous 3/4/5 days as training sets, marked as **3-day**, **4-day**, and **5-day**, respectively. The details of the datasets are shown in Table 9.5.

Table 9.5 The statistics of the datasets

Dataset	3-day	4-day	5-day
#Train $\langle u, i, v \rangle$	3,324,367	4,443,996	5,611,531
#Train *Users*	1,064,426	1,315,126	1,546,017
#Train *Items*	537,048	679,784	818,290
#Valid $\langle u, i, v \rangle$	1,401,395		
#Valid *Users*	539,959		
#Valid *Items*	247,907		

Baselines We select feature based models (i.e., LR, DNN, and XGBoost) and GNN models (i.e., GraphSAGE, IGC, and MEIRec) as baselines. Since IGC and MEIRec cannot handle ternary recommendation, we also provide tripartite versions (i.e., IGC+ and MEIRec+) for share recommendation. To validate delicate designs in HGSRec, we also test two variants of HGSRec (HGSRec$_{\setminus att}$ and HGSRec$_{\setminus tra}$).

Evaluation Metrics and Hyper-parameter Settings We select AUC as the evaluation metric and RMSProp as optimizer. We uniformly set feature embedding to 8, node embedding to 128, batch size to 1024, learning rate to 0.01, and dropout rate to 0.6 for deep models. For XGBoost, we set tree depth to 6 and tree number to 10. For LR, we set the L1 reg to 1. For HeteGNNs, we sample 5, 10, 2 neighbors via *U-s-U*, *U-v-I-v-U*, *U-b-I-b-U* to learn multiple user embeddings and sample 50 neighbors via *U-b-I* to learn item embedding.

9.4.4.2 Offline Performance Evaluation

As shown in Table 9.6, we have the following observations: (1) HGSRec consistently performs better than all baselines with significant improvements. Compared to the best baseline, the improvements are up to 11.7%–14.5%, indicating the superiority of HGSRec. (2) Most of GNNs (i.e., GraphSAGE, IGC, and MEIRec) outperform feature based methods (i.e., LR, DNN, and XGBoost), indicating the importance of structure information. When deeper insight into these methods, we can find, if employing ternary interactions, the tripartite versions (i.e., IGC+ and MEIRec+) significantly outperform the original versions. It further confirms the benefits of modeling ternary interaction for share recommendation. (3) Comparing the performance of HGSRec with its variants, we can find HGSRec achieves the best performance. The degradation of HGSRec$_{\setminus att}$ indicates the effectiveness of the dual co-attention mechanism, while the degradation of HGSRec$_{\setminus tra}$ validates the superiority of transitive triplet representation. Note that the degradation of HGSRec$_{\setminus tra}$ is much more significant than that of HGSRec$_{\setminus att}$, which implies that transitive triple representation may make higher contribution than dual co-attention mechanism.

9.4.4.3 Attention Analysis

The dual co-attention mechanism can dynamically fuse multiple embeddings of *User* and *Friend* with regard to different *Items* and improve the share suitabilities. We first present the macro-level analysis via the box-plot figure of attention distributions over *User* on 3-day dataset in Fig. 9.11a. Note that attention value distributions over *Friend* also show similar phenomena. As can be seen, the attention distributions of meta-paths are different, and the attention values of *U-b-I-b-U* are the largest with a higher variance, which illustrates that this meta-path is the most important for most users. The reason is that *U-b-I-b-U* is related to

Table 9.6 The AUC comparisons of different methods. The best results of all methods are indicated in bold

Model	3-day				4-day				5-day			
	40%	60%	80%	100%	40%	60%	80%	100%	40%	60%	80%	100%
LR	67.56	67.62	67.26	67.69	67.58	67.65	67.68	67.72	67.62	67.67	67.72	67.74
XGBoost	72.04	72.14	72.13	72.18	72.08	72.11	72.15	72.49	72.72	72.54	71.78	72.14
DNN	71.30	71.20	71.67	72.03	71.04	71.33	71.48	71.80	70.96	71.12	71.46	71.51
SAGE	70.55	70.97	70.86	70.89	69.82	69.69	70.46	71.03	69.11	69.66	71.25	71.06
IGC	62.23	61.78	62.20	62.25	61.87	62.30	63.11	63.17	62.60	62.91	63.11	63.15
IGC+	73.15	73.37	73.92	74.34	73.87	73.99	74.22	74.51	74.14	74.22	74.53	74.79
MEIRec	64.94	65.10	65.30	65.53	65.45	65.55	65.66	65.72	65.19	65.58	66.20	65.63
MEIRec+	76.82	77.40	77.06	78.29	76.97	77.75	76.87	76.36	76.58	77.29	76.63	77.66
HGSRec$_{\backslash att}$	86.63	86.95	87.16	87.26	87.00	87.27	87.31	87.51	87.11	87.23	87.34	87.59
HGSRec$_{\backslash tra}$	78.17	79.10	79.50	79.95	76.40	79.12	77.09	79.63	78.22	78.89	78.83	81.37
HGSRec	**86.84**	**87.20**	**87.36**	**87.45**	**87.05**	**87.39**	**87.43**	**87.69**	**87.27**	**87.53**	**87.72**	**87.92**
Impro(%).	13.0	12.7	13.4	11.7	13.1	12.4	13.7	14.8	14.0	13.2	14.5	13.2

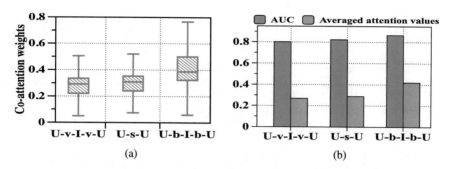

Fig. 9.11 The attention analysis on 3-day dataset. (**a**) Attention distributions. (**b**) Performance and averaged attention values

user purchasing behavior that reflects the strongest user preference. The higher variance of *U-b-I-b-U* also implies its importances vary greatly for different users. We further test HGSRec with single meta-path and show their performances with the corresponding averaged attention values in Fig. 9.11b. Consistent with attention distribution, *U-b-I-b-U* is the most useful meta-path that achieves the highest AUC and gets the largest attention value.

9.4.4.4 Online Experiments

We deploy HGSRec on Taobao APP for online share recommendation and compare HGSRec with XGBoost via online A/B testing. Online service needs to satisfy the following requirements: (1) Storage and processing for massive data. Share recommendation system is stored on MaxCompute as adjacency list for memory efficiency. (2) Abnormal share action. We filter abnormal share actions (e.g., a user shares more than thousands of items with his friend within 24 h). (3) New feature and missing feature. New features come every day, so we leverage hash function to map all features, leading a slight loss of performance when hash collision happens. Missing features are padded with a specific *token*.

The online results range from 2020/01/08 to 2020/02/02 (25 days) are shown in Fig. 9.12. Here we select UCTR (UCTR = Unique Click/Unique Visitor) for online evaluation. The larger UCTR, the better performance. The long-term observations show that HGSRec consistently outperforms XGBoost with a significant gap, demonstrating the high industrial practicability and stability of HGSRec.

The more detailed method description and experiment validation can be seen in [12].

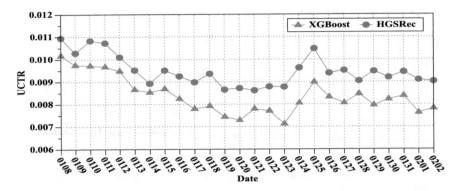

Fig. 9.12 The results of online A/B testing

9.5 Friend-Enhanced Recommendation

9.5.1 Overview

Nowadays, with the thriving of online social networks, people are more willing to actively express their opinions and share information with friends on social platforms. Friends become essential information sources and high-quality information filters. Impressed by the great successes of social influence in recommendation, a novel scenario named **Friend-Enhanced Recommendation (FER)** is proposed, which multiplies the influence of friends in social recommendation. FER has two major differences from the classical social recommendation: (1) FER only recommends to the user what his/her friends have interacted with, regarding friends as high-quality information filters to provide more high-quality items. (2) All friends who have interacted with the item are explicitly displayed to the user attached to the recommended item, which highlights the critical importance of explicit social factors and improves the interpretability for user behaviors.

In recent years, FER systems are blooming and have been widely used by hundreds of millions of users. Figure 9.13 gives a typical illustration of a real-world FER. For each user–item pair, FER explicitly shows the friend set having interacted with the item, which is defined as the **friend referral circle (FRC)** of the user to the item. In FER, multiple factors contribute to user clicks. The reasons for a user clicking an article may come from (1) his interests in item contents (item), (2) the recommendation of an expert (item–friend combination), or even (3) the concerns on his friends themselves (friend). In FER, users have the tendency to see *what their friends have read*, rather than to merely see *what themselves are interested in*. It could even say that social recommendation focuses on bringing social information to better recommend items, while FER aims to recommend the combination of both items and friend referrals.

As the critical characteristic of FER, the explicit FRC brings in two challenges: (1) *How to extract key information from multifaceted heterogeneous factors?* (2)

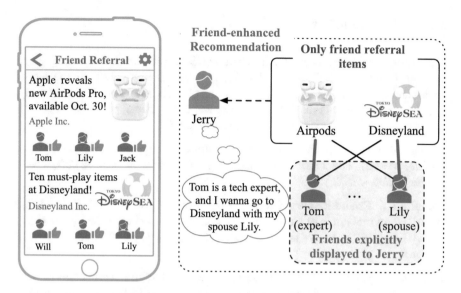

Fig. 9.13 A typical illustration of the friend-enhanced recommendation. The left shows the scenario that *Jerry* is recommended two articles, with friends (e.g., *Tom*) who have interacted with (shared, liked, etc.) them explicitly shown underneath. The right shows the formalization of the FER problem, where only friend referral items will be recommended and friends who interacted with the item are explicitly displayed to user

How to exploit explicit friend referral information? To solve these issues, we propose a novel **S**ocial **I**nfluence **A**ttentive **N**eural network (named **SIAN**). Specifically, we define the FER as a user–item interaction prediction task on a heterogeneous social graph, which flexibly integrates rich information in heterogeneous objects and their interactions. First, we design an attentive feature aggregator with both node- and type-level aggregations to learn user and item representations, without being restricted to pre-defined meta-paths in some previous efforts [6, 28]. Next, we implement a social influence coupler to model the coupled influence diffusing through the explicit friend referral circles, which combines the influences of multiple factors (e.g., friends and items) with an attentive mechanism. Overall, SIAN captures valuable multifaceted factors in FER, which successfully distills the most essential preferences of users from a heterogeneous graph and friend referral circles. In experiments, SIAN significantly outperforms all competitive baselines in multiple metrics on three large, real-world datasets. Further quantitative analyses on attentive aggregation and social influence also reveal impressive sociological discoveries.

9.5.2 Preliminaries

Definition 5 Heterogeneous Social Graph (HSG). A heterogeneous social graph is denoted as $\mathcal{G} = (\mathcal{V}, \mathcal{E})$, where $\mathcal{V} = \mathcal{V}_U \cup \mathcal{V}_I$ and $\mathcal{E} = \mathcal{E}_F \cup \mathcal{E}_R$ are the sets of nodes and edges. Here $\mathcal{V}_U$ and $\mathcal{V}_I$ are the sets of users and items. For $u, v \in \mathcal{V}_U$, $\langle u, v \rangle \in \mathcal{E}_F$ represents the friendship between users. For $u \in \mathcal{V}_U$ and $i \in \mathcal{V}_I$, $\langle u, i \rangle \in \mathcal{E}_R$ is the interaction relation between u and i.

Example 4 Figure 9.13 shows an HSG containing three types of nodes, i.e., {*User, Article, Media*}, and multiple relations, e.g., {*User–User, User–Article, User–Media, Article–Media*}.

Definition 6 Friend Referral Circle (FRC). Given an HSG $\mathcal{G} = (\mathcal{V}, \mathcal{E})$, we define the friend referral circle of a user u w.r.t. a non-interacting item i (i.e., $\langle u, i \rangle \notin \mathcal{E}_R$) as $\mathcal{C}_u(i) = \{v | \langle u, v \rangle \in \mathcal{E}_F \cap \langle v, i \rangle \in \mathcal{E}_R\}$. Here v is called an **influential friend** of user u.

Example 5 Taking Fig. 9.13 as an example, the friend referral circle of *Jerry* w.r.t. the non-interacting article about *AirPods* is {*Tom, Lily, Jack*}, while the FRC in terms of the article about *Disneyland* is $\mathcal{C}_{Jerry}(Disneyland) = \{Will, Tom, Lily\}$.

Definition 7 Friend-Enhanced Recommendation (FER). Given an HSG $\mathcal{G} = (\mathcal{V}, \mathcal{E})$ and the FRC $\mathcal{C}_u(i)$ of a user u w.r.t. a non-interacting item i, the FER aims to predict whether user u has a potential preference to item i. That is, a prediction function $\hat{y}_{ui} = \mathcal{F}(\mathcal{G}, \mathcal{C}_u(i); \Theta)$ is to be learned, where $\hat{y}_{ui}$ is the probability that user u will interact with item i, and Θ is the model parameters.

9.5.3 The SIAN Model

9.5.3.1 Model Framework

As illustrated in Fig. 9.14, SIAN models the FER with an HSG. In addition to the user and item representations (e.g., $\mathbf{h}_u$ for *Jerry* and $\mathbf{h}_i$ for the *Disneyland* article), SIAN learns a social influence representation (e.g., $\mathbf{h}_{ui}$) by coupling each influential friend (e.g., *Tom*) with the item. They are jointly responsible for predicting the probability $\hat{y}_{ui}$ of interaction between user u and item i. First, each user or item node is equipped with an attentive feature aggregator with node- and type-level aggregations, which is designed to exploit multifaceted information. Second, the influence from an influential friend (e.g., *Tom*) and an item (e.g., the *Disneyland* article) is jointly captured with a social influence coupler, which quantifies the degree of their coupled influence.

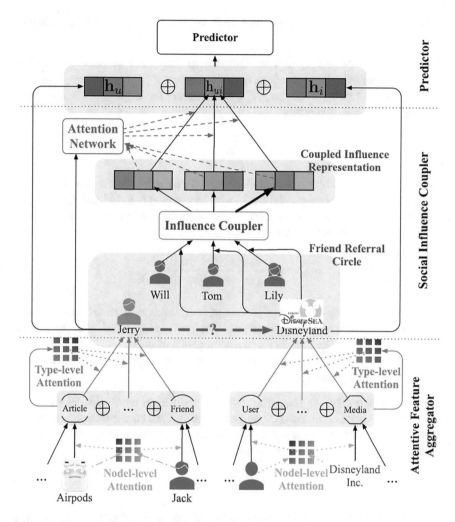

Fig. 9.14 The overall architecture of SIAN. The attentive feature aggregator hierarchically aggregates heterogeneous neighbor features with node- and type-level attention and outputs the representations of users and items (i.e., $\mathbf{h}_u$ and $\mathbf{h}_i$). The social influence coupler couples the influence of each influential friend and the item, to encode the explicit social influence into the representation (i.e., $\mathbf{h}_{ui}$)

9.5.3.2 Attentive Feature Aggregator

Given an HSG $\mathcal{G} = \{\mathcal{V}, \mathcal{E}\}$, attentive feature aggregator aims to learn user and item representations (i.e., $\mathbf{h}_u$ and $\mathbf{h}_i$, $u, i \in \mathcal{V}$). Considering that different neighbors of the same type might not equally contribute to the feature aggregation, and different types entail multifaceted information, we design a hierarchical node- and type-level attentive aggregation. Node-level aggregation separately models user/item features in a fine-grained manner, while type-level aggregations capture heterogeneous information.

Node-Level Attentive Aggregation Formally, given a user u, let $\mathcal{N}_u = \mathcal{N}_u^{t_1} \cup \mathcal{N}_u^{t_2} \cup \cdots \cup \mathcal{N}_u^{t_{|\mathcal{T}|}}$ denote his/her neighbors, which is a union of $|\mathcal{T}|$ types of neighbor sets. For neighbors of type $t \in \mathcal{T}$ (i.e., $\mathcal{N}_u^t$), we represent the aggregation in the t type space as the following function:

$$\mathbf{p}_u^t = \text{ReLU}(\mathbf{W}_p(\sum_{k \in \mathcal{N}_u^t} \alpha_{ku} \mathbf{x}_k) + \mathbf{b}_p), \tag{9.32}$$

where $\mathbf{p}_u^t \in \mathbb{R}^d$ is the aggregated embeddings of user u in t type space. $\mathbf{x}_k \in \mathbb{R}^d$ is the initial embedding of the neighbor k, which is randomly initialized. Here $\mathbf{W}_p \in \mathbb{R}^{d \times d}$ and $\mathbf{b}_p \in \mathbb{R}^d$ are the weight and bias of a neural network. α_{ku} is the attentive contribution of neighbor k to the feature aggregation of u,

$$\alpha_{ku} = \frac{\exp(f([\mathbf{x}_k \oplus \mathbf{x}_u]))}{\sum_{k' \in \mathcal{N}_u^t} \exp(f([\mathbf{x}_{k'} \oplus \mathbf{x}_u]))}, \tag{9.33}$$

where $f(\cdot)$ is a two-layer neural network activated with ReLu function and $\oplus$ denotes the concatenation operation. Obviously, the larger α_{ku}, the greater contribution of neighbor k to the feature aggregation of user u.

Given multiple types of neighbors, we can get multiple embeddings for u in various type spaces, denoted as $\{\mathbf{p}_u^{t_1}, \cdots, \mathbf{p}_u^{t_{|\mathcal{T}|}}\}$.

Type-Level Attentive Aggregation Intuitively, different types of neighbors indicate various aspects of information, and a node is likely to have different preferences for multiple aspects. Given a user u and his/her node-level aggregated embeddings in different type spaces, we aggregate them as follows:

$$\mathbf{h}_u = \text{ReLU}(\mathbf{W}_h \sum_{t \in \mathcal{T}} \beta_{tu} \mathbf{p}_u^t + \mathbf{b}_h), \tag{9.34}$$

where $\mathbf{h}_u \in \mathbb{R}^d$ is the latent representation of user u. $\{\mathbf{W}_h \in \mathbb{R}^{d \times d}, \mathbf{b}_h \in \mathbb{R}^d\}$ are parameters of a neural network. β_{tu} is the attentive preferences of type t w.r.t. the feature aggregation of user u, as various types of neighbors contain multifaceted information and are expected to collaborate with each other. For user u, we concatenate the aggregated representations of all neighbor types and define the following weight:

$$\beta_{tu} = \frac{\exp(\mathbf{a}_t^\top [\mathbf{p}_u^{t_1} \oplus \mathbf{p}_u^{t_2} \oplus \cdots \oplus \mathbf{p}_u^{t_{|\mathcal{T}|}}])}{\sum_{t' \in \mathcal{T}} \exp(\mathbf{a}_{t'}^\top [\mathbf{p}_u^{t_1} \oplus \mathbf{p}_u^{t_2} \oplus \cdots \oplus \mathbf{p}_u^{t_{|\mathcal{T}|}}])}, \tag{9.35}$$

where $\mathbf{a}_t \in \mathbb{R}^{|\mathcal{T}|d}$ is a type-aware attention vector shared by all users. With Eq. 9.35, the concatenation of various neighbor types captures multifaceted information for a user, and $\mathbf{a}_t$ encodes the global preference of each type.

Similarly, for each item i, the attentive feature aggregator takes the neighbors of i as input and outputs the latent representation of i, denoted as $\mathbf{h}_i$.

9.5.3.3 Social Influence Coupler

To exploit the FRCs and capture the effects of influential friends, we propose a social influence coupler. The different impact of the influential friends and the item on social behaviors is first coupled together, and then we attentively represent the overall influence on the FRC.

Coupled Influence Representation Following [13], human behaviors are affected by various factors. In FER, whether u interacts with i is not simply driven by only the item itself or only the friends. More likely, the co-occurrence of friends and the item has a significant impact. As in the previous example (Fig. 9.13), when it is technology-related, the coupling between the expert (e.g., *Tom*) and the item (e.g., *AirPods*) has a greater impact than the coupling between the spouse and a tech-item, but the opposite scenario may happen for entertainment-related items. Hence, given user u, item i, and the FRC $C_u(i)$, we couple the influence of each friend $v \in C_u(i)$ and item i as the following:

$$\mathbf{c}_{\langle v,i \rangle} = \sigma(\mathbf{W}_c \phi(\mathbf{h}_v, \mathbf{h}_i) + \mathbf{b}_c), \qquad (9.36)$$

where $\mathbf{h}_v$ and $\mathbf{h}_i$ are aggregated representations of user v and item i. $\phi(\cdot, \cdot)$ serves as a fusion function, which can be element-wise product or concatenation (here we adopt concatenation). σ is the ReLU function. Obviously, Eq. 9.36 couples the features of item i and the influential friend v, capturing the influence of both.

Attentive Influence Degree With the coupled influence representation $\mathbf{c}_{\langle v,i \rangle}$, our next goal is to obtain the influence degree of $\mathbf{c}_{\langle v,i \rangle}$ on the user u. Since the influence score depends on user u, we incorporate the representation of user u (i.e., $\mathbf{h}_u$) into the influence score calculation with a two-layer neural network parameterized by $\{\mathbf{W}_1, \mathbf{W}_2, \mathbf{b}_1, b_2\}$:

$$d'_{u \leftarrow \langle v,i \rangle} = \sigma(\mathbf{W}_2(\sigma(\mathbf{W}_1 \phi(\mathbf{c}_{v,i}, \mathbf{h}_u) + \mathbf{b}_1)) + b_2). \qquad (9.37)$$

Then, the attentive influence degree is obtained by normalizing $d'_{u \leftarrow \langle v,i \rangle}$, which can be interpreted as the impact of the influential friend v on the user behavior:

$$d_{u \leftarrow \langle v,i \rangle} = \frac{\exp(d'_{u \leftarrow \langle v,i \rangle})}{\sum_{v' \in C_u(i)} \exp(d'_{u \leftarrow \langle v',i \rangle})}. \qquad (9.38)$$

Since the influences of friends propagate from the FRC, we attentively sum the coupled influences of the influential friends and item v on user u:

$$\mathbf{h}_{ui} = \sum_{v \in \mathcal{C}_u(i)} d_{u \leftarrow \langle v,i \rangle} \mathbf{c}_{\langle v,i \rangle}. \tag{9.39}$$

As the coupled influence representation $\mathbf{c}_{\langle v,i \rangle}$ incorporates the latent factors of the influential friend and the item, Eq. 9.39 guarantees that the social influence propagating among them can be encoded into the latent representation $\mathbf{h}_{ui}$.

9.5.3.4 Behavior Prediction and Model Learning

With the representations of user, item, and the coupled influence (i.e., $\mathbf{h}_u$, $\mathbf{h}_i$, and $\mathbf{h}_{ui}$), we concatenate them and then feed it into a two-layer neural network:

$$\mathbf{h}_o = \sigma(\mathbf{W}_{o_2}(\sigma(\mathbf{W}_{o_1}([\mathbf{h}_u \oplus \mathbf{h}_{ui} \oplus \mathbf{h}_i]) + \mathbf{b}_{o_1}) + \mathbf{b}_{o_2}). \tag{9.40}$$

Then, the predicted probability of a user–item pair is obtained via a regression layer with a weight vector $\mathbf{w}_y$ and bias b_y:

$$\hat{y}_{ui} = \text{sigmoid}(\mathbf{w}_y^\top \mathbf{h}_o + b_y). \tag{9.41}$$

Finally, to estimate model parameters Θ of SIAN, we optimize the following cross-entropy loss, where y_{ui} is the ground truth and λ is the L2 regularization parameter for reducing over-fitting:

$$-\sum_{\langle u,i \rangle \in \mathcal{E}_R} \left(y_{ui} \log \hat{y}_{ui} + (1 - y_{ui}) \log \left(1 - \hat{y}_{ui}\right)\right) + \lambda ||\Theta||_2^2. \tag{9.42}$$

9.5.4 Experiments

9.5.4.1 Experimental Settings

Datasets Yelp and Douban are classical open datasets widely used in recommendation, for which we build FRCs for each user–item pair to simulate the FER scenarios. FWD is extracted from a deployed live FER system with real FRCs displayed to users. The detailed statistics of datasets are shown in Table 9.7:

- **Yelp**[5] is a business review dataset containing both interactions and social relations. We first sample a set of users. For each user u, we construct a set of FRCs based on the given user–user relations and user–item interactions.

---

[5] https://www.yelp.com/dataset/challenge.

Table 9.7 Statistics of datasets

Datasets	Nodes	#Nodes	Relations	#Relations
Yelp	User (U)	8163	User–User	92, 248
	Item (I)	7900	User–Item	36, 571
Douban	User (U)	12, 748	User–User	169, 150
	Book (B)	13, 342	User–Book	224, 175
FWD	User (U)	72, 371	User–User	8, 639, 884
	Article (A)	22, 218	User–Article	2, 465, 675
	Media (M)	218, 887	User–Media	1, 368, 868
			Article–Media	22, 218

- **Douban**[6] is a social network related to sharing books, which includes friendships between users and interaction records between users and items. As preprocesses done for Yelp, we construct a set of FRCs based on the given user–user relations and user–item interactions.
- **Friends Watching Data (FWD)** is extracted from a real-world live FER system named WeChat Top Stories after data masking, where FRCs are explicitly displayed. Based on FWD, we construct an HSG containing nearly 313 thousand nodes and 12 million edges.

Baselines We compare the proposed SIAN against four types of methods, including feature/structure based methods (i.e., MLP, DeepWalk [19], node2vec [10], and metapath2vec [6]), fusion of feature/structure based methods (i.e., Deep-Walk+fea, node2vec+fea and metapath2vec+fea), graph neural network methods (i.e., GCN [15], GAT [25] and HAN [28]), and social recommendation methods (i.e., TrustMF [30] and DiffNet [29]).

Hyper-Parameter Settings For each dataset, the ratio of training, validation, and test set is 7:1:2. We adopt Adam optimizer [14] with the PyTorch implementation. The learning rate, batch size, and regularization parameter are set to 0.001, 1, 024, and 0.0005 using grid search [3], determined by optimizing AUC on the validation set. For random walk-based baselines, we set the walk number, walk length, and window size as 10, 50, and 5, respectively. For graph neural network-based methods, the number of layers is set to 2. For DiffNet, we set the regularization parameter as 0.001. The depth parameter is set to 2 as recommended in [29]. For other parameters of baselines, we optimize them empirically under the guidance of the literature. Finally, for all methods except MLP, we set the size of feature vector as 64 and report performances under different embedding dimensions {32, 64}.

---

[6] https://book.douban.com.

9.5.4.2   Experimental Results

We adopt three widely used metrics AUC, F1, and Accuracy to evaluate performance. The results w.r.t. the dimension of latent representation are reported in Table 9.8, from which we have the following findings:

(1) SIAN outperforms all baselines in all metrics on three datasets with statistical significance ($p < 0.01$) under paired t-test. It indicates that SIAN can well capture user core concerns from multifaceted factors in FER. The improvements derive from both high-quality node representations generated from node- and type-level attentive aggregations, and the social influence coupler that digs out what users are socially inclined to.

(2) Compared with the graph neural network methods, the impressive improvements of SIAN prove the effectiveness of the node- and type-level attentive aggregations. Especially, SIAN achieves better performances than HAN that is also designed for heterogeneous graphs with a two-level aggregation. It is because that the type-level attentive aggregation in SIAN captures heterogeneous information in multiple aspects, without being limited by the pre-defined meta-paths used in HAN. Moreover, the improvements also indicate the significance of our social influence coupler in FER.

(3) Social recommendation baselines also achieve promising performances, which further substantiates the importance of social influence in FER. Compared with baselines that only treat social relations as side information, the improvements imply that the friend referral factor may take the dominating position in FER, which should be carefully modeled. In particular, our SIAN achieves the best performance, reconfirming the capability of our social influence coupler in encoding diverse social factors for FER.

9.5.4.3   Analysis on Social Influence in FER

We have verified that FRC is the most essential factor in FER. However, a friend could impact user from different aspects (e.g., authority or similarity). Next, we show how different user attributes affect user behaviors in FER.

Evaluation Protocol The attention in social influence coupler reflects the importance of different friends. We assume that the friend v having the highest attention value (i.e., $d_{u \leftarrow \langle v, i \rangle}$ in Eq. 9.37) is the most influential friend w.r.t. item i for user u, and all of v's attribute values are equally regarded as contributing to the influence. Given a user attribute and a user group, we define the *background distribution* by counting the attribute values of all friends in FRCs of users in this group and also define the *influence distribution* by counting the attribute values of the most influential friends of users in the group. Thus, the background distribution represents the characteristics of general friends of this user group, while the influence distribution represents the characteristics of the most influential friends of

Table 9.8 Results on three datasets. The best method is bolded, and the second best is underlined. * indicates the significance level of 0.01. The best results of all methods are indicated in bold

Dataset	Model	AUC		F1		Accuracy	
		$d = 32$	$d = 64$	$d = 32$	$d = 64$	$d = 32$	$d = 64$
Yelp	MLP	0.6704	0.6876	0.6001	0.6209	0.6589	0.6795
	DeepWalk	0.7693	0.7964	0.6024	0.6393	0.7001	0.7264
	node2vec	0.7903	0.8026	0.6287	0.6531	0.7102	0.7342
	metapath2vec	0.8194	0.8346	0.6309	0.6539	0.7076	0.7399
	DeepWalk+fea	0.7899	0.8067	0.6096	0.6391	0.7493	0.7629
	node2vec+fea	0.8011	0.8116	0.6634	0.6871	0.7215	0.7442
	metapath2vec+fea	0.8301	0.8427	0.6621	0.6804	0.7611	0.7856
	GCN	0.8022	0.8251	0.6779	0.6922	0.7602	0.7882
	GAT	0.8076	0.8456	0.6735	0.6945	0.7783	0.7934
	HAN	0.8218	0.8476	0.7003	0.7312	0.7893	0.8102
	TrustMF	0.8183	0.8301	0.6823	0.7093	0.7931	0.8027
	DiffNet	<u>0.8793</u>	<u>0.8929</u>	<u>0.8724</u>	<u>0.8923</u>	<u>0.8698</u>	<u>0.8905</u>
	SIAN	**0.9486***	**0.9571***	**0.8976***	**0.9128***	**0.9096***	**0.9295***
Douban	MLP	0.7689	0.7945	0.7567	0.7732	0.7641	0.7894
	DeepWalk	0.8084	0.8301	0.7995	0.8054	0.8295	0.8464
	node2vec	0.8545	0.8623	0.8304	0.8416	0.8578	0.8594
	metapath2vec	0.8709	0.8901	0.8593	0.8648	0.8609	0.8783
	DeepWalk+fea	0.8535	0.8795	0.8347	0.8578	0.8548	0.8693
	node2vec+fea	0.8994	0.9045	0.8732	0.8958	0.8896	0.8935
	metapath2vec+fea	0.9248	0.9309	0.8998	0.9134	0.8975	0.9104
	GCN	0.9032	0.9098	0.8934	0.9123	0.9032	0.9112
	GAT	0.9214	0.9385	0.8987	0.9103	0.8998	0.9145
	HAN	0.9321	0.9523	<u>0.9096</u>	0.9221	<u>0.9098</u>	0.9205
	TrustMF	0.9034	0.9342	0.8798	0.9054	0.9002	0.9145
	DiffNet	<u>0.9509</u>	<u>0.9634</u>	0.9005	<u>0.9259</u>	0.9024	<u>0.9301</u>
	SIAN	**0.9742***	**0.9873***	**0.9139***	**0.9429***	**0.9171***	**0.9457***
FWD	MLP	0.5094	0.5182	0.1883	0.1932	0.2205	0.2302
	DeepWalk	0.5587	0.5636	0.2673	0.2781	0.1997	0.2056
	node2vec	0.5632	0.5712	0.2674	0.2715	0.2699	0.2767
	metapath2vec	0.5744	0.5834	0.2651	0.2724	0.4152	0.4244
	DeepWalk+fea	0.5301	0.5433	0.2689	0.2799	0.2377	0.2495
	node2vec+fea	0.5672	0.5715	0.2691	0.2744	0.3547	0.3603
	metapath2vec+fea	0.5685	0.5871	0.2511	0.2635	0.4698	0.4935
	GCN	0.5875	0.5986	0.2607	0.2789	0.4782	0.4853
	GAT	0.5944	0.6006	0.2867	0.2912	0.4812	0.4936
	HAN	0.5913	0.6025	0.2932	0.3011	0.4807	0.4937
	TrustMF	0.6001	0.6023	0.3013	0.3154	0.5298	0.5404
	DiffNet	<u>0.6418</u>	<u>0.6594</u>	<u>0.3228</u>	<u>0.3379</u>	<u>0.6493</u>	<u>0.6576</u>
	SIAN	**0.6845***	**0.6928***	**0.3517***	**0.3651***	**0.6933***	**0.7018***

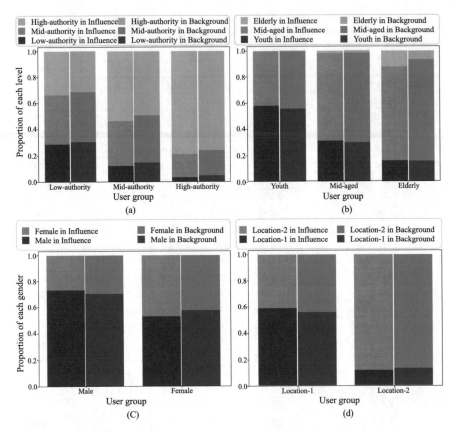

Fig. 9.15 Social influence analysis w.r.t. user attributes. For each attribute and user group (e.g., the authority and the low-authority group in (a)), the left is the influence distribution, while the right is the background distribution. In each bar, the height of each different colored segment means the proportion of an attribute value in the influence or background distribution. Best read in color. (**a**) Authority. (**b**) Age. (**c**) Gender. (**d**) Location

this user group. If the two distributions perfectly agree with each other, this attribute is not a key social factor in influencing this user group. In contrast, the differences between the two distributions imply how much this attribute is a key social factor, and how its different values affect user behaviors.

Results and Analysis As shown in Fig. 9.15, we find out the following:

(1) In Fig. 9.15a, we observe that user behaviors are more influenced by their friends who are more authoritative, regardless of what authority the user him/herself has. In all three user groups of varying authority, the proportion of high-authority in the influence distribution is larger than that in the background distribution. For instance, in the mid-authority user group, the top red block (high-authority influence) is larger than the top blue one (high-authority

background), which implies that high-authority friends are more influential for mid-authority users. The result is not surprising as users are usually more susceptible and easy to be affected by authoritative persons, which is consistent with common sense.

(2) We also conduct several analyses on influences w.r.t. other user attributes. We find that *users are easy to be influenced by their friends who are similar to themselves*. Specifically, Fig. 9.15b shows that people like items recommended by their peers, especially for the youth and the elderly; meanwhile, Fig. 9.15c and d shows that users tend to watch articles recommended by their friends with the same gender or location. Recommendation with user similarity, which has been widely assumed in collaborative filtering, is still classical even in FER.

The more detailed method description and experiment validation can be seen in [18].

9.6 Conclusions

As HG representation has a great power to fuse heterogeneous information, it has become one of the major techniques to apply HG analysis to real-world applications. This chapter presents several advanced HG representation methods applying to E-commercial systems and online social networks. Particularly, we first study the cash-out user detection problem and propose the HACUD, which is a hierarchical heterogeneous GNN method. The model could extract the user's structural features through a hierarchical attention mechanism. In addition, we study a newly emerged problem in E-commercial system, named intent recommendation, and propose a novel meta-path guided heterogeneous GNN method, called MEIRec. The MEIRec model learns users' and queries' embeddings through multiple pre-defined meta-paths. Furthermore, we study the share recommendation problem, which is a unique recommendation paradigm in social e-commerce, and propose a tripartite heterogeneous GNN, named HGSRec. Different from MEIRec model concatenating multiple learned embeddings, the proposed method aggregates multiple embeddings through a co-attention mechanism. Apart from e-commerce, we also study the friend-enhanced recommendation problem in online social networks and propose a novel social influence attentive neural network method. The experiments of these methods solidly validate the effectiveness of HG embedding methods on real-world applications.

More interesting future applications are worth being exploited on HG representation. For example, in the biological area, there are multiple relations between gene expression and phenotype, which can be naturally constructed as an HG. Besides, in the software engineering area, there are complex relations among test sample, requisition form, and problem form, which can be naturally modeled as an HG. Therefore, HG representation is expected to open up broad prospects for these new application areas and become a promising analytical tool.

References

1. Antoine, B., Nicolas, U., Alberto, G.D., Jason, W., Oksana, Y.: Translating embeddings for modeling multi-relational data. In: Advances in Neural Information Processing Systems, pp. 2787–2795 (2013)
2. Bahdanau, D., Cho, K., Bengio, Y.: Neural machine translation by jointly learning to align and translate. In: Proceedings of the Conference ICLR (2015)
3. Bergstra, J.S., Bardenet, R., Bengio, Y., Kégl, B.: Algorithms for hyper-parameter optimization. In: Advances in Neural Information Processing Systems, pp. 2546–2554 (2011)
4. Chen, T., Guestrin, C.: XGBoost: a scalable tree boosting system. In: Proceedings of the 22nd ACM SIGKDD International Conference on Knowledge Discovery and Data Mining (KDD), pp. 785–794 (2016)
5. Dai, H., Dai, B., Song, L.: Discriminative embeddings of latent variable models for structured data. In: International Conference on Machine Learning (ICML), pp. 2702–2711 (2016)
6. Dong, Y., Chawla, N.V., Swami, A.: metapath2vec: scalable representation learning for heterogeneous networks. In: Proceedings of the 23rd ACM SIGKDD International Conference on Knowledge Discovery and Data Mining (KDD), pp. 135–144 (2017)
7. Fan, S., Zhu, J., Han, X., Shi, C., Hu, L., Ma, B., Li, Y.: Metapath-guided heterogeneous graph neural network for intent recommendation. In: Proceedings of the 25th ACM SIGKDD International Conference on Knowledge Discovery and Data Mining (KDD), pp. 2478–2486 (2019)
8. Friedman, J.H.: Greedy function approximation: a gradient boosting machine. Ann. Stat. **29**(5), 1189–1232 (2001)
9. Glorot, X., Bengio, Y.: Understanding the difficulty of training deep feedforward neural networks. In: Proceedings of the Thirteenth International Conference on Artificial Intelligence and Statistics. JMLR Workshop and Conference Proceedings (AISTATS), pp. 249–256 (2010)
10. Grover, A., Leskovec, J.: node2vec: scalable feature learning for networks. In: Proceedings of the 22nd ACM SIGKDD International Conference on Knowledge Discovery and Data Mining (KDD), pp. 855–864 (2016)
11. Hu, B., Zhang, Z., Shi, C., Zhou, J., Li, X., Qi, Y.: Cash-out user detection based on attributed heterogeneous information network with a hierarchical attention mechanism. In: Proceedings of the AAAI Conference on Artificial Intelligence (AAAI), pp. 946–953 (2019)
12. Ji, H., Zhu, J., Wang, X., Shi, C., Wang, B., Tan, X., Li, Y., He, S.: Who you would like to share with? A study of share recommendation in social e-commerce. In: Proceedings of the AAAI Conference on Artificial Intelligence (AAAI) (2021)
13. Jolly, A.: Lemur social behavior and primate intelligence. Science **153**(3735), 501–506 (1966)
14. Kingma, D., Ba, J.: Adam: a method for stochastic optimization. In: International Conference on Learning Representations (ICLR) (2015)
15. Kipf, T.N., Welling, M.: Semi-supervised classification with graph convolutional networks. In: International Conference on Learning Representations (ICLR) (2017)
16. Liang, J., Jacobs, P., Sun, J., Parthasarathy, S.: Semi-supervised embedding in attributed networks with outliers. In: Proceedings of the 2018 SIAM International Conference on Data Mining. Society for Industrial and Applied Mathematics (SDM), pp. 153–161 (2018)
17. Lobo, J.M., Jiménez-Valverde, A., Real, R.: AUC: a misleading measure of the performance of predictive distribution models. Glob. Ecol. Biogeogr. **17**(2), 145–151 (2008)
18. Lu, Y., Xie, R., Shi, C., Fang, Y., Wang, W., Zhang, X., Lin, L.: Social influence attentive neural network for friend-enhanced recommendation. In: Proceedings of The European Conference on Machine Learning and Principles and Practice of Knowledge Discovery in Databases, Ghent (ECML-PKDD) (2020)
19. Perozzi, B., Al-Rfou, R., Skiena, S.: DeepWalk: online learning of social representations. In: Proceedings of the 20th ACM SIGKDD International Conference on Knowledge Discovery and Data Mining (KDD) (2014)

20. Qu, M., Tang, J., Shang, J., Ren, X., Zhang, M., Han, J.: An attention-based collaboration framework for multi-view network representation learning. In: Proceedings of the 2017 ACM on Conference on Information and Knowledge Management (CIKM), pp. 1767–1776 (2017)
21. Shi, C., Li, Y., Zhang, J., Sun, Y., Philip, S.Y.: A survey of heterogeneous information network analysis. IEEE Trans. Knowl. Data Eng. **29**(1), 17–37 (2017)
22. Shi, C., Hu, B., Zhao, W.X., Philip, S.Y.: Heterogeneous information network embedding for recommendation. IEEE Trans. Knowl. Data Eng. **31**(2), 357–370 (2019)
23. Sun, Y., Han, J., Yan, X., Yu, P.S., Wu, T.: PathSim: meta path-based top-k similarity search in heterogeneous information networks. In: Proceedings of the VLDB Endowment, pp. 992–1003 (2011)
24. Tieleman, T., Hinton, G.: Lecture 6.5-rmsprop: divide the gradient by a running average of its recent magnitude. COURSERA Neural Netw. Mach. Learn. **4**(2), 26–31 (2012)
25. Veličković, P., Cucurull, G., Casanova, A., Romero, A., Liò, P., Bengio, Y.: Graph attention networks. In: Proceedings of the Conference ICLR (2018)
26. Wang, Z., Liao, J., Cao, Q., Qi, H., Wang, Z.: Friendbook: a semantic-based friend recommendation system for social networks. IEEE Trans. Mob. Comput. **14**(3), 538–551 (2014)
27. Wang, X., He, X., Wang, M., Feng, F., Chua, T.S.: Neural graph collaborative filtering. In: Proceedings of the 42nd International ACM SIGIR Conference on Research and Development in Information Retrieval (SIGIR), pp. 165–174 (2019)
28. Wang, X., Ji, H., Shi, C., Wang, B., Ye, Y., Cui, P., Yu, P.S.: Heterogeneous graph attention network. In: The World Wide Web Conference (WWW), pp. 2022–2032 (2019)
29. Wu, L., Sun, P., Fu, Y., Hong, R., Wang, X., Wang, M.: A neural influence diffusion model for social recommendation. In: Proceedings of the 42nd International ACM SIGIR Conference on Research and Development in Information Retrieval (SIGIR), pp. 235–244 (2019)
30. Yang, B., Lei, Y., Liu, J., Li, W.: Social collaborative filtering by trust. In: IEEE Transactions on Pattern Analysis and Machine Intelligence, pp. 1633–1647 (2016)
31. Zhang, Z., Yang, H., Bu, J., Zhou, S., Yu, P., Zhang, J., Ester, M., Wang, C.: ANRL: attributed network representation learning via deep neural networks. In: Proceedings of the Twenty-Seventh International Joint Conference on Artificial Intelligence (IJCAI), pp. 3155–3161 (2018)

Chapter 10
Platforms and Practice of Heterogeneous Graph Representation Learning

Abstract It is challenging to build a Heterogeneous Graph (HG) representation learning model because HG is heterogeneous, irregular, and sparse. An easy-to-use and friendly framework is important for a beginner to make an understanding and get deep into this field. In this chapter, we are going to introduce OpenHGNN, a toolkit that can help to build HG models in a predesigned pipeline. And we will present the procedures with three well-known HG models that are HAN, the model first introduces attention mechanism to heterogeneous graph neural networks, RGCN, a model used to model multi-relational graphs with GCN, and HERec, a heterogeneous graph embedding method for recommendation.

10.1 Introduction

Graph is a kind of irregular structure data. Compared to regular structure data like images, graph cannot be directly computed by traditional deep learning platforms (e.g., TensorFlow [1], PyTorch [6], etc.). And graph is usually very large-scale and highly sparse, which also increases the difficulty to apply these frameworks. To tackle these problems, some libraries (e.g., DGL [9], PyG [3], etc.) have been developed to support tensor computation over graphs with the principle of message passing followed by most Graph Neural Network (GNN). These frameworks greatly facilitate the implementation of GNN for engineers and researchers.

However, because a Heterogeneous Graph (HG) can be more complex than a homogeneous graph and Heterogeneous Graph Neural Networks (HGNN) are designed with different paradigms, most of the existing graph learning libraries are not friendly enough to implement a HGNN or even do not support HG. Here, we are going to introduce OpenHGNN,[1] an open-source toolkit for HGNN based on DGL and PyTorch, developed by BUPT GAMMA Lab.[2] OpenHGNN is specifically

---

[1] https://github.com/BUPT-GAMMA/OpenHGNN.

[2] https://github.com/BUPT-GAMMA.

© The Author(s), under exclusive license to Springer Nature Singapore Pte Ltd. 2022   285
C. Shi et al., *Heterogeneous Graph Representation Learning and Applications*,
Artificial Intelligence: Foundations, Theory, and Algorithms,
https://doi.org/10.1007/978-981-16-6166-2_10

designed for heterogeneous graph representation learning, integrating many popular models and datasets, and it is easy-to-use, extensible, and efficient.

In this chapter, we will show how to build a HGNN in practice based on OpenHGNN. We will first give a brief introduction to the mainstream deep learning platforms and graph leaning libraries. Then we introduce the pipeline of developing on OpenHGNN, and the following part is three heterogeneous graph neural network model instances (HAN [10], HERec [8], RGCN [7]) that developed on OpenHGNN.

10.2   Foundation Platforms

The platforms to be introduced below have three types: deep learning platforms, platforms of graph machine learning, and platforms of heterogeneous graph representation learning. These three types of platforms usually have a hierarchical relation, i.e., platforms of graph machine learning are built upon deep learning platforms and the platforms of heterogeneous graph representation learning are built upon platforms of graph machine learning. From the first to the third type of platforms, their application scenarios are more and more focused, and they are more and more friendly and convenient for developing models of heterogeneous graphs.

10.2.1   Deep Learning Platforms

Deep learning platforms are collections of software which abstracts the underlying hardware and software stacks to expose simple APIs to deep learning developers. They usually support GPU devices and have efficiency optimizations on common computation operations like matrix multiplication, which significantly accelerate the computation speed of machine learning programs. Platforms of graph machine learning are also usually built based on these deep learning platforms.

10.2.1.1   TensorFlow

TensorFlow [1] is an end-to-end open-source machine learning platform. It was originally developed by the Google Brain team in Google's Machine Intelligence Research organization to conduct machine learning and deep neural networks research. It is now general enough to be applicable in a wide variety of other domains. It is one of the earliest platforms and is still thriving nowadays.

TensorFlow has a comprehensive, flexible ecosystem of tools, libraries, and community resources that lets researchers push the state-of-the-art in ML and developers easily build and deploy ML-powered applications. And it provides stable Python and C++ APIs, as well as non-guaranteed backward compatible API for other languages like Java and JavaScript. The modules of TensorFlow can be

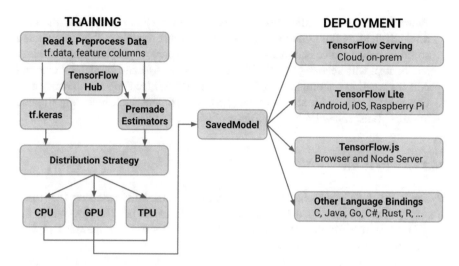

Fig. 10.1 TensorFlow architecture

classified into modules for training and modules for deployment, which can be seen in Fig. 10.1.

Beginners of deep learning platforms may be confused by the concept of the graph mode in TensorFlow. Actually, many other deep learning platforms also have similar concepts and hence it is worth elaborating. The graphs, also known as computational graphs, are kinds of data structure and they can be saved, executed, and restored without the original Python code. Graph execution enables portability outside Python and tends to offer better performance, while you can also run a TensorFlow program *eagerly*, i.e., operation by operation like a normal python program. The eager mode has high flexibility and is easy to learn while graph mode is more efficient.

TensorFlow has the following features:

- **Easy Model Building**. TensorFlow offers multiple levels of abstraction so you can choose the right one according to your needs. Building and training models by using the high-level Keras API can help beginners of TensorFlow and machine learning. For large ML training tasks, TensorFlow also provides the Distribution Strategy API for distributed training on different hardware configurations without changing the model definition.
- **Robust ML Production**. TensorFlow also provides a direct and robust path to production. Whether on servers, edge devices, or the web, training and deploying models using TensorFlow can be easy, no matter what language or platform you use. If you require a full production ML pipeline, TensorFlow Extended (TFX) is needed. For running inference on mobile and edge devices, use TensorFlow Lite. For training and deploying models in JavaScript environments, use TensorFlow.js.

- **Powerful Experimentation for Research**. As a frequently used platform in the research field, building and training state-of-the-art models without sacrificing speed or performance is a matter of course. TensorFlow provides the flexibility and control with features like the Keras Functional API and Model Subclassing API for creation of complex topologies. TensorFlow also supports an ecosystem of powerful add-on libraries and models to experiment with, including Ragged Tensors, TensorFlow Probability, Tensor2Tensor, and BERT.

10.2.1.2   PyTorch

PyTorch [6] is an open-source machine learning framework based on the Torch library, primarily developed by Facebook's AI Research lab (FAIR). While the initial release of PyTorch is about one year later than TensorFlow, there is a rapidly rising trend of its users in recent years.

PyTorch designs tensor computation (like NumPy) with strong GPU acceleration. Hence it can be utilized as a replacement for NumPy to use the power of GPUs. Besides, it uses a technique called reverse-mode auto-differentiation, which allows users to change the way network behaves arbitrarily with zero lag or overhead. Based on these characteristics, it has grown as a deep learning research platform that provides both flexibility and speed.

PyTorch and TensorFlow are usually considered the two most famous deep learning platforms. They may be very different at the beginning while they have similar features later. For example, in TensorFlow1.x, there was no eager mode which made debugging on TensorFlow codes very difficult. Nowadays, both PyTorch and TensorFlow have graph mode and eager mode.

PyTorch has the following features:

- **Production Ready**. With TorchScript, PyTorch provides ease-of-use and flexibility in eager mode, while seamlessly transitioning to graph mode for speed, optimization, and functionality in C++ runtime environments. For deploying PyTorch models at scale, TorchServe is an easy-to-use tool. It is cloud and environment agnostic and supports features such as multi-model serving, logging, metrics, and the creation of RESTful endpoints for application integration. PyTorch also supports asynchronous execution of collective operations and peer-to-peer communication accessible from both Python and C++, which are helpful for distributed training.
- **Robust Ecosystem**. An active community of researchers and developers have built a rich ecosystem of tools and libraries, which can extend PyTorch and support development in nearly all machine learning fields from computer vision to reinforcement learning. PyTorch also supports exporting models in the standard ONNX (Open Neural Network Exchange) format for direct access to ONNX-compatible platforms, runtimes, visualizers, etc. Furthermore, PyTorch is well supported on mainstream cloud platforms, providing frictionless development

and easy scaling through prebuilt images, large-scale training on GPUs, ability to run models in a production scale environment, etc.

- **C++ Frontend**. Except Python, PyTorch also provides a C++ frontend. It is a pure C++ interface that follows the design and architecture of the Python frontend. It is intended to enable research in high-performance, low latency, and bare metal C++ applications.

10.2.1.3 MXNet

Apache MXNet [2] is an open-source deep learning software framework, developed by Apache software foundation. Unlike the rest three deep learning platforms, MXNet does not have a background of a single company or institution.

MXNet is designed for both efficiency and flexibility, which is similar to the design philosophy of PyTorch. It allows users to mix symbolic and imperative programming to maximize both efficiency and productivity. At its core, MXNet contains a dynamic dependency scheduler that automatically parallelizes both symbolic and imperative operations on the fly. A graph optimization layer on top of that makes symbolic execution fast and memory efficient.

MXNet has the following features:

- **Hybrid Frontend** MXNet provides hybridize functionality to switch to symbolic mode from imperative mode simply.
- **Distributed Training**. MXNet supports multi-gpu or multi-host training with near-linear scaling efficiency. MXNet recently introduced support for Horovod, the distributed learning framework developed by Uber.
- **8 Language Bindings**. MXNet supports 8 languages. It is integrated into Python deeply and supports Scala, Julia, Clojure, Java, C++, R, and Perl. Combined with the hybridization feature, this allows a very smooth transition from Python training to deployment in the language of your choice to shorten the time to production.

10.2.1.4 PaddlePaddle

PaddlePaddle[3] is an open-source deep learning platform which is derived from industry practice. It is developed by Baidu based on its research and industrial application experience of deep learning technologies.

A prominent characteristic of PaddlePaddle is that it has a close relation with the industry field. For example, there are abundant open-source algorithms, especially pretrained models integrated into the platform, which can accelerate the application to industrial scenarios.

---

[3] https://paddlepaddle.org.cn.

PaddlePaddle has the following features:

- **Convenient Development**. PaddlePaddle also supports both declarative and imperative programming. The network structures can be designed automatically to some degree, and even in some scenarios, the auto-designed model can outperform human experts.
- **Large-Scale Training**. PaddlePaddle supports hundreds of billions of features and trillions of parameters, which is necessary for industrial development.
- **Deployment on Multiple Terminals**. PaddlePaddle is compatible with models trained by multiple open-source frameworks and can be deployed to multiple kinds of terminals with high inference speed.

10.2.2   Platforms of Graph Machine Learning

In the early years, without platforms of graph machine learning, developers and researchers had to design graph machine learning models on deep learning platforms mentioned above. However, there are significant semantic gaps between the tensor-centric perspective of the platforms above and that of a graph. The performance gaps between the computation and memory-access patterns, which are induced by the sparse nature of graphs and the underlying parallel hardware optimized for dense tensor operations, also exist. Hence platforms of graph machine learning thrives in recent years to abstract and integrate these to provide a succinct programming interface.

10.2.2.1   DGL

DGL [9], Deep Graph Library, is an easy-to-use, high-performance, and scalable Python package for deep learning on graphs. It is built for easy implementation of graph neural network model family, on the top of existing deep learning frameworks. The main sponsors of DGL include AWS, NSF, NVIDIA, and Intel.

DGL is framework agnostic, meaning if a deep graph model is a component of an end-to-end application, the rest of the logics can be implemented in any major frameworks, such as PyTorch, Apache MXNet, or TensorFlow. DGL offers a versatile control of message passing, speed optimization via auto-batching and highly tuned sparse matrix kernels, and multi-GPU/CPU training to scale to graphs of hundreds of millions of nodes and edges.

DGL has the following features:

- **Framework Agnostic**. The backend platform of DGL can be anyone of PyTorch, TensorFlow, and MXNet. Users can choose their familiar deep learning platforms and get used to DGL in a short time.
- **Supporting GPU Well** DGL provides a powerful graph class that can reside on either CPU or GPU. It bundles structural data as well as features for a better

control. DGL provides a variety of functions for computing with graph objects including efficient and customizable message passing primitives for graph neural networks.

- **Models, Modules, and Benchmarks** DGL collects a rich set of example implementations of popular GNN models of a wide range of topics. Researchers can search for related models to innovate new ideas from or use them as baselines for experiments. Moreover, DGL provides many state-of-the-art GNN layers and modules for users to build new model architectures based on them.
- **Scalable and Efficient** DGL distills the computational patterns of GNNs into a few generalized sparse tensor operations suitable for extensive parallelization. It also extensively optimizes the whole stack to reduce the overhead in communication, memory consumption, and synchronization. As a result, it is convenient to train models using DGL on large-scale graphs across multiple GPUs or multiple machines and relatively easy to scale to billion-sized graphs.

10.2.2.2 PyG

PyG [3], PyTorch Geometric, is a library for deep learning on irregularly structured input data such as graphs, point clouds, and manifolds built upon PyTorch.

PyG contains various methods for deep learning on graphs and other irregular structures, also known as geometric deep learning, from a variety of published papers. It also contains easy-to-use mini-batch loaders for operating on many small and single giant graphs, multi-GPU support, a large number of common benchmark datasets, the GraphGym experiment manager, and helpful transforms for learning on arbitrary graphs, 3D meshes, or point clouds. Unlike DGL, PyG can only use PyTorch as its backend.

PyG has the following features:

- **Friendly to PyTorch Users** PyG is especially friendly to PyTorch users. It utilizes a tensor-centric API and keeps design principles close to vanilla PyTorch.
- **Comprehensive and Well-maintained GNN Models** PyG also contains Comprehensive and well-maintained GNN models. Most of the state-of-the-art Graph Neural Network architectures have been implemented by library developers or authors of research papers.
- **High Flexibility** Existing PyG models can easily be extended for conducting your own research with GNNs. Making modifications to existing models or creating new architectures is simple, thanks to its easy-to-use message passing API, and a variety of operators and utility functions.
- **GraphGym Integration** GraphGym lets users easily reproduce GNN experiments, and is able to launch and analyze thousands of different GNN configurations. It is also customizable by registering new modules to a GNN learning pipeline.

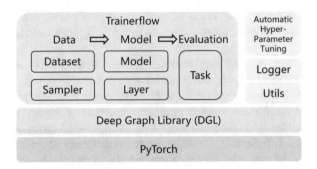

Fig. 10.2 OpenHGNN architecture

10.2.3 Platforms of Heterogeneous Graph Representation Learning

It may seem that platforms of graph machine learning are strong enough after having a tour on them, while heterogeneous graph representation learning has its unique characteristics. To handle heterogeneous node types and relation types and mine the abundant semantic information behind them, heterogeneous graph models are usually more complicated than usual graph models, while many mainstream heterogeneous models are not integrated into the platforms of graph machine learning mentioned above. Furthermore, the training procedures of heterogeneous models are also usually more complicated, which may be an obstacle for beginners to comprehend.

OpenHGNN[4] is an open-source toolkit for heterogeneous graph neural networks based on DGL and PyTorch, developed by BUPT GAMMA Lab.

The main modules of OpenHGNN are dataset, model, sampler, layer, and task, which are shown in Fig. 10.2. The functions and relations of these modules will be elaborated on in the next section.

OpenHGNN is helpful to its users in every aspect in the field of heterogeneous graph models. Reading the well-organized model codes provided by the library makes users more clear to details of models, comprehending the trainerflow architectures makes users more familiar with heterogeneous graph tasks and modifying on these models makes users build their new models conveniently and flexibly. OpenHGNN is also convenient for researchers to test model performances as baselines in one command. Here is an example:

```
python main.py -m GTN -d imdb4GTN -t node_classification -g 0
```

---

[4] https://github.com/BUPT-GAMMA/OpenHGNN.

OpenHGNN has the following features:

- **Various Models** OpenHGNN provides many open-source mainstream heterogeneous models, such as HAN, HetGNN [12], and GTN [11]. It is more convenient to learn heterogeneous models by virtue of well-organized and well-commented model codes. Modification on these existing models is also easy.
- **High Extensibility** OpenHGNN is highly extensible. It unifies all these models into a single framework composed of tasks and models, which will be elaborated on later. Datasets, trainerflows, and models are decoupled well, and therefore users can customize one of them without affecting others. For example, if a user wants to implement his own model, he does not have to implement the trainer flow and only needs to implement the bare model class.
- **High Efficiency** The efficiency of OpenHGNN is guaranteed by the efficient APIs of the backend DGL.

10.3   Practice of Heterogeneous Graph Representation Learning

In this part, we will show how to build a heterogeneous graph learning model based on OpenHGNN in practice. OpenHGNN mainly contains three parts, **trainerflow**, **model**, and **task**. The relations between them are shown in Fig. 10.3. The trainer-flow, containing the model and the task, is an abstraction of a predesigned workflow that trains and evaluates a model on a given dataset for a specific use case. The

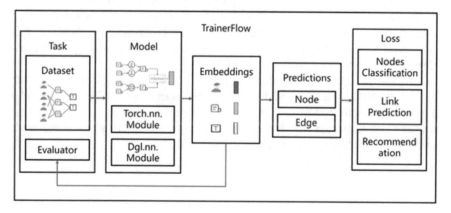

Fig. 10.3 Pipeline of OpenHGNN

model plays the role of an encoder and it will output the node embedding for a given heterogeneous graph. The task makes a connection between the embeddings and downstream tasks.

In the next part of the section, we will show how to build a new model from scratch with source codes. The rest of this section is organized in the following order:

- **Construct a dataset**. This part will introduce how to construct a new dataset OpenHGNN not integrated. We will first introduce how to create a `dgl.heter- ograph` object and then how to integrate it to OpenHGNN.
- **Build a trainerflow**. Not all the models follow a universal training procedure, especially for HGNN. In this part, we will show how to implement a model and how to build a new trainerflow for it if needed.
- **Examples in practice**. The last part is the implementation of three models (i.e., HAN [10], RGCN [7], HERec [8]).

10.3.1   Build a New Dataset

In order to better introduce the pipeline and practice of OpenHGNN, we first introduce the data structure and data processing in OpenHGNN. We have processed a number of heterogeneous graph dataset in advance and integrated into the OpenHGNN. Several types of datasets including academic networks (e.g., ACM, DBLP, Aminer), information networks (e.g., IMDB, LastFM), and recommendation graphs (e.g., Amazon, MovieLens, Yelp) are the most wildly used datasets which could be specified directly when running experiments with the command line.

Before implementing a model or testing an existing model performance, we need to check if the dataset is supported by OpenHGNN. If not, what we should do first is to construct a new dataset. In this section, we use the ACM dataset [10] for node classification as an example to show the procedures.

First we will show an example about how to build a node classification dataset. In DGL, an HG is specified with a series of relation subgraphs as below. Each relation name is a string (`source node type, edge type, destination node type`) associated with a relation subgraph. The following code snippet is an example of creating a heterogeneous graph in DGL.

```
>>> import dgl
>>> import torch as th

>>> # Create a heterograph with 3 node types and 3 edge types.
>>> graph_data = {
...        ('drug','interacts','drug'): (th.tensor([0,1]), th.tensor([1,2])),
...        ('drug','interacts','gene'): (th.tensor([0,1]), th.tensor([2,3])),
...        ('drug','treats','disease'): (th.tensor([1]), th.tensor([2]))
... }
>>> g = dgl.heterograph(graph_data)
>>> g.ntypes
['disease', 'drug', 'gene']
>>> g.etypes
['interacts', 'interacts', 'treats']
>>> g.canonical_etypes
[('drug', 'interacts', 'drug'),
 ('drug', 'interacts', 'gene'),
 ('drug', 'treats', 'disease')]
```

We recommend to set the feature name as 'h'.

```
>>> g.nodes['drug'].data['h'] = th.ones(3, 1)
```

DGL provides `dgl.save_graphs` and `dgl.load_graphs` respectively for saving and loading heterogeneous graphs in binary format. So we can use `dgl.save_graphs` to store graphs into the disk.

```
>>> dgl.save_graphs("demo_graph.bin", g)
```

Now a binary file containing the dataset has been created, and we should move it to the directory *openhgnn/dataset/*. In *NodeClassificationDataset.py*, the dataset will be loaded and some extra information will also be needed. For example, the `category`, `num_classes` and `multi_label` (if necessary) should be set with "drug", 3, True, representing the node type to predict, the number of classes, and whether the task is multi-label classification, respectively.

```
>>> if name_dataset == 'demo_graph':
...        data_path = './openhgnn/dataset/demo_graph.bin'
...        g, _ = load_graphs(data_path)
...        g = g[0].long()
...        self.category = 'drug'
...        self.num_classes = 3
...        self.multi_label = False
```

We have described how to construct a heterogeneous graph in DGL. In the following part, we will introduce how to build the ACM dataset. We will first provide the description of the dataset and then we will show how to process the original file to DGL heterogeneous graph data structure.

The ACM dataset is a citation network that is commonly used on HGNN. There are four node types and eight link types in the original ACM dataset. Node types contain **paper**, **author**, **subject**, and **term**. Each node has a unique ID and a type ID used to identify the node type. Taking **paper** as an example, it contains a unique ID, the type ID of the node (e.g., 0 for paper), and the feature of this node. In the original txt file, each row represents a node and its format is as follows:

```
0      'Influence and correlation...'  0    "1,1,1..."
1      'Efficient semi-streaming...'   0    "0,1,0..."
...
3025 'IMohammad Mahdian'              1    "1,1,1..."
3026 'Ravi Kumar'                     1    "1,1,1..."
...
```

The first column is node ID, the second column is node name, the third column is node type ID, and the last column is node feature, and the feature dimensions of nodes with the same type are consistent. The meta information of node type is as follows:

```
{
    "node type":
    {
        "0": "paper",
        "1": "author",
        "2": "subject",
        "3": "term"
    }
}
```

Note that node IDs are continuous. Taking **paper-cite-paper** as an edge example, it contains the source node ID, the target node ID, the edge type ID (e.g., 0 for paper-cite-paper and 7 for term-paper), and the weight of each edge (In ACM dataset, all weights are 1.0). In the original file, each row represents an edge and its format is as follows:

```
0 179  0 1.0
0 2697 0 1.0
1 2523 0 1.0
1 2589 0 1.0
...
```

For example, the first column means an edge from node 0 to node 179, the edge type ID is 0, and the edge weight is 1.0. The meta information of edge type is as follows:

```
{
    "link type": {
        "0": {
            "start": 0,
            "end": 0,
            "meaning": "paper-cite-paper"
        },
        [...]
        "7": {
            "start": 3,
            "end": 0,
            "meaning": "term-paper"
        }
    }
}
```

Finally, there is a file about node label. Each row represents the target node ID, the node type and its category, and the format is as follows:

```
622   0 2
1923  0 0
951   0 0
1957  0 1
```

Note that only the target node has category (paper is the target node type here), so each element in the second column has the same value 0. The meta information of node category is as follows:

```
{
    "node type": {
        "0": {
            "0": "database",
            "1": "wireless communication",
            "2": "data mining"
        }
    }
}
```

For example, the paper node 622 belongs to the data mining field and the paper node 1923 is published in the database field.

In the next part, we first show how to generate a dictionary which can be transformed into DGL heterogeneous graph data structure. This can be done by the following code:

```
>>> meta_graphs = {}
>>> for i in range(8):
...       edge = edges[edges[2] == i]
...       source_node = edge.iloc[:,0].values -
...                     np.min(edge.iloc[:,0].values)
...       target_node = edge.iloc[:,1].values -
...                     np.min(edge.iloc[:,1].values)
...       meta_graphs[(node_info[str(link_info[str(i)]['start'])],
...                   link_info[str(i)]['meaning'],
...                   node_info[str(link_info[str(i)]['end'])])]
...          = (torch.tensor(source_node), torch.tensor(target_node))
```

Then a heterogeneous graph object can be constructed with function `dgl.hete- rograph`:

```
>>> g = dgl.heterograph(meta_graphs)
```

The following code can store node features directly on the graph:

```
>>> g.nodes['paper'].data['h'] = torch.FloatTensor(paper_feature)
>>> g.nodes['author'].data['h'] = torch.FloatTensor(author_feature)
>>> g.nodes['subject'].data['h'] = torch.FloatTensor(subject_feature)
>>> dgl.save_graphs("acm.bin", g)
```

In the end, we can output the specific details of the processed graph through the following code:

```
>>> import dgl
>>> g = dgl.load_graphs('acm.bin')
>>> print(g)
Graph(num_nodes={'author':5959, 'paper':3025, 'subject':56, 'term':1902},
      num_edges={('author', 'author-paper', 'paper'): 9949,
                ('paper', 'paper-author', 'author'): 9949,
                ('paper', 'paper-cite-paper', 'paper'): 5343,
                ('paper', 'paper-ref-paper', 'paper'): 5343,
                ('paper', 'paper-subject', 'subject'): 3025,
                ('paper', 'paper-term', 'term'): 255619,
                ('subject', 'subject-paper', 'paper'): 3025,
                ('term', 'term-paper', 'paper'): 255619},
      metagraph=[('author', 'paper', 'author-paper'),
                ('paper', 'author', 'paper-author'),
                ('paper', 'paper', 'paper-cite-paper'),
                ('paper', 'paper', 'paper-ref-paper'),
                ('paper', 'subject', 'paper-subject'),
                ('paper', 'term', 'paper-term'),
                ('subject', 'paper', 'subject-paper'),
                ('term', 'paper', 'term-paper')])
```

By now, the binary graph file you stored can be used directly by OpenHGNN to train your model.

10.3.2 Build a New Model

10.3.2.1 Model

The model acts as a graph encoder. Given a heterogeneous graph and node features, the model will return a dictionary of node embeddings. It also allows to output the embedding of target nodes which are participated in loss calculation.

The Model mainly consists of two parts, a model builder and a forward propagator. Each model inherits the BaseModel class and must implement the classmethod `build_model_from_args`. With that, two parameters named **args** and **hg** can be used to build up a custom model with model-specific hyper-parameters. So it is necessary to implement the function `build_model_from_args` in the model. Here is an example:

```
class RGAT(BaseModel):
    @classmethod def build_model_from_args(cls, args, hg):
    return cls(in_dim=args.hidden_dim,
               out_dim=args.hidden_dim,
               h_dim=args.out_dim,
               etypes=hg.etypes,
               num_heads=args.num_heads,
               dropout=args.dropout)
```

The function `forward` is the core of the model. It is noted that in OpenHGNN, we prepossess the feature of dataset outside of model. Specifically, we use a linear layer with a bias for each node type to map all node features to a shared feature space. So the parameter **h_dict** of `forward` in model is not original, and the model need not feature prepossessing.

10.3.2.2 Trainerflow

The trainerflow, containing the model and the task, is an abstraction of a predesigned-workflow that trains and evaluates a model on a given dataset for specific utilization.

As shown in Fig. 10.3, the trainerflow contains the whole training process. A Trainerflow object includes **task, model, optimizer**, and **dataloader** in order to execute the training process. The main training process is done by function `train` and the single training step is done by `_full_train_step` and its mini-batch version `_mini_train_step`. After one or several training steps, the function `_test_step` can be applied to test the validation performance of model to select the best model parameters.

There are three main trainers named **node_classification_flow** which is designed for a semi-supervised node classification model, **link_prediction** which is designed for link prediction task and **recommendation** for recommendation scenario.

Some models (e.g., HetGNN [12], HeGAN [4], etc.) demanding special training processes have their own trainerflow. Generally, if the paper has a fancy loss function or an accelerated sampling method, it needs a customized trainerflow.

The following demo may give a sight of how a trainerflow organized and worked in OpenHGNN.

```python
class Trainerflow(BaseFlow):
    def train():
        # _full_train_step()
        # _test_step()
    def _full_train_step():
        # output = model(input)
        # loss = loss_fn(output, label)
        # loss.backward()
        # optimizer.step()
    def _test_step():
        # task.evaluate()
```

10.3.2.3   Task

The Task module encapsulates several objects related to the task, including **Dataset**, **Evaluator**, **Labels**, and **dataset_split**. The dataset contains the heterogeneous graph as DGLGraph, as well as node/edge features and additional dataset-specific information. The evaluator takes the predictions of the model and the labels to give the evaluation results, and this can be done by the function `evaluate`. The function `get_graph` will return the DGLGraph, and the function `get_labels` will return the ground truth labels. There are three kinds of supported tasks in the OpenHGNN, including **Node classification**, **Link prediction**, and **Recommendation**.

10.3.2.4   Register in OpenHGNN

Before training the model in OpenHGNN, what we should do first is to register the model with `@register_model(New_model)`. Here is an example:

```python
from openhgnn.models import BaseModel, register_model
@register_model('New_model')
class New_model(BaseModel):
    # Implementation details
```

The training process of models differs from each other, which needs a model trainer that can be applied. Similar to the implementation of a model, we need to register the model trainer with `@register_trainer("New_trainer")`. Here is an example:

```python
from openhgnn.trainerflow import BaseFlow, register_flow
@register_flow('New_trainer')
class New_trainer(BaseFlow):
    # Implementation details
```

For more details about the developing process based on OpenHGNN, please refer to the official documents.[5] In the following section, we will introduce several examples on how to implement a model based on OpenHGNN.

10.3.3   Practice of HAN

HAN [10] firstly proposes a heterogeneous graph neural network based on the hierarchical attention, including node-level and semantic-level attentions. It enables the graph neural network to be directly applied to the heterogeneous graph and further facilitates the heterogeneous graph based applications.

The node-level attention aims to learn the importance between a node and its meta-path based neighbors. The mathematical formulas are as follows:

$$\alpha_{ij}^{\Phi} = \text{softmax}_j\left(e_{ij}^{\Phi}\right) = \frac{\exp\left(\sigma\left(\mathbf{a}_{\Phi}^{\mathrm{T}} \cdot \left[\mathbf{h}_i' \| \mathbf{h}_j'\right]\right)\right)}{\sum_{k \in \mathcal{N}_i^{\Phi}} \exp\left(\sigma\left(\mathbf{a}_{\Phi}^{\mathrm{T}} \cdot \left[\mathbf{h}_i' \| \mathbf{h}_k'\right]\right)\right)},$$

$$\mathbf{z}_i^{\Phi} = \prod_{k=1}^{K} \sigma\left(\sum_{j \in \mathcal{N}_i^{\Phi}} \alpha_{ij}^{\Phi} \cdot \mathbf{h}_j'\right).$$

The semantic-level attention is able to learn the importance of different meta-paths. The mathematical formula is as follows:

$$\mathbf{Z} = \mathcal{F}_{att}(Z^{\Phi_1}, Z^{\Phi_2}, \ldots, Z^{\Phi_p}).$$

With the importance score learned from both node level and semantic level, the proposed model can generate node embedding by aggregating features from neighbors based on meta-path in a hierarchical manner. In the following part, we will show how to use OpenHGNN to implement it.

---

[5] https://openhgnn.readthedocs.io/en/latest/.

10.3.3.1   HAN Basemodel

To create a model HAN, we could imitate the case of 10.3.2. In this section, we will show each functional part of the HAN source code and explain the message passing process over HG and its mathematical definition. We implement the class method build_model_from_args, other functions like __init__. Here is our code.

```python
@register_model('HAN')
class HAN(BaseModel):
    @classmethod
    def build_model_from_args(cls, args, hg):
        if args.meta_paths is None:
            meta_paths = extract_metapaths(args.category,
                                           hg.canonical_etypes)
        else:
            meta_paths = args.meta_paths

        return cls(meta_paths=meta_paths,
                   category=args.category,
                   in_size=args.hidden_dim,
                   hidden_size=args.hidden_dim,
                   out_size=args.out_dim,
                   num_heads=args.num_heads,
                   dropout=args.dropout)

    def __init__(self, meta_paths, category, in_size, hidden_size,
                 out_size, num_heads, dropout):
        super(HAN, self).__init__()
        self.category = category
        self.layers = nn.ModuleList()
        self.layers.append(HANLayer(meta_paths, in_size, hidden_size,
        num_heads[0], dropout))
        for l in range(1, len(num_heads)):
            self.layers.append(
                HANLayer(meta_paths, hidden_size * num_heads[l-1],
                         hidden_size, num_heads[l], dropout)
            )
        self.linear = nn.Linear(hidden_size * num_heads[-1], out_size)

    def forward(self, g, h_dict):
        h = h_dict[self.category]
        for gnn in self.layers:
            h = gnn(g, h)

        return {self.category: self.linear(h)}
```

This piece of code is the overall structure about HAN model. Function
__init__ will initialize the model and the function `forward` will encoder
original features into new features. In general, the model should output all the node
embeddings in a form of dictionary. It is allowed to output only the embedding of
the target node participating in the loss calculation.

10.3.3.2 HANlayer

HAN is mainly composed of two parts, **Node-level Attention** and **Semantic-level
Attention**. The model which we implement in OpenHGNN incorporates these two
parts in HANLayer.

```python
class HANLayer(nn.Module):
    def __init__(self, meta_paths, in_size, out_size,
                 layer_num_heads, dropout):
        super(HANLayer, self).__init__()
        self.meta_paths = meta_paths
        # One GAT layer for each meta path based adjacency matrix
        self.gat_layers = nn.ModuleList()
        semantic_attention = SemanticAttention(in_size=out_size
                                               * layer_num_heads)
        self.model = MetapathConv(
            meta_paths,
            [GATConv(in_size, out_size, layer_num_heads, dropout,
                     activation=F.elu, allow_zero_in_degree=True)
             for _ in meta_paths],
            semantic_attention
        )
        self._cached_graph = None
        self._cached_coalesced_graph = {}
    def forward(self, g, h):
        if self._cached_graph is None or self._cached_graph is not g:
            self._cached_graph = g
            self._cached_coalesced_graph.clear()
            for meta_path in self.meta_paths:
                self._cached_coalesced_graph[meta_path] = \
                dgl.metapath_reachable_graph(g, meta_path)
        h = self.model(self._cached_coalesced_graph, h)

        return h
```

Meta-path Subgraph Extraction

We use `dgl.metapath_reachable_graph(g,meta_path)` API encapsulated by DGL to extract homogeneous graph from heterogeneous graph according to meta-path. For example:

```
g = dgl.heterograph({
    ('A', 'AB', 'B'): ([0, 1, 2], [1, 2, 3]),
    ('B', 'BA', 'A'): ([1, 2, 3], [0, 1, 2])})
new_g = dgl.metapath_reachable_graph(g, ['AB', 'BA'])
new_g.edges(order='eid')
# (tensor([0, 1, 2]), tensor([0, 1, 2]))
```

This part of code is the core component of HAN. `HANLayer` is inherited from `nn.Module` and acts as a complete aggregation of HAN. We use `GATConv`, which is implemented by DGL, to implement the node-level attention mechanism. And semantic-level attention is implemented by `MetapathConv`, which we will introduce later.

After the aggregation of node-level attention mechanism, we will get node embeddings based on different meta-paths, as much as the predetermined meta-paths. At the same time, we extend node-level attention to multi-head attention and concatenate the learned embeddings to enhance the representation power of the model. Then based on different meta-paths we perform semantic-level attention fusion to learn different importance weights of each meta-path. The code of `MetapathConv` is as follows:

```
class MetapathConv(nn.Module):
    def __init__(self, meta_paths, mods, macro_func, **kargs):
        super(MetapathConv, self).__init__()
        # One GAT layer for each meta path based adjacency matrix
        self.mods = nn.ModuleList(mods)
        self.meta_paths = meta_paths
        self.SemanticConv = macro_func

    def forward(self, g_list, h):
        for i, meta_path in enumerate(self.meta_paths):
            new_g = g_list[meta_path]
            semantic_embeddings.append(self.mods[i](new_g, h).flatten(1))

        return self.SemanticConv(semantic_embeddings)
```

`self.SemanticConv` is an aggregation function fusing meta-paths and we could get the final embeddings by it.

The code of `SemanticConv` is as follows:

```
class SemanticAttention(nn.Module):
    def __init__(self, in_size, hidden_size=128):
        super(SemanticAttention, self).__init__()

        self.project = nn.Sequential(
            nn.Linear(in_size, hidden_size),
            nn.Tanh(),
            nn.Linear(hidden_size, 1, bias=False)
        )

    def forward(self, z, nty=None):
        if len(z) == 0:
            return None
        z = torch.stack(z, dim=1)
        w = self.project(z).mean(0)                        # (M, 1)
        beta = torch.softmax(w, dim=0)                     # (M, 1)
        beta = beta.expand((z.shape[0],) + beta.shape)     # (N, M, 1)

        return (beta * z).sum(1)                           # (N, D * K)
```

After that, we will get the final embedding which aggregates all meta-paths information by Attention mechanism.

10.3.4   Practice of RGCN

RGCN [7] is the first to show that the GCN [5] framework can be applied to modeling relational data, which contributes to the application of neural network in heterogeneous graph. In RGCN, we execute graph convolution on subgraph of every relation to produce relation-specific representations. Then we can sum these relation-specific nodes representation with a single self-connection of each node. We summarize the above as the following formula:

$$h_i^{(l+1)} = \sigma \left(\sum_{r \in \mathcal{R}} \sum_{j \in \mathcal{N}_i^r} \frac{1}{c_{i,r}} W_r^{(l)} h_j^{(l)} + W_0^{(l)} h_i^{(l)} \right).$$

OpenHGNN also uses DGL method which is based on message passing mechanism to implement RGCN layer. In this part, we focus on describing the interface of RGCN and the code of convolution.

Here is the interface of RGCN model in OpenHGNN:

```
def __init__(self, in_dim, h_dim, out_dim, etypes, num_bases,
              num_hidden_layers=1, dropout=0, use_self_loop=False)
```

Next is the code of RelGraphConvLayer:

```
self.conv = dglnn.HeteroGraphConv({
        rel: dglnn.GraphConv(in_feat, out_feat, norm='right',
        weight=False, bias=False) for rel in rel_names
    })
```

The function `HeteroGraphConv` receives a dictionary containing all the subgraph convolution models as a parameter. DGL gives a pseudo-code for implementation of `HeteroGraphConv`:

```
outputs = {nty : [] for nty in g.dsttypes}
# Apply submodules on their associating relation graphs in parallel
for relation in g.canonical_etypes:
    stype, etype, dtype = relation
    dstdata = relation_submodule(g[relation], ...)
    outputs[dtype].append(dstdata)


# Aggregate the results for each destination node type
rsts = {}
for ntype, ntype_outputs in outputs.items():
    if len(ntype_outputs) != 0:
        rsts[ntype] = aggregate(ntype_outputs)


return rsts
```

The code of `self.conv` used in `forward` function:

```
def forward(self, g, inputs):
    g = g.local_var()
    if self.use_weight:
        weight = self.basis() if self.use_basis else self.weight
        wdict = {self.rel_names[i]: {'weight': w.squeeze(0)}
                 for i, w in enumerate(th.split(weight, 1, dim=0))}
    else:
        wdict = {}

    if g.is_block:
        inputs_src = inputs
        inputs_dst = {k: v[:g.number_of_dst_nodes(k)]
                      for k, v in inputs.items()}
    else:
        inputs_src = inputs_dst = inputs

    hs = self.conv(g, inputs_src, mod_kwargs=wdict)
```

The function `forward` receives a parameter `inputs` which contains the input of function `HeteroGraphConv`. The dict `inputs_src` contains node features for each node type, and the dict `wdict` is the weight dict of every subrelation which is used in a linear or non-linear transformation.

10.3.5   Practice of HERec

HERec [8] is the first attempt which adopts the network embedding approach to extract useful information from heterogeneous information network and leverage such information for rating prediction. HERec mainly contains two key components: meta-path based random walk and Skip-Gram model. As a practice of HERec, we will focus on the former component meta-path based random walk. As the earliest model, HERec utilizes a similar random walk with meatapth2vec model. And we will introduce how to implement the two ways of random walk.

10.3.5.1   HERec Random Walk

As mentioned in 10.3.3.2, we use the `dgl.metapath_reachable_graph(g, meta_path)` API to extract meta-path subgraph which is a homogeneous graph.

For a homogeneous graph, we can use the random walk applied in the homogeneous graph. `dgl.sampling.random_walk` provides an API to generate random walk traces.

Example of normal random walk, g1 will generate four traces from the source nodes [0, 1, 2, 0] and each trace has five nodes.

```
>>> g1 = dgl.graph(([0, 1, 1, 2, 3], [1, 2, 3, 0, 0]))
>>> dgl.sampling.random_walk(g1, [0, 1, 2, 0], length=4)
(tensor([[0, 1, 2, 0, 1],
         [1, 3, 0, 1, 3],
         [2, 0, 1, 3, 0],
         [0, 1, 2, 0, 1]]), tensor([0, 0, 0, 0, 0]))
```

Taking ACM as an example, we extract two meta-path subgraphs with meta-path PAP and PSP. PAP-subgraph can generate traces by random walk.

```
>>> PAP = ['paper-author', 'author-paper']
>>> PSP = ['paper-subject', 'subject-paper']
>>> pap_subgraph = dgl.metapath_reachable_graph(hg, PAP)
Graph(num_nodes=3025, num_edges=29767,
        ndata_schemes={'test_mask': Scheme(shape=(), dtype=torch.uint8),
        'train_mask': Scheme(shape=(), dtype=torch.uint8),
        'label': Scheme(shape=(), dtype=torch.float32),
        'h': Scheme(shape=(1902,), dtype=torch.float32)}
        edata_schemes={})
>>> psp_subgraph = dgl.metapath_reachable_graph(hg, PSP)
Graph(num_nodes=3025, num_edges=2217089,
        ndata_schemes={'test_mask': Scheme(shape=(), dtype=torch.uint8),
        'train_mask': Scheme(shape=(), dtype=torch.uint8),
        'label': Scheme(shape=(), dtype=torch.float32),
        'h': Scheme(shape=(1902,), dtype=torch.float32)}
        edata_schemes={})
>>> pap_traces = dgl.sampling.random_walk(pap_subgraph,
...     torch.arange(pap_subgraph.num_nodes()), length=4)
(tensor([[   0,    0,   20, 1807,  734],
        [   1,  773,    5,    1, 2576],
        [   2, 2519, 2701,  616,  616],
        ...,
        [3022, 1678, 3022,  275,  275],
        [3023, 3023, 3023, 3023, 3023],
        [3024, 3024, 3024, 3024, 3024]]), tensor([0, 0, 0, 0, 0]))
>>> psp_traces = dgl.sampling.random_walk(psp_subgraph,
...     torch.arange(psp_subgraph.num_nodes()), length=4)
(tensor([[   0,   75,   75,  586,  716],
        [   1, 1764, 2512, 1468, 1641],
        [   2,  189, 2786,  737, 2743],
        ...,
        [3022, 2641,  347,  684, 2923],
        [3023,  722, 2707, 2394, 1380],
        [3024, 1450,  235,  803, 1884]]), tensor([0, 0, 0, 0, 0]))
```

And traces could generate embedding through skipgram model. Finally, we can get the final embedding by fusing the two meta-path-types embedding of paper nodes.

10.3.5.2   Meta-Path Random Walk

We also generate random walk traces from an array of starting nodes directly based on the given meta-path.

```
>>> g2 = dgl.heterograph({
...      ('user','follow','user'): ([0, 1, 1, 2, 3], [1, 2, 3, 0, 0]),
...      ('user','view','item'): ([0, 0, 1, 2, 3, 3], [0, 1, 1, 2, 2, 1]),
...      ('item','viewed-by','user'): ([0, 1, 1, 2, 2, 1],
...                                     [0, 0, 1, 2, 3, 3])
... })
>>> dgl.sampling.random_walk(g2, [0, 1, 2, 0],
...      metapath=['follow', 'view', 'viewed-by'] * 2)
(tensor([[0, 1, 1, 1, 2, 2, 3],
         [1, 3, 1, 1, 2, 2, 2],
         [2, 0, 1, 1, 3, 1, 1],
         [0, 1, 1, 0, 1, 1, 3]]), tensor([0, 0, 1, 0, 0, 1, 0]))
```

Taking ACM as an example too, we will generate traces through meta-path PAPAP.

```
>>> dgl.sampling.random_walk(
...      hg, torch.arange(hg.num_nodes('paper')),
...      metapath=['paper-author', 'author-paper']*2)
(tensor([[   0,    1,  731,    1,   20],
         [   1,    3,    1,    5,    1],
         [   2,    7,  229,   12,    4],
         ...,
         [3022, 3774, 1678, 3775, 1678],
         [3023, 5915, 2998, 5915, 2998],
         [3024, 5956, 3024, 5955, 3024]]), tensor([1, 0, 1, 0, 1]))
```

10.4 Conclusion

The implementation of HGNN is a major obstacle for a beginner getting started. Although the source codes of some models are released, it is not easy to understand and apply to new tasks. With the help of OpenHGNN, we can easily implement a new model and test the performance among various datasets. This chapter introduces the HGNN toolkit OpenHGNN and the pipeline of implementing a new model on it. We also provide three classical model implementation instances. The first model is HAN, which is the first model to introduce attention mechanism to HGNN. The second is RGCN, the first model to use graph convolution network to model multi-relational graph. The last one is HERec, which applies heterogeneous graph embedding to recommendation scenarios. These models are the most representative for learning on heterogeneous graphs. For the built-in models and datasets, we can run the model in just one command. For non-built-in models or datasets, implementation and testing are also very simple. OpenHGNN is under development and more features and more models (including graph embedding methods) will be incorporated in the future version. For more details about OpenHGNN, please visit the website *https://github.com/BUPT-GAMMA/OpenHGNN.*

References

1. Abadi, M., Agarwal, A., Barham, P., Brevdo, E., Chen, Z., Citro, C., Corrado, G.S., Davis, A., Dean, J., Devin, M., Ghemawat, S., Goodfellow, I., Harp, A., Irving, G., Isard, M., Jia, Y., Jozefowicz, R., Kaiser, L., Kudlur, M., Levenberg, J., Mané, D., Monga, R., Moore, S., Murray, D., Olah, C., Schuster, M., Shlens, J., Steiner, B., Sutskever, I., Talwar, K., Tucker, P., Vanhoucke, V., Vasudevan, V., Viégas, F., Vinyals, O., Warden, P., Wattenberg, M., Wicke, M., Yu, Y., Zheng, X.: TensorFlow: large-scale machine learning on heterogeneous systems (2015). https://www.tensorflow.org/. Software available from tensorflow.org
2. Chen, T., Li, M., Li, Y., Lin, M., Wang, N., Wang, M., Xiao, T., Xu, B., Zhang, C., Zhang, Z.: Mxnet: A flexible and efficient machine learning library for heterogeneous distributed systems. Preprint. arXiv:1512.01274 (2015)
3. Fey, M., Lenssen, J.E.: Fast graph representation learning with PyTorch Geometric. In: ICLR Workshop on Representation Learning on Graphs and Manifolds (2019)
4. Hu, B., Fang, Y., Shi, C.: Adversarial learning on heterogeneous information networks. In: Proceedings of the 25th ACM SIGKDD International Conference on Knowledge Discovery & Data Mining, pp. 120–129 (2019)
5. Kipf, T.N., Welling, M.: Semi-supervised classification with graph convolutional networks. In: ICLR (2017)
6. Paszke, A., Gross, S., Massa, F., Lerer, A., Bradbury, J., Chanan, G., Killeen, T., Lin, Z., Gimelshein, N., Antiga, L., Desmaison, A., Kopf, A., Yang, E., DeVito, Z., Raison, M., Tejani, A., Chilamkurthy, S., Steiner, B., Fang, L., Bai, J., Chintala, S.: Pytorch: An imperative style, high-performance deep learning library. In: Wallach, H., Larochelle, H., Beygelzimer, A., d'Alché-Buc, F., Fox, E., Garnett, R. (eds.) Advances in Neural Information Processing Systems, vol. 32. Curran Associates, Inc., Red Hook (2019). https://proceedings.neurips.cc/paper/2019/file/bdbca288fee7f92f2bfa9f7012727740-Paper.pdf
7. Schlichtkrull, M., Kipf, T.N., Bloem, P., Van Den Berg, R., Titov, I., Welling, M.: Modeling relational data with graph convolutional networks. In: European Semantic Web Conference, pp. 593–607. Springer, Berlin (2018)
8. Shi, C., Hu, B., Zhao, W.X., Yu, P.S.: Heterogeneous information network embedding for recommendation. IEEE Trans. Knowl. Data Eng. **31**(2), 357–370 (2018)
9. Wang, M., Zheng, D., Ye, Z., Gan, Q., Li, M., Song, X., Zhou, J., Ma, C., Yu, L., Gai, Y., Xiao, T., He, T., Karypis, G., Li, J., Zhang, Z.: Deep graph library: a graph-centric, highly-performant package for graph neural networks. Preprint. arXiv:1909.01315 (2019)
10. Wang, X., Ji, H., Shi, C., Wang, B., Ye, Y., Cui, P., Yu, P.S.: Heterogeneous graph attention network. In: The World Wide Web Conference, pp. 2022–2032 (2019)
11. Yun, S., Jeong, M., Kim, R., Kang, J., Kim, H.J.: Graph transformer networks. In: Advances in Neural Information Processing Systems, pp. 11960–11970 (2019)
12. Zhang, C., Song, D., Huang, C., Swami, A., Chawla, N.V.: Heterogeneous graph neural network. In: KDD '19: Proceedings of the 25th ACM SIGKDD International Conference on Knowledge Discovery & Data Mining, pp. 793–803 (2019)

Chapter 11
Future Research Directions

Abstract Heterogeneous graph (HG) representation has made great progress in recent years, which clearly shows that it is a powerful and promising graph analysis paradigm. However, it is still a young and promising research field. In this chapter, we first make a summarization of this book and then illustrate some advanced topics, including challenging research issues, and explore a series of possible future research directions. One major potential direction is exploring fundamental ways to keep intrinsic structures or properties in HG. And another direction is to integrate the techniques widely used or newly emerged in machine learning to further enhance the applicability of HG on more key fields. We will illustrate more fine-grained potential works along with these two directions.

11.1 Introduction

Heterogeneous graph (HG) representation has significantly facilitated the HG analysis and related applications. This book conducts a comprehensive study of the state-of-the-art HG representation methods. Thorough discussions and summarization of the reviewed methods, along with the widely used benchmarks and resources, are systematically presented. Then, in part one of this book, we present the advanced HG representation techniques. Particularly, we first introduce several classical structure-preserved HG methods. These methods preserve the heterogeneous structure by most fundamental elements in HG, including meta-path, relation, and network schema. Then attribute information is introduced to enrich the characteristics of nodes. Heterogeneous graph neural networks (HGNNs) naturally provide an alternative way to integrate attributes with structural information. Besides static heterogeneous graphs, we introduce dynamic HGNN methods, which mainly focus on updating node representation in an efficient way or learning node representations while considering sequential evolution. In part two, the necessity of HG to fuse abundant heterogeneous interactions is comprehensively displayed in several prevalent applications. The recommendation is one of such prevalent applications, as the interactions of users and items can be naturally built as an HG.

C. Shi et al., *Heterogeneous Graph Representation Learning and Applications*,
Artificial Intelligence: Foundations, Theory, and Algorithms,
https://doi.org/10.1007/978-981-16-6166-2_11

311

In particular, three advanced HG-based recommendation methods are presented to show the effectiveness of integrating heterogeneous information. Another interesting application is using HG to overcome data sparsity problems in text mining. We summarize the methods, which utilize the powerful capabilities of HG to integrate extra information, to demonstrate the superiority of HG representation methods in text mining. More importantly, one of the unique characteristics of this book is that we summarize not only the methods invented based on public academic data but also the methods deployed in real-world systems. These methods further promote the application of the HG methods towards industrial production. We hope that this book can provide a clean sketch and key technique summarization on HG representation, which could help both the interested readers and the researchers who wish to continue working in this area.

In this chapter, we point out some promising research directions on HG representation. Preserving HG structures and properties is deemed as one of the most fundamental ways to encode heterogeneous information. More fundamental but largely ignored methods are pointed out by us, such as motif or network-schema preserved methods, and dynamic and uncertainty properties of HG captured methods, etc. Besides shallow methods, deep GNN is a developing topic in recent years. Self-supervised learning and pretraining are emerged topics in GNN. And we point out that they are also worth exploring in HGNN. Moreover, to further deepen the reliability of HG representation methods in more key fields, it is important to integrate extra knowledge to make HG representation methods more fair, robust, explainable, and stable. Last but not least, we believe that exploring more potential industrial applications of HG representation methods holds great promise in the future.

11.2   Preserving HG Structures

The basic success of HG representation builds on the HG structure preservation. This also motivates many HG representation methods to exploit different HG structures, where the most typical one is meta-path [8, 32]. Following this line, meta-graph structure is naturally considered [42]. However, the fundamental elements to characterize the HG are far more these structures. Selecting the most appropriate meta-path is still very challenging in the real world. An improper meta-path will fundamentally hinder the performance of HG representation method. Whether we can explore other techniques, e.g., motif [15, 44] or network schema [46], to capture HG structure is worth pursuing. Moreover, if we rethink the goal of traditional graph representation, i.e., replacing the structure information with the distance or similarity in a metric space, a research direction to explore is whether we can design an HG representation method that can naturally learn such distance or similarity rather than using pre-defined meta-path or meta-graph.

11.3   Capturing HG Properties

As mentioned before, many current HG representation methods mainly take the structures into account. However, some properties, which usually provide additional useful information to model HG, have not been fully considered. One typical property is the dynamics of HG, i.e., a real-world HG always evolves over time. Despite that the incremental learning on dynamic HG is proposed [40], dynamic HG representation is still facing big challenges. For example, [1] is only proposed with a shallow model, which greatly limits its representation ability. How can we learn dynamic HG representation in deep learning framework is worth pursuing. The other property is the uncertainty of HG, i.e., the generation of HG is usually multifaceted and the node in an HG contains different semantics. Traditionally, learning a vector representation usually cannot well-capture such uncertainty. Gaussian distribution may innately represent the uncertainty property [17, 47], which is largely ignored by current HG representation methods. This suggests a huge potential direction for improving HG representation.

11.4   Deep Graph Learning on HG Data

We have witnessed the great success and large impact of GNNs, where most of the existing GNNs are proposed for homogeneous graph [18, 35]. Recently, HGNNs have attracted considerable attention [6, 10, 38, 43].

One natural question may arise: What is the essential difference between GNNs and HGNNs? More theoretical analyses on HGNNs are seriously lacking. For example, it is well-accepted that the GNNs suffer from over-smoothing problem [19], so will heterogeneous GNNs also have such a problem? If the answer is yes, what factor causes the over-smoothing problem in HGNNs since they usually contain multiple aggregation strategies [38, 43]. Moreover, some researchers have derived the generalization bounds for GNNs [20, 21] and analyzed the key factors dominating the generalization error. Hence, a natural question arises. What is the key factor influencing the generalization ability of HG representation methods? Meta-path or the aggregation function?

In addition to theoretical analysis, new technique design is also important. One of the most important directions is the self-supervised learning. It uses the pretext tasks to train the neural networks, thus reducing the dependence on manual labels [23]. Considering the actual demand that label is insufficient, self-supervised learning can greatly benefit the unsupervised and semi-supervised learning and has shown remarkable performance on homogeneous graph representation [27, 33, 36, 41]. Therefore, exploring self-supervised learning on HG representation is expected to further facilitate the development of this area.

Another important direction is the pretraining of HGNNs [14, 29]. Nowadays, HGNNs are designed independently, i.e., the proposed method usually works well for some certain tasks, but the transfer ability across different tasks is ill-considered. When dealing with a new HG or task, we have to train an HG representation method from scratch, which is time consuming and requires large amounts of labels. In this situation, if there is a well pretrained HGNN with strong generalization that can be fine-tuned with few labels, the time and label consumption can be reduced.

11.5   Making HG Representation Reliable

Except from the properties and techniques in HG, we are also concerned about the ethical issues in HG representation, such as fairness, robustness, and interpretability. Considering that most methods are black boxes, making HG representation reliable is an important future work.

Fair HG Representation   The representations learned by methods are sometimes highly related to certain attributes, e.g., age or gender, which may amplify the societal stereotypes in the prediction results [3, 9]. Therefore, learning fair or de-biased representations is an important research direction. There are some researches on the fairness of homogeneous graph representation [3, 30]. However, the fairness of HG is still an unsolved problem, which is an important research direction in the future.

Robust HG Representation   Also, the robustness of HG representation, especially the adversarial attacking, is always an important problem [25]. Since many real-world applications are built based on HG, the robustness of HG representation becomes an urgent yet unsolved problem. What is the weakness of HG representation and how to enhance it to improve the robustness need to be further studied.

Explainable HG Representation   Moreover, in some risk-aware scenarios, e.g., fraud detection [13] and biomedicine [5], the explanation of models or representations is important. A significant advantage of HG is that it contains rich semantics, which may provide eminent insight to promote the explanation of heterogeneous GNNs. Besides, the emerging disentangled learning [24, 26], which divides the representation into different latent spaces to improve the interpretability, can also be considered. Learning post-explanation model for GNNs has attracted attention in the recent years [28]. Then it is necessary to develop a post-explanation model for HGNNs to explain the prediction mechanism of these methods.

Stable HG Representation   Furthermore, most HG representation methods assume that the training graph and testing graph are drawn from the same distribution. However, this assumption is easy to be violated, as distribution shifts may arise from different environments that are common in real-world data collection pipelines, such as locations, times, experimental conditions, etc. [12]. For considering the generalization ability of HG representation methods, it is necessary to improve the stability of HG representation methods under unknown testing environments. Causal variables and relations are deemed to be invariant across environments.

Recently, some literature aimed to discovery such variables in representation learning [31]. It will be a promising direction to marry causal learning with HG representation methods to enhance the stability of HG representation methods on agnostic environments.

11.6 Technique Deployment in Real-World Applications

Many HG-based applications have stepped into the era of graph representation. This survey has demonstrated the strong performance of HG representation methods on E-commerce and cybersecurity. Exploring more capacity of HG representation on other areas holds great potential in the future. For example, in software engineering area, there are complex relations among test sample, requisition form, and problem form, which can be naturally modeled as HG. Therefore, HG representation is expected to open up broad prospects for these new areas and become promising analytical tool. Another area is the biological systems, which can also be naturally modeled as an HG. A typical biological system contains many types of objects, e.g., gene expression, chemical, phenotype, and microbe. There are also multiple relations between gene expression and phenotype [34]. HG structure has been applied to biological system as an analytical tool, implying that HG representation is expected to provide more promising results. For another area, e.g., transportation prediction, the data usually consists of heterogeneous objects, such car, traffic light, etc., and exists in a spatiotemporal format, so it is natural to model such complex data with HG while considering the spatiotemporal information.

In addition, since the complexity of HGNNs is relatively large and the techniques are difficult to parallelize, it is difficult to apply the existing HGNNs to large-scale industrial scenarios. For example, the number of nodes in E-commerce recommendation may reach one billion [45]. Therefore, successful technique deployment in various applications while resolving the scalability and efficiency challenges will be very promising.

11.7 Others

Last but not least, there are also some important future works that cannot be summarized in the previous sections. Therefore, we carefully discuss them in this section.

Hyperbolic Heterogeneous Graph Representation Some recent researches point out that the underlying latent space of graph may be non-Euclidean, but in hyperbolic space [4]. Some attempts have been made towards hyperbolic graph or heterogeneous graph representation, and the results are rather promising [7, 22, 39].

However, how to design an effective hyperbolic heterogeneous GNNs is still challenging, which can be another research direction.

Heterogeneous Graph Structure Learning Under the current HG representation framework, HG is usually constructed beforehand, which is independent of the HG representation. This may result in that the input HG is not suitable for the final task. HG structure learning can be further integrated with HG representation, so that they can promote each other.

Heterophily Heterogeneous Graph Representation Current HG representation methods focus on the leverage of network homophily. Due to the recent research on homogeneous networks that study learning network representation on Heterophily network [2, 48], it would be interesting to find heterophily HG and explore how to generalize design principles and paradigms used in heterophily homogeneous network representation to HG representation.

Connections with Knowledge Graph Knowledge graph representation has great potential on knowledge reasoning [16]. However, knowledge graph representation and HG representation are usually investigated separately. Recently, knowledge graph representation has been successfully applied to other areas, e.g., recommender system [11, 37]. It is worth studying that how to combine knowledge graph representation with HG representation and incorporate knowledge into HG representation.

References

1. Bian, R., Koh, Y.S., Dobbie, G., Divoli, A.: Network embedding and change modeling in dynamic heterogeneous networks. In: Proceedings of the 42nd International ACM SIGIR Conference on Research and Development in Information Retrieval (SIGIR), pp. 861–864. ACM, New York (2019)
2. Bo, D., Wang, X., Shi, C., Shen, H.: Beyond low-frequency information in graph convolutional networks. In: Proceedings of the AAAI Conference on Artificial Intelligence (AAAI) (2021)
3. Bose, A.J., Hamilton, W.L.: Compositional fairness constraints for graph embeddings. In: International Conference on Machine Learning (ICML), pp. 715–724 (2019)
4. Bronstein, M.M., Bruna, J., LeCun, Y., Szlam, A., Vandergheynst, P.: Geometric deep learning: Going beyond Euclidean data. IEEE Signal Process. Mag. **34**(4), 18–42 (2017)
5. Cao, Y., Peng, H., Philip, S.Y.: Multi-information source HIN for medical concept embedding. In: Pacific-Asia Conference on Knowledge Discovery and Data Mining (PAKDD), pp. 396–408 (2020)
6. Cen, Y., Zou, X., Zhang, J., Yang, H., Zhou, J., Tang, J.: Representation learning for attributed multiplex heterogeneous network. In: Proceedings of the 25th ACM SIGKDD International Conference on Knowledge Discovery and Data Mining (KDD). ACM, New York (2019)
7. Chami, I., Ying, Z., Ré, C., Leskovec, J.: Hyperbolic graph convolutional neural networks. In: Advances in Neural Information Processing Systems (NeurIPS), pp. 4869–4880 (2019)
8. Dong, Y., Chawla, N.V., Swami, A.: metapath2vec: Scalable representation learning for heterogeneous networks. In: Proceedings of the 23rd ACM SIGKDD International Conference on Knowledge Discovery and Data Mining (SIGKDD), pp. 135–144 (2017)
9. Du, M., Yang, F., Zou, N., Hu, X.: Fairness in deep learning: A computational perspective. CoRR abs/1908.08843 (2019)

10. Fu, X., Zhang, J., Meng, Z., King, I.: MAGNN: Metapath aggregated graph neural network for heterogeneous graph embedding. In: Proceedings of the Web Conference 2020 (WWW), pp. 2331–2341 (2020)

11. Guo, Q., Zhuang, F., Qin, C., Zhu, H., Xie, X., Xiong, H., He, Q.: A survey on knowledge graph-based recommender systems. CoRR abs/2003.00911 (2020)

12. Hendrycks, D., Dietterich, T.: Benchmarking neural network robustness to common corruptions and perturbations. arXiv preprint:1903.12261 (2019)

13. Hu, B., Zhang, Z., Shi, C., Zhou, J., Li, X., Qi, Y.: Cash-out user detection based on attributed heterogeneous information network with a hierarchical attention mechanism. In: Proceedings of the AAAI Conference on Artificial Intelligence (AAAI), vol. 33(1), pp. 946–953 (2019)

14. Hu, Z., Dong, Y., Wang, K., Chang, K., Sun, Y.: GPT-GNN: generative pre-training of graph neural networks. In: Proceedings of the 26th ACM SIGKDD International Conference on Knowledge Discovery and Data Mining (KDD), pp. 1857–1867 (2020)

15. Huang, Z., Zheng, Y., Cheng, R., Sun, Y., Mamoulis, N., Li, X.: Meta structure: computing relevance in large heterogeneous information networks. In: Proceedings of the 22nd ACM SIGKDD International Conference on Knowledge Discovery and Data Mining (KDD), pp. 1595–1604 (2016)

16. Ji, S., Pan, S., Cambria, E., Marttinen, P., Yu, P.S.: A survey on knowledge graphs: representation, acquisition and applications. CoRR abs/2002.00388 (2020)

17. Kipf, T.N., Welling, M.: Variational graph auto-encoders. CoRR abs/1611.07308 (2016)

18. Kipf, T.N., Welling, M.: Semi-supervised classification with graph convolutional networks. In: Proceedings of the Conference ICLR (2017)

19. Li, Q., Han, Z., Wu, X.: Deeper insights into graph convolutional networks for semi-supervised learning. In: Thirty-Second AAAI Conference on Artificial Intelligence (AAAI), pp. 3538–3545 (2018)

20. Liao, R., Urtasun, R., Zemel, R.S.: Generalization and representational limits of graph neural networks. In: International Conference on Machine Learning (ICML), pp. 3419–3430 (2020)

21. Liao, R., Urtasun, R., Zemel, R.S.: A PAC-Bayesian approach to generalization bounds for graph neural networks. In: Proceedings of the Conference ICLR (2021)

22. Liu, Q., Nickel, M., Kiela, D.: Hyperbolic graph neural networks. In: 33rd Conference on Neural Information Processing Systems (NeurIPS 2019), pp. 8228–8239 (2019)

23. Liu, X., Zhang, F., Hou, Z., Wang, Z., Mian, L., Zhang, J., Tang, J.: Self-supervised learning: Generative or contrastive. CoRR abs/2006.08218 (2020)

24. Ma, J., Zhou, C., Cui, P., Yang, H., Zhu, W.: Learning disentangled representations for recommendation. In: Advances in Neural Information Processing Systems (NeurIPS), pp. 5712–5723 (2019)

25. Madry, A., Makelov, A., Schmidt, L., Tsipras, D., Vladu, A.: Towards deep learning models resistant to adversarial attacks. In: Proceedings of the Conference ICLR (2018)

26. Narayanaswamy, S., Paige, B., van de Meent, J., Desmaison, A., Goodman, N.D., Kohli, P., Wood, F.D., Torr, P.H.S.: Learning disentangled representations with semi-supervised deep generative models. In: Proceedings of the 31st International Conference on Neural Information Processing Systems (NeurIPS), pp. 5925–5935 (2017)

27. Peng, Z., Dong, Y., Luo, M., Wu, X., Zheng, Q.: Self-supervised graph representation learning via global context prediction. CoRR abs/2003.01604 (2020)

28. Pope, P.E., Kolouri, S., Rostami, M., Martin, C.E., Hoffmann, H.: Explainability methods for graph convolutional neural networks. In: Proceedings of the IEEE/CVF Conference on Computer Vision and Pattern Recognition, pp. 10772–10781 (2019)

29. Qiu, J., Chen, Q., Dong, Y., Zhang, J., Yang, H., Ding, M., Wang, K., Tang, J.: GCC: graph contrastive coding for graph neural network pre-training. In: Proceedings of the 26th ACM SIGKDD International Conference on Knowledge Discovery and Data Mining (KDD), pp. 1150–1160 (2020)

30. Rahman, T.A., Surma, B., Backes, M., Zhang, Y.: Fairwalk: towards fair graph embedding. In: Proceedings of the Twenty-Eighth International Joint Conference on Artificial Intelligence (IJCAI), pp. 3289–3295 (2019)

31. Schölkopf, B., Locatello, F., Bauer, S., Ke, N.R., Kalchbrenner, N., Goyal, A., Bengio, Y.: Towards causal representation learning. In: Special Issue of Proceedings of the IEEE— Advances in Machine Learning and Deep Neural Networks (2021)

32. Shi, C., Li, Y., Zhang, J., Sun, Y., Yu, P.S.: A survey of heterogeneous information network analysis. IEEE Trans. Knowl. Data Eng. **29**(1), 17–37 (2017)

33. Sun, K., Lin, Z., Zhu, Z.: Multi-stage self-supervised learning for graph convolutional networks on graphs with few labeled nodes. In: Proceedings of the AAAI Conference on Artificial Intelligence (AAAI), pp. 5892–5899 (2020)

34. Tsuyuzaki, K., Nikaido, I.: Biological systems as heterogeneous information networks: a mini-review and perspectives. CoRR abs/1712.08865 (2017)

35. Veličković, P., Cucurull, G., Casanova, A., Romero, A., Lio, P., Bengio, Y.: Graph attention networks. In: Proceedings of the Conference ICLR (2018)

36. Velickovic, P., Fedus, W., Hamilton, W.L., Liò, P., Bengio, Y., Hjelm, R.D.: Deep graph infomax. In: Proceedings of the Conference ICLR (2019)

37. Wang, H., Zhao, M., Xie, X., Li, W., Guo, M.: Knowledge graph convolutional networks for recommender systems. In: The World Wide Web Conference (WWW), pp. 3307–3313 (2019)

38. Wang, X., Ji, H., Shi, C., Wang, B., Ye, Y., Cui, P., Yu, P.S.: Heterogeneous graph attention network. In: The World Wide Web Conference (WWW), pp. 2022–2032 (2019)

39. Wang, X., Zhang, Y., Shi, C.: Hyperbolic heterogeneous information network embedding. In: Proceedings of the AAAI Conference on Artificial Intelligence (AAAI), pp. 5337–5344 (2019)

40. Wang, X., Lu, Y., Shi, C., Wang, R., Cui, P., Mou, S.: Dynamic heterogeneous information network embedding with meta-path based proximity. IEEE Trans. Knowl. Data Eng. (2020)

41. You, Y., Chen, T., Wang, Z., Shen, Y.: When does self-supervision help graph convolutional networks? CoRR abs/2006.09136 (2020)

42. Zhang, D., Yin, J., Zhu, X., Zhang, C.: Metagraph2vec: complex semantic path augmented heterogeneous network embedding. In: Pacific-Asia Conference on Knowledge Discovery and Data Mining (PAKDD), pp. 196–208. Springer, Berlin (2018)

43. Zhang, C., Song, D., Huang, C., Swami, A., Chawla, N.V.: Heterogeneous graph neural network. In: Proceedings of the 25th ACM SIGKDD International Conference on Knowledge Discovery and Data Mining (KDD), pp. 793–803 (2019)

44. Zhao, H., Zhou, Y., Song, Y., Lee, D.L.: Motif enhanced recommendation over heterogeneous information network. In: Proceedings of the 28th ACM International Conference on Information and Knowledge Management (CIKM), pp. 2189–2192 (2019)

45. Zhao, J., Zhou, Z., Guan, Z., Zhao, W., Ning, W., Qiu, G., He, X.: IntentGC: a scalable graph convolution framework fusing heterogeneous information for recommendation. In: Proceedings of the 25th ACM SIGKDD International Conference on Knowledge Discovery and Data Mining (KDD), pp. 2347–2357 (2019)

46. Zhao, J., Wang, X., Shi, C., Liu, Z., Ye, Y.: Network schema preserving heterogeneous information network embedding. In: IJCAI (2020)

47. Zhu, D., Cui, P., Wang, D., Zhu, W.: Deep variational network embedding in Wasserstein space. In: Proceedings of the 24th ACM SIGKDD International Conference on Knowledge Discovery and Data Mining (KDD), pp. 2827–2836 (2018)

48. Zhu, J., Yan, Y., Zhao, L., Heimann, M., Akoglu, L., Koutra, D.: Generalizing graph neural networks beyond homophily. In: Proceedings of the 34th Conference on Neural Information Processing Systems (NeurIPS) (2020)